T0205665

Electroporation Protocols
for Microorganisms

Methods in Molecular Biology™

John M. Walker, SERIES EDITOR

Methods in Molecular Biology • 47

Electroporation Protocols for Microorganisms

Edited by

Jac A. Nickoloff

Harvard University, Boston, MA

Humana Press ✳ Totowa, New Jersey

This publication is printed on acid-free paper. ∞
ANSI Z39.48-1984 (American National Standards Institute)
Permanence of Paper for Printed Library Materials

Printed in the United States of America. 10 9 8 7 6 5 4 3 2 1

Library of Congress Cataloging in Publication Data

Electroporation protocols for microorganisms/edited by Jac A. Nickoloff.
    p.   cm.—(Methods in molecular biology™; 47)
Includes index.
ISBN 0-89603-310-4
   1. Electroporation.   2. Microbial genetics—Methodology.   I. Nickoloff, Jack A. II. Series:
Methods in molecular biology™ (Totowa, NJ); 47
QH5895.5.E48E45 1995
576'.1390'028—dc20
                                           95-14992
                                             CIP

Preface

Electroporation is one of the most widespread techniques used in modern molecular genetics. It is most commonly used to introduce DNA into cells for investigations of gene structure and function, and in this regard, electroporation is both highly versatile, being effective with nearly all species and cell types, and highly efficient. For many cell types, electroporation is either the most efficient or the only means known to effect gene transfer. However, exposure of cells to brief, high-intensity electric fields has found broad application in other aspects of biological research, and is now routinely used to introduce other types of biological and analytic molecules into cells, to induce cell–cell fusion, and to transfer DNA directly between different species.

The first seven chapters of *Electroporation Protocols for Microorganisms* describe the underlying theory of electroporation, the commercially available instrumentation, and a number of specialized electroporation applications, such as cDNA library construction and interspecies DNA electrotransfer. Each of the remaining chapters presents a well developed method for electrotransformation of a particular bacterial, fungal, or protist species. These chapters also serve to introduce those new to the field the important research questions that are currently being addressed with particular organisms, highlighting both the major advantages and limitations of each species as a model organism, and explaining the roles that electroporation has played in the development of the molecular genetic systems currently in use. Microorganisms continue to play key roles in the development of our understanding of basic biological processes, as well as being important human, plant, and animal disease vectors. Because electroporation has such broad application, protocols for all microorganisms that have been successfully electrotransformed could not be included in this volume. However, protocols are included for a diverse array of bacterial, fungal, and protist species, including many that are important in human disease, and most chapters provide literature references for electroporation protocols of related species.

Although many of the procedures for electrotransformation of different organisms are similar, subtle differences are often important, especially when an experimental design requires optimum transformation. For example, electroporation efficiency is often strongly affected by growth conditions and growth phase at the time of cell harvest. Therefore, each chapter provides detailed information about growth conditions for the particular organism. Because specific approaches are not always successful, comparisons of procedures used with similar (or even quite different) organisms might provide valuable insight to researchers working to solve a particular problem. In addition, the chapter on electroporation theory can be used to develop new protocols or modify existing ones. In sum, I feel that this volume will be an especially valuable resource for molecular geneticists working with the widest variety of cell systems, both with respect to technical hints and troubleshooting advice, which are presented as "Notes" at the end of each protocol, and as a guide to the various applications of electroporation in different model systems.

I want to express my gratitude to all of the contributors, for both their timely submissions and their many suggestions. I also want to thank Debra Horensky for clarifying many fine points of microbial taxonomy, and for assistance in the selection of topics relevant to human disease. And I thank John Walker for his considerable assistance and thoughtful advice.

Jac A. Nickoloff

Contents

Contents

CONTENTS FOR THE COMPANION VOLUME

Animal Cell Electroporation and Electrofusion Protocols

CONTENTS FOR THE COMPANION VOLUME

Plant Cell Electroporation
and Electrofusion Protocols

Contributors

TAKETOSHI ARAI • *Laboratory of Microbiology, Showa College of Pharmaceutical Sciences, Tokyo, Japan*

THEA W. AUKRUST • *MATFORSK, Norwegian Food Research Institute, Ås, Norway*

JULIET BAILEY • *McArdle Laboratory, Madison, WI*

GERALD S. BARON • *Department of Biochemistry and Microbiology, University of Victoria, Canada*

HERMANN BERG • *Institute of Molecular Biotechnology, Jena, Germany*

DIETMAR BECHER • *Institute of Genetics and Biochemistry, E. M. Arendt University, Greifswald, Switzerland*

PAT A. BLUNDELL • *London School of Hygiene and Tropical Medicine, London, UK*

STEPHEN M. BOYLE • *Center for Molecular Medicine and Infectious Diseases, Virginia–Maryland Regional College of Veterinary Medicine, Virginia Polytechnic Institute and State University, Blacksburg, VA*

MARTIN BRENDEL • *Institute for Microbiology, J. W. Goethe University, Frankfurt-am-Main, Germany*

MAY B. BRURBERG • *Laboratory of Microbial Gene Technology Agricultural University of Norway, Ås, Norway*

TIMOTHY G. BURLAND • *McArdle Laboratory, Madison, WI*

RICHARD DANA • *BIO 101, Inc., La Jolla, CA*

JONATHAN J. DENNIS • *Department of Microbiology and Infectious Diseases, University of Calgary Health Sciences Centre, Calgary, Canada*

JOSEPH J. FERRETTI • *Department of Microbiology and Immunology, University of Oklahoma Health Sciences Center, Oklahoma City, OK*

JACEK GAERTIG • *Department of Zoology, The University of Georgia, Athens, GA*

DANIEL R. GALLIE • *Department of Biochemistry, University of California, Riverside, CA*

JAMES GAUTSCH • *BIO 101, Inc., La Jolla, CA*

MICHAEL S. GILMORE • *Department of Microbiology and Immunology, University of Oklahoma Health Sciences Center, Oklahoma City, OK*

CARLOS F. GONZALEZ • *Department of Plant Pathology and Microbiology, Texas A & M University, College Station, TX*

MARTIN A. GOROVSKY • *Department of Zoology, The University of Georgia, Athens, GA*

MARTIN GREY • *Institute for Microbiology, J. W. Goethe University, Frankfurt-am-Main, Germany*

CHRISTIAN E. GRUBER • *Research and Development, Life Technologies, Gaithersburg, MD*

LAURA GUNN • *Department of Cancer Biology, Harvard University School of Public Health, Boston, MA*

HAJIME HAMASHIMA • *Laboratory of Microbiology, Showa College of Pharmaceutical Sciences, Tokyo, Japan*

ANDREW HESSEL • *Salmonella Genetic Stock Centre, Department of Biological Sciences, University of Calgary, Canada*

MERL F. HOEKSTRA • *ICOS Corporation, Bothell, WA*

GUNTER A. HOFMANN • *Genetronics, Inc., San Diego, CA*

HELGE HOLO • *Laboratory of Microbial Gene Technology, Agricultural University of Norway, Ås, Norway, and Norwegian Dairies Association, Oslo, Norway*

MARK T. HOOD • *Department of Biology, Boston College, Chestnut Hill, MA*

MAKOTO IWASAKI • *Laboratory of Microbiology, Showa College of Pharmaceutical Sciences, Tokyo, Japan*

M. KAPOOR • *Department of Biological Sciences, University of Calgary, Canada*

JOHN M. KELLY • *London School of Hygiene and Tropical Medicine, London, UK*

DAVID KNECHT • *Department of Molecular and Cell Biology, The University of Connecticut, Storrs, CT*

JEAN C. LEE • *Department of Medicine, Brigham and Women's Hospital and Harvard Medical School, Boston, MA*

JHY-JHU LIN • *Agriculture Biotechnology and Molecular Biology R & D, Life Technologies Inc., Gaithersburg, MD*

P. RONALD MACLACHLAN • *Salmonella Genetic Stock Centre, Department of Biological Sciences, University of Calgary, Canada. Present Address: Veterinary Infectious Disease Organization, University of Saskatchewan, Saskatoon, Canada*

ROBERT E. MCLAUGHLIN • *Department of Microbiology and Immunology, University of Oklahoma Health Sciences Center, Oklahoma City, OK*

JOHN R. MCQUISTON • *Center for Molecular Medicine and Infectious Diseases, Virginia–Maryland Regional College of Veterinary Medicine, Virginia Polytechnic Institute and State University, Blacksburg, VA*

ELIZABETH M. MILLER • *Department of Cancer Biology, Harvard University School of Public Health, Boston, MA*

SVETLANA V. MYLTSEVA • *Department of Biochemistry and Microbiology, University of Victoria, Canada*

FRANCIS E. NANO • *Department of Biochemistry and Microbiology, University of Victoria, Canada*

INGOLF F. NES • *Laboratory of Microbial Gene Technology Agricultural University of Norway, Ås, Norway*

JAC A. NICKOLOFF • *Department of Cancer Biology, Harvard University School of Public Health, Boston, MA*

STEPHEN G. OLIVER • *Institute of Genetics and Biochemistry, E. M. Arendt University, Greifswald, Switzerland*

KA MING PANG • *Department of Molecular and Cell Biology, The University of Connecticut, Storrs, CT*

T. PARISH • *Bacterial Molecular Genetics Unit, Department of Clinical Sciences, London School of Hygiene and Tropical Medicine, London, UK*

MARY K. PHILLIPS-JONES • *Krebs Institute for Biomolecular Research, Department of Molecular Biology and Biotechnology, University of Sheffield, UK*

GLORIA RUDENKO • *London School of Hygiene and Tropical Medicine, London, UK*

D. SCOTT SAMUELS • *Bacterial Pathogenesis Section, Rocky Mountain Laboratories Microscopy Branch, National Institute of Allergy and Infectious Diseases, Hamilton, MT. Present Address: Division of Biological Sciences, University of Montana, Missoula, MT*

KENNETH E. SANDERSON • *Salmonella Genetic Stock Centre, Department of Biological Sciences, University of Calgary, Canada*

GERHARDT G. SCHURIG • *Center for Molecular Medicine and Infectious Diseases, Virginia–Maryland Regional College of Veterinary Medicine, Virginia Polytechnic Institute and State University, Blacksburg, VA*

ELLYN D. SEGAL • *Department of Microbiology and Immunology, Digestive Disease Center, Stanford University, Stanford, CA*

TIINA SEPP • *Department of Pharmaceutical Chemistry, University of California, San Francisco, CA*

BRETT D. SHEPARD • *Department of Microbiology and Immunology, University of Oklahoma Health Sciences Center, Oklahoma City, OK*

C. JEFFREY SMITH • *Department of Microbiology and Immunology, East Carolina University, Greenville, NC*

PAMELA A. SOKOL • *Department of Microbiology and Infectious Diseases, University of Calgary Health Sciences Centre, Calgary, Canada*

NAMMALWAR SRIRANGANATHAN • *Center for Molecular Medicine and Infectious Diseases, Virginia–Maryland Regional College of Veterinary Medicine, Virginia Polytechnic Institute and State University, Blacksburg, VA*

C. S. STACHOW • *Department of Biology, Boston College, Chestnut Hill, MA*

N. G. STOKER • *Bacterial Molecular Genetics Unit, Department of Clinical Sciences, London School of Hygiene and Tropical Medicine, London, UK*

LISA STOWERS • *BIO 101, Inc., La Jolla, CA*

Martin C. Taylor • *London School of Hygiene and Tropical Medicine, London, UK*

C. C. Wang • *Department of Pharmaceutical Chemistry, University of California, San Francisco, CA*

A. L. Wang • *Department of Pharmaceutical Chemistry, University of California, San Francisco, CA*

James C. Weaver • *Harvard-MIT Division of Health Sciences and Technology, Massachusetts Institute of Technology, Cambridge, MA*

Herbert Weber • *Jena Inc., Jena, Germany*

Jennifer Whelden • *Department of Cancer Biology, Harvard University School of Public Health, Boston, MA*

Teresa J. White • *Department of Plant Pathology and Microbiology, Texas A & M University, College Station, TX*

Helen L. Withers • *Department of Genetics, Cambridge University, Cambridge, UK. Present Address: Department of Microbiology, Biomedical Center, Uppsala College, Uppsala, Sweden*

Chapter 1

Electroporation Theory

Concepts and Mechanisms

James C. Weaver

1. Introduction

Application of strong electric field pulses to cells and tissue is known to cause some type of structural rearrangement of the cell membrane. Significant progress has been made by adopting the hypothesis that some of these rearrangements consist of temporary aqueous pathways ("pores"), with the electric field playing the dual role of causing pore formation and providing a local driving force for ionic and molecular transport through the pores. Introduction of DNA into cells in vitro is now the most common application. With imagination, however, many other uses seem likely. For example, in vitro electroporation has been used to introduce into cells enzymes, antibodies, and other biochemical reagents for intracellular assays; to load larger cells preferentially with molecules in the presence of many smaller cells; to introduce particles into cells, including viruses; to kill cells purposefully under otherwise mild conditions; and to insert membrane macromolecules into the cell membrane itself. Only recently has the exploration of in vivo electroporation for use with intact tissue begun. Several possible applications have been identified, viz. combined electroporation and anticancer drugs for improved solid tumor chemotherapy, localized gene therapy, transdermal drug delivery, and noninvasive extraction of analytes for biochemical assays.

The present view is that electroporation is a universal bilayer membrane phenomenon (*1–7*). Short (μs to ms) electric field pulses that cause

From: *Methods in Molecular Biology, Vol. 47: Electroporation Protocols for Microorganisms*
Edited by: J. A. Nickoloff Humana Press Inc., Totowa, NJ

the transmembrane voltage, U(t), to rise to about 0.5–1.0 V cause electroporation. For isolated cells, the necessary single electric field pulse amplitude is in the range of 10^3–10^4 V/cm, with the value depending on cell size. Reversible electrical breakdown (REB) then occurs and is accompanied by greatly enhanced transport of molecules across the membrane. REB also results in a rapid membrane discharge, with U(t) returning to small values after the pulse ends. Membrane recovery is often orders of magnitude slower. Cell stress probably occurs because of relatively nonspecific chemical exchange with the extracellular environment. Whether or not the cell survives probably depends on the cell type, the extracellular medium composition, and the ratio of intra- to extracellular volume. Progress toward a mechanistic understanding has been based mainly on theoretical models involving transient aqueous pores. An electric field pulse in the extracellular medium causes the transmembrane voltage, U(t), to rise rapidly. The resulting increase in electric field energy within the membrane and ever-present thermal fluctuations combine to create and expand a heterogeneous population of pores. Scientific understanding of electroporation at the molecular level is based on the hypothesis that pores are microscopic membrane perforations, which allow hindered transport of ions and molecules across the membrane.

These pores are presently believed to be responsible for the following reasons:

1. Dramatic electrical behavior, particularly REB, during which the membrane rapidly discharges by conducting small ions (mainly Na^+ and Cl^-) through the transient pores. In this way, the membrane protects itself from destructive processes;
2. Mechanical behavior, such as rupture, a destructive phenomenon in which pulses too small or too short cause REB and lead to one or more supracritical pores, and these expand so as to remove a portion of the cell membrane; and
3. Molecular transport behavior, especially the uptake of polar molecules into the cell interior.

Both the transient pore population, and possibly a small number of metastable pores, may contribute. In the case of cells, relatively nonspecific molecular exchange between the intra- and extracellular volumes probably occurs, and can lead to chemical imbalances. Depending on the ratio of intra- and extracellular volume, the composition of the extracellular medium, and the cell type, the cell may not recover from the associated stress and will therefore die.

2. Basis of the Cell Bilayer
Membrane Barrier Function

It is widely appreciated that cells have membranes in order to separate the intra- and extracellular compartments, but what does this really mean? Some molecules utilized by cells have specific transmembrane transport mechanisms, but these are not of interest here. Instead, we consider the relatively nonspecific transport governed by diffusive permeation. In this case, the permeability of the membrane to a molecule of type "s" is $P_{m,s}$, which is governed by the relative solubility (partition coefficient), $g_{m,s}$, and the diffusion constant, $D_{m,s}$, within the membrane. In the simple case of steady-state transport, the rate of diffusive, nonspecific molecular transport, N_s, is:

$$N_s = A_m P_{m,s} \Delta C_s = A_m [g_{m,s} D_{m,s}/d] \Delta C_s \tag{1}$$

where N_s, is the number of molecules of type "s" per unit time transported, ΔC_s is the concentration difference across the membrane, $d \approx 6$ nm is the bilayer membrane thickness, and A_m is the area of the bilayer portion of the cell membrane. As discussed below, for charged species, the small value of $g_{m,s}$ is the main source of the large barrier imposed by a bilayer membrane.

Once a molecule dissolves in the membrane, its diffusive transport is proportional to Δc_s and $D_{m,s}$. The dependence on $D_{m,s}$ gives a significant, but not tremendously rapid, decrease in molecular transport as size is increased. The key parameter is $g_{m,s}$, which governs entry of the molecule into the membrane. For electrically neutral molecules, $g_{m,s}$ decreases with molecular size, but not dramatically. In the case of charged molecules, however, entry is drastically reduced as charge is increased. The essential features of a greatly reduced $g_{m,s}$ can be understood in terms of electrostatic energy considerations.

The essence of the cell membrane is a thin (≈ 6 nm) region of low dielectric constant ($K_m \approx 2$–3) lipid, within which many important proteins reside. Fundamental physical considerations show that a thin sheet of low dielectric constant material should exclude ions and charged molecules. This exclusion is owing to a "Born energy" barrier, i.e., a significant cost in energy that accompanies movement of charge from a high dielectric medium, such as water (dielectric constant $K_w \approx 80$), into a low dielectric medium, such as the lipid interior of a bilayer membrane (dielectric constant $K_m \approx 2$) *(8)*.

The Born energy associated with a particular system of dielectrics and charges, W_{Born}, is the electrostatic energy needed to assemble that system of dielectric materials and electric charge. W_{Born} can be computed by specifying the distribution of electrical potential and the distribution of charge, or it can be computed by specifying the electric field, E, and the permittivity $\varepsilon = K\varepsilon_0$ (K is the dielectric constant and $\varepsilon_0 = 8.85 \times 10^{-12}$ F/m) *(9)*. Using the second approach:

$$W_{Born} \equiv \int_{\substack{\text{all space} \\ \text{except ion}}} 1/2\ \varepsilon E^2 dV \tag{2}$$

The energy cost for insertion of a small ion into a membrane can now be understood by estimating the maximum change in Born energy, $\Delta W_{Born,max}$, as the ion is moved from water into the lipid interior of the membrane. It turns out that W_{Born} rises rapidly as the ion enters the membrane, and that much of the change occurs once the ion is slightly inside the low dielectric region. This means that it is reasonable to make an estimate based on treating the ion as a charged sphere of radius r_s and charge $q = ze$ with $z = \pm 1$ where $e = 1.6 \times 10^{-19}$ C. The sphere is envisioned as surrounded by water when it is located far from the membrane, and this gives ($W_{Born,i}$). When it is then moved to the center of the membrane, there is a new electrostatic energy, ($W_{Born,f}$). The difference in these two energies gives the barrier height, $\Delta W_{Born} \equiv W_{Born,f} - W_{Born,i}$. Even for small ions, such as Na^+ and Cl^-, this barrier is substantial (Fig. 1). More detailed, numerical computations confirm that ΔW_{Born} depends on both the membrane thickness, d, and ion radius, r_s.

Here we present a simple estimate of ΔW_{Born}. It is based on the recognition that if the ion diameter is small, $2r_s \approx 0.4$ nm, compared to the membrane thickness, $d \approx 3$–6 nm, then ΔW_{Born} can be estimated by neglecting the finite size of the membrane. This is reasonable, because the largest electric field occurs near the ion, and this in turn means that the details of the membrane can be replaced with bulk lipid. The resulting estimate is:

$$\Delta W_{Born} \approx e^2/8\pi\varepsilon_0 r_s[1/K_m - 1/K_w] \approx 65\ kT \tag{3}$$

where T = 37°C = 310 K. A complex numerical computation for a thin low dielectric constant sheet immersed in water confirms this simple estimate (Fig. 1). This barrier is so large that spontaneous ion transport

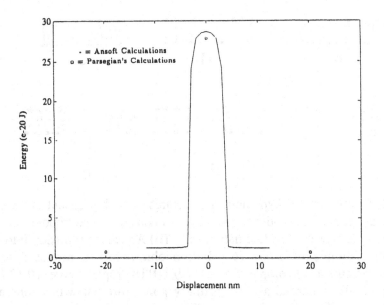

Fig. 1. Numerical calculation of the Born energy barrier for transport of a charged sphere across a membrane (thickness $d = 4$ nm). The numerical solution was obtained by using commercially available software (Ansoft, Inc., Pittsburgh, PA) to solve Poisson's equation for a continuum model consisting of a circular patch of a flow dielectric constant material ($K_m = 2$) immersed in water ($K_w = 80$). The ion was represented by a charged sphere of radius ($r_s = 0.2$ nm), and positioned at a number of different displacements on the axis of rotation of the disk. No pore was present. The electric field and the corresponding electrostatic energy were computed for each case to obtain the values plotted here as a solid line ("- Ansoft Calculations"). The single value denoted by o ("Parsegian's Calculations;" 8) is just under the Ansoft peak. As suggested by the simple estimate of Eq. (2), the barrier is large, viz. $\Delta W \approx 2.8 \times 10^{-19}$ J ≈ 65 kT. As is well appreciated, this effectively rules out significant spontaneous ion transport. The appearance of aqueous pathways ("pores"; Fig. 2) provides a large reduction in this barrier. Reproduced with permission (47).

resulting from thermal fluctuations is negligible. For example, a large transmembrane voltage, U_{direct}, would be needed to force an ion directly across the membrane. The estimated value is $U_{\text{direct}} \approx 65kT/e = 1.7$ V for $z = \pm1$. However, 1.7 V is considerably larger than the usual "resting values" of the transmembrane voltage (about 0.1 ± 0.05 V). The scientific literature on electroporation is consistent with the idea that some sort of membrane structural rearrangement occurs at a smaller voltage.

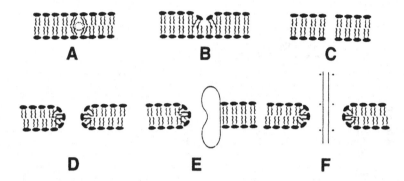

Fig. 2. Illustrations of hypothetical structures of both transient and meta-stable membrane conformations that may be involved in electroporation *(4)*. **(A)** Membrane-free volume fluctuation *(62)*, **(B)** Aqueous protrusion into the membrane ("dimple") *(12,63)*, **(C)** Hydrophobic pore first proposed as an immediate precursor to hydrophilic pores *(10)*, **(D)** Hydrophilic pore *(10,17,18)*; that is generally regarded as the "primary pore" through which ions and molecules pass, **(E)** Composite pore with one or more proteins at the pore's inner edge *(20)*, and **(F)** Composite pore with "foot-in-the-door" charged macromolecule inserted into a hydrophilic pore *(31)*. Although the actual transitions are not known, the transient aqueous pore model assumes that transitions from A → B → C or D occur with increasing frequency as U is increased. Type E may form by entry of a tethered macromolecule during the time that U is significantly elevated, and then persist after U has decayed to a small value because of pore conduction. These hypothetical structures have not been directly observed. Instead, evidence for them comes from interpretation of a variety of experiments involving electrical, optical, mechanical, and molecular transport behavior. Reproduced with permission *(4)*.

3. Aqueous Pathways ("Pores") Reduce the Membrane Barrier

A significant reduction in ΔW_{Born} occurs if the ion (1) is placed into a (mobile) aqueous cavity or (2) can pass through an aqueous channel *(8)*. Both types of structural changes have transport function based on a local aqueous environment, and can therefore be regarded as aqueous pathways. Both allow charged species to cross the membrane much more readily. Although both aqueous configurations lower ΔW_{Born}, the greater reduction is achieved by the pore *(8)*, and is the basis of the "transient aqueous pore" theory of electroporation.

Why should the hypothesis of pore formation be taken seriously? As shown in Fig. 2, it is imagined that some types of prepore structural changes can occur in a microscopic, fluctuating system, such as the bilayer membrane. Although the particular structures presented there are plausible, there is no direct evidence for them. In fact, it is unlikely that transient pores can be visualized by any present form of microscopy, because of the small size, short lifetime, and lack of a contrast-forming interaction. Instead, information regarding pores will probably be entirely indirect, mainly through their involvement in ionic and molecular transport *(4)*. Without pores, a still larger voltage would be needed to move multivalent ions directly across the membrane. For example, if $z = \pm 2$, then $U_{direct} \approx 7$ V, which for a cell membrane is huge.

Qualitatively, formation of aqueous pores is a plausible mechanism for transporting charged molecules across the bilayer membrane portion of cell membranes. The question of how pores form in a highly interactive way with the instantaneous transmembrane voltage has been one of the basic challenges in understanding electroporation.

4. Large U(t) Simultaneously Causes Increased Permeability and a Local Driving Force

Electroporation is more than an increase in membrane permeability to water-soluble species owing to the presence of pores. The temporary existence of a relatively large electric field within the pores also provides an important, local driving force for ionic and molecular transport. This is emphasized below, where it is argued that massive ionic conduction through the transient aqueous pores leads to a highly interactive membrane response. Such an approach provides an explanation of how a planar membrane can rupture at small voltages, but exhibits a protective REB at large voltages. At first this seems paradoxical, but the transient aqueous pore theory predicts that the membrane is actually protected by the rapid achievement of a large conductance. The large conductance limits the transmembrane voltage, rapidly discharges the membrane after a pulse, and thereby saves the membrane from irreversible breakdown (rupture). The local driving force is also essential to the prediction of an approximate plateau in the transport of charged molecules.

5. Membrane-Level and Cell-Level Phenomena

For applications, electroporation should be considered at two levels: (1) the membrane level, which allows consideration of both artificial and cell membranes, and (2) the cellular level, which leads to consideration of secondary processes that affect the cell. The distinction of these two levels is particularly important to the present concepts of reversible and irreversible electroporation. A key concept at the membrane level is that molecular transport occurs through a dynamic pore population. A related hypothesis is that electroporation itself can be reversible at the membrane level, but that large molecular transport can lead to significant chemical stress of a cell, and it is this secondary, cell-level event that leads to irreversible cell electroporation. This will be brought out in part of the presentation that follows.

6. Reversible and Irreversible Electroporation at the Membrane Level

Put simply, reversible electroporation involves creation of a dynamic pore population that eventually collapses, returning the membrane to its initial state of a very few pores. As will be discussed, reversible electroporation generally involves REB, which is actually a temporary high conductance state. Both artificial planar bilayer membranes and cell membranes are presently believed capable of experiencing reversible electroporation. In contrast, the question of how irreversible electroporation occurs is reasonably well understood for artificial planar bilayer membranes, but significantly more complicated for cells.

7. Electroporation in Artificial Planar and in Cell Membranes

Artificial planar bilayer membrane studies led to the first proposals of a theoretical mechanism for electroporation *(10–16)*. However, not all aspects of planar membrane electroporation are directly relevant to cell membrane electroporation. Specifically, quantitative understanding of the stochastic rupture ("irreversible breakdown") in planar membranes was the first major accomplishment of the pore hypothesis. Although cell membranes can also be damaged by electroporation, there are two possible mechanisms. The first possibility is lysis resulting from a secondary result of reversible electroporation of the cell membrane. According to this hypothesis, even though the membrane recovers (the dynamic pore population returns to the initial state), there can be so much molecular transport that the cell is chemically or osmotically stressed, and this secondary event leads to cell destruction through lysis. The second possibility is that rupture of an isolated portion of a cell membrane occurs, because one or more bounded portions of the membrane behave like small planar membranes. If this is the case, the mechanistic understanding of planar membrane rupture is relevant to cells.

8. Energy Cost to Create a Pore
at Zero Transmembrane Voltage ($U = 0$)

The first published descriptions of pore formation in bilayer membranes were based on the idea that spontaneous (thermal fluctuation driven) structural changes in the membrane could create pores. A basic premise was that the large pores could destroy a membrane by rupture, which was suggested to occur as a purely mechanical event, i.e., without electrical assistance *(17,18)*. The energy needed to make a pore was considered to involve two contributions. The first is the "edge energy," which relates to the creation of a stressed pore edge, of length $2\pi r$, so that if the "edge energy" (energy cost per length) was γ, then the cost to make the pore's edge was $2\pi r\gamma$. The second is the "area energy" change associated with removal of a circular patch of membrane, $-\pi r^2\Gamma$. Here Γ is the energy per area (both sides of the membrane) of a flat membrane.

Put simply, this process is a "cookie cutter" model for a pore creation. The free energy change, $\Delta W_p(r)$, is based on a gain in edge energy and a simultaneous reduction in area energy. The interpretation is simple: a pore-free membrane is envisioned, then a circular region is cut out of the membrane, and the difference in energy between these two states calculated, and identified as ΔW_p. The corresponding equation for the pore energy is:

$$\Delta W_p(r) = 2\pi\gamma r - \pi\Gamma r^2 \text{ at } U = 0 \qquad (4)$$

A basic consequence of this model is that $\Delta W_p(r)$ describes a parabolic barrier for pores. In its simplest form, one can imagine that pores might be first made, but then expanded at the cost of additional energy. If the barrier peak is reached, however, then pores moving over the barrier can expand indefinitely, leading to membrane rupture. In the initial models (which did not include the effect of the transmembrane voltage), spontaneous thermal fluctuations were hypothesized to create pores, but the probability of surmounting the parabolic barrier was thought to be small. For this reason, it was concluded that spontaneous rupture of a red blood cell membrane by spontaneous pore formation and expansion was concluded to be negligible *(17)*. At essentially the same time, it was independently suggested that pores might provide sites in the membrane where spontaneous translocation of membrane lipid molecules ("flip flop") should preferentially occur *(18)*.

9. Energy Cost to Create a Pore at U > 0

In order to represent the electrical interaction, a pore is regarded as having an energy associated with the change of its specific capacitance, C_p. This was first presented in a series of seven back-to-back papers *(10–16)*. Early on, it was recognized that it was unfavorable for ions to enter small pores because of the Born energy change discussed previously. For this reason, a relatively small number of ions will be available within small pores to contribute to the electrical conductance of the pore. With this justification, a pore is represented by a water-filled, rather than electrolyte-filled, capacitor. However, for small hydrophilic pores, even if bulk electrolyte exists within the pores, the permittivity would be $\varepsilon \approx 70\varepsilon_0$, only about 10% different from that of pure water.

In this case, the pore resistance is still large, $R_p = \rho_e h/\pi r^2$, and is also large in comparison to the spreading resistance discussed below. If so, the voltage across the pore is approximately U. With this in mind, in the presence of a transmembrane electric field, the free energy of pore formation should be *(10)*:

$$\Delta W_p(r,\text{U}) = 2\pi\gamma r - \pi\Gamma r^2 - 0.5C_p\text{U}^2 r^2 \qquad (5)$$

Here U is the transmembrane voltage spatially averaged over the membrane. A basic feature is already apparent in the above equation: as *U* increases, the pore energy, ΔW_p, decreases, and it becomes much more favorable to create pores. In later versions of the transient aqueous pore model, the smaller, local transmembrane voltage, U_p, for a conducting pore is used. As water replaces lipid to make a pore, the capacitance of the membrane increases slightly.

10. Heterogeneous Distribution of Pore Sizes

A spread in pore sizes is fundamentally expected *(19–22)*. The origin of this size heterogeneity is the participation of thermal fluctuations along with electric field energy within the membrane in making pores. The basic idea is that these fluctuations spread out the pore population as pores expand against the barrier described by $\Delta W_p(r, U)$. Two extreme cases illustrate this point: (1) occasional escape of large pores over the barrier described by $\Delta W_p(r, U)$ leads to rupture, and (2) the rapid creation of many small pores ($r \approx r_{\min}$) causes the large conductance that is responsible for REB. In this sense, rupture is a large-pore phenomenon, and REB is a small-pore phenomenon. The moderate value of U(t) associated with rupture leads to only a modest conductance, so that there is ample time for the pore population to evolve such that one or a small number of large pores appear and diffusively pass over the barrier, which is still fairly large. The pore population associated with REB is quite different; at larger voltages, a great many more small pores appear, and these discharge the membrane before the pore population evolves any large "critical" pores that lead to rupture.

11. Quantitative Explanation of Rupture

As the transmembrane voltage increases, the barrier $\Delta W_p(r, U)$ changes its height, ΔW_{max} and the location of its peak. The latter is associated with a critical pore radius, r_c, such that pores with $r > r_c$ tend to expand without limit. A property of $\Delta W_p(r, U)$ is that both ΔW_{max} and r_c decrease as U increases. This provides a readily visualized explanation of planar membrane rupture: as U increases, the barrier height decreases, and this increases the probability of the membrane acquiring one or more pores with $r > r(U)_c$. The appearance of even one supracritical pore is, however, sufficient to rupture the membrane. Any pore with $r > r_c$ tends to expand until it reaches the macroscopic aperture that defines the planar membrane. When this occurs, the membrane material has all collected at the aperture, and it makes no sense to talk about a membrane being present. In this case, the membrane is destroyed.

The critical pore radius, r_c, associated with the barrier maximum, $\Delta W_{p,max} = \Delta W_p(r_c, U)$, is *(10)*:

$$r_c = (\gamma/\Gamma + 0.5C_pU^2) \text{ and } \Delta W_{p,max} = \pi\gamma^2/\Gamma + 0.5C_pU^2) \qquad (6)$$

The associated pore energy, $\Delta W_{p,max}$ also decreases. Overcoming energy barriers generally depends nonlinearly on parameters, such as *U*, because Boltzmann factors are involved. For this reason, a nonlinear dependence on *U* was expected.

The electrical conductance of the membrane increases tremendously because of the appearance of pores, but the pores, particularly the many small ones, are not very good conductors. The reason for this relatively poor conduction of ions by small pores is again the Born energy change; conduction within a pore can be suppressed over bulk electrolyte conduction because of Born energy exclusion owing to the nearby low dielectric constant lipid. The motion of ions through a pore only somewhat larger than the ion itself can be sterically hindered. This has been accounted for by using the Renkin equation to describe the essential features of hindrance *(23)*. This function provides for reduced transport of a spherical ion or molecule of radius r_s through cylindrical pathway of radius r (representing a pore) *(20,21,24)*.

12. Planar Membrane Destruction by Emergence of Even One "Critical Pore"

As a striking example of the significance of heterogeneity within the pore population, it has been shown that one or a small number of large pores can destroy the membrane by causing rupture *(11)*. The original approach treated the diffusive escape of pores over an energy barrier. Later, an alternative, simpler approach for theoretically estimating the average membrane lifetime against rupture, $\bar{\tau}$, was proposed *(25)*. This approach used an absolute rate estimate for critical pore appearance in which a Boltzmann factor containing $\Delta W_p/kT$ and an order of magnitude estimate for the prefactor was used. The resulting estimate for the rate of critical pore appearance is:

$$\bar{\tau} \approx (1/\nu_0 V_m) \exp(+\Delta W_{p,c}/kT) \tag{7}$$

This estimate used an attempt rate density, ν_0, which is based on a collision frequency density within the fluid bilayer membrane. The order of magnitude of ν_0 was obtained by estimating the volume density of collisions per time in the fluid membrane. The factor $V_m = hA_m$ is the total volume of the membrane. By choosing a plausible value (e.g., 1 s), the value of $\Delta W_{p,c}$, and hence of U_c, can be found. This is interpreted as the critical voltage for rupture. Because of the strong nonlinear behavior of Eq. (7), using values, such as 0.1 or 10 s, results in only small differences in the predicted $U_c \approx 0.3$–0.5 V.

13. Behavior of the Transmembrane Voltage During Rupture

Using this approach, reasonable (but not perfect) agreement for the behavior of $U(t)$ was found. Both the experimental and theoretical behaviors of U exhibit a sigmoidal decay during rupture, but the duration of the decay phase is longer for the experimental values. Both are much longer than the rapid discharge found for REB. Many experiments have shown that both artificial planar bilayer membranes and cell membranes exhibit REB, and its occurrence coincides with tremendously enhanced molecular transport across cell membranes. However, the term "breakdown" is misleading, because REB is now believed to be a protective behavior, in which the membrane acquires a very large conductance in the form of pores. In planar membranes challenged by short pulses (the "charge injection" method mentioned above), a characteristic of REB is the progressively faster membrane discharge as larger and larger pulses are used *(26)*.

14. Reversible Electroporation

Unlike reversible electroporation (rupture) of planar membranes, in which the role of one or a small number of critical pores is dominant, reversible electroporation is believed to involve the rapid creation of so many small pores that membrane discharge occurs before any critical pores can evolve from the small pores. The transition in a planar membrane from rupture to REB can be qualitatively understood in terms of a competition between the kinetics of pore creation and of pore expansion. If only a few pores are present owing to a modest voltage pulse, the membrane discharges very slowly (e.g., ms) and there is time for evolution of critical pores. If a very large number of pores are present because of a large pulse, then the high conductance of these pores discharges the membrane rapidly, before rupture can occur. One basic challenge in a mechanistic understanding is to find a quantitative description of the transition from rupture to REB, i.e., to show that a planar membrane can experience rupture for modest pulses, but makes a transition to REB as the pulse amplitude is increased *(19–22)*. This requires a physical model for both pore creation and destruction, and also the behavior of a dynamic, heterogeneous pore population.

15. Conducting Pores Slow Their Growth

An important aspect of the interaction of conducting pores with the changing transmembrane voltage is that pores experience a progressively smaller expanding force as they expand *(21,27)*. This occurs because there are inhomogeneous electric fields (and an associated "spreading resistance") just outside a pore's entrance and exit, such that as the pore grows, a progressively greater fraction of U appears across this spreading resistance. This means that less voltage appears across the pore itself, and therefore, the electrical expanding pressure is less. For this reason, pores tend to slow their growth as they expand. The resistance of the internal portion of the pore is also important, and as already mentioned, has a reduced internal resistance because $\sigma_p < \sigma_e$ because of Born energy "repulsion." The voltage divider effect means simply that the voltage across the pore is reduced to:

$$U_p = U \, [R_p/(R_p + R_s)] \leq U \tag{8}$$

Here R_p is the electrical resistance associated with the pore interior, and R_s is the resistance associated with the external inhomogeneous electric field near the entrance and exit to the pore. The fact that U_p becomes less than U means that the electrical expanding force owing to the gradient of ΔW_p in pore radius space is reduced. In turn, this means that pores grow more slowly as they become larger, a basic pore response that contributes to reversibility (21,27).

16. Reversible Electroporation and "Reversible Electrical Breakdown"

For planar membranes, the transition from irreversible behavior ("rupture") to reversible behavior ("REB" or incomplete reversible electrical breakdown) can be explained by the evolution of a dynamic, heterogeneous pore population (20–22,24). One prediction of the transient aqueous pore model is that a planar membrane should also exhibit incomplete reversible electrical breakdown, i.e., a rapid discharge that does not bring U down to zero. Indeed, this is predicted to occur for somewhat smaller pulses than those that produce REB. Qualitatively, the following is believed to occur. During the initial rapid discharge, pores rapidly shrink and some disappear. As a result, the membrane conductance, G(t), rapidly reaches such a small value that further discharge occurs very slowly. On the time scale (μs) of the experiment, discharge appears to stop, and the membrane has a small transmembrane voltage, e.g., $U \approx 50$ mV.

Although irreversible electroporation of planar membranes now seems to be reasonably accounted for by a transient aqueous pore theory, the case of irreversibility in cells is more complicated and still not fully understood. The rupture of planar membranes is explained by recognizing that expansion of one or more supracritical pores can destroy the membrane. When it is created, the planar membrane covers a macroscopic aperture, but also connects to a meniscus at the edge of the aperture. This meniscus also contains phospholipids, and can be thought of as a reservoir that can exchange phospholipid molecules with the thinner bilayer membrane. As a result of this connection to the meniscus, the bilayer membrane has a total surface tension (both sides of the membrane), Γ, which favors expansion of pores. Thus, during rupture, the membrane material is carried by pore expansion into the meniscus, and the membrane itself vanishes.

However, there is no corresponding reservoir of membrane molecules in the case of the closed membrane of a vesicle or cell. For this reason, if the osmotic pressure difference across the cell membrane is zero, the cell membrane effectively has $\Gamma = 0$. For this reason, a simple vesicle cannot rupture *(28)*. Although a cell membrane has the same topology as a vesicle, the cell membrane is much more complicated, and usually contains other, membrane-connecting structures. With this in mind, suppose that a portion of a cell membrane is bounded by the cytoskeleton or some other cellular structure, such that membrane molecules can accumulate there if pores are created (Fig. 2). If so, these bounded portions of the cell membrane may be able to rupture, since a portion of the cell membrane would behave like a microscopic planar bilayer membrane. This localized but limited rupture would create an essentially permanent hole in the cell membrane, and would lead to cell death. Another possibility is that reversible electroporation occurs, with REB and a large, relatively nonspecific molecular transport (*see* Section 21.) across the cell membrane.

17. Tremendous Increase in Membrane Conductance, G(t) During REB

Creation of aqueous pathways across the membrane is, of course, the phenomenon of interest. This is represented by the total membrane conductance, $G(t) = 1/R(t)$. As pores appear during reversible electroporation, R changes by orders of magnitude. A series of electrical experiments using a planar bilayer membrane provided conditions and results that motivated the choice of particular parameters, including the use of a very short (0.4 μs) square pulse *(26)*. In these experiments, a current pulse of amplitude I_i passes through R_N, thereby creating a voltage pulse, V_0 (Fig. 2). For $0 < t < t_{\text{pulse}}$ current flows into and/or across the membrane, and at $t = t_{\text{pulse}}$, the pulse is terminated by opening the switch. Because the generator is then electronically disconnected, membrane discharge can occur only through the membrane for a planar membrane (not true for a cell). Predictions of electroporation behavior were obtained by generating self-consistent numerical solutions to these equations.

18. Evidence for Metastable Pores

Pores do not necessarily disappear when U returns to small values. For example, electrical experiments with artificial planar bilayer membranes have shown that small pores remain after U is decreased. Other experiments with cells have examined the response of cells to dyes supplied after electrical pulsing, and find that a subpopulation of cells takes up these molecules *(29,30)*. Although not yet understood quantitatively in terms of an underlying mechanism, it is qualitatively plausible that some type of complex, metastable pores can form. Such pores may involve other components of a cell, e.g., the cytoskeleton or tethered cytoplasmic molecules (Fig. 2), that lead to metastable pores. For example, entry of a portion of a tethered, charged molecule should lead to a "foot-in-the-door" mechanism in which the pore cannot close *(31)*. However, pore destruction is not well understood. Initial theories assumed that pore disappearance occurs independently of other pores. This is plausible, since pores are widely spaced even when the total (aqueous) area is maximum *(22)*. Although this approximate treatment has contributed to reasonable theoretical descriptions of some experimental behavior, a complete, detailed treatment of pore disappearance remains an unsolved problem.

19. Interaction of the Membrane with the External Environment

It is not sufficient to describe only the membrane. Instead, an attempt to describe an experiment should include that part of the experimental apparatus that directly interacts with the membrane. Specifically, the electrical properties of the bathing electrolyte, electrodes, and output characteristics of the pulse generator should be included. Otherwise, there is no possibility for including the limiting effects of this part of the experiment. Clearly there is a pathway by which current flows in order to cause interfacial polarization, and thereby increase $U(t)$.

An initial attempt to include membrane–environment interactions used a simple circuit model to represent the most important aspects of the membrane and the external environment, which shows the relationship among the pulse generator, the charging pathway resistance, and the membrane *(19,21)*. The membrane is represented as the membrane capacitance, C, connected in parallel with the membrane resistance, $R(t)$. As pores begin to appear in the membrane, the membrane conductance $G(t) = 1/R(t)$ starts to increase, and therefore $R(t)$ drops. The membrane does not experience the applied pulse immediately, however, since the membrane capacitance has to charge through the external resistance of the electrolyte, which baths the membrane, the electrode resistance, and the output resistance of the pulse generator. This limitation is represented by a single resistor, R_E. This explicit, but approximate, treatment of the membrane's environment provides a reasonable approach to achieving theoretical descriptions of measurable quantities that can be compared to experimental results.

20. Fractional Aqueous Area of the Membrane During Electroporation

The membrane capacitance is treated as being constant, which is consistent with experimental data *(32)*. It is also consistent with the theoretical model, as shown by computer simulations that use the model to predict correctly basic features of the transmembrane voltage, $U(t)$. The simulation allows the slight change in C to be predicted simultaneously, and finds that only a small fraction ($F_{w,max} \approx 5 \times 10^{-4}$) of the membrane becomes aqueous through the appearance of pores. The additional capacitance owing to this small amount of water leads to a slight (on the order of 1%) change in the capacitance *(22)*, which is consistent with experimental results *(32)*.

The fractional aqueous area, $F_w(t)$, changes rapidly with time as pores appear, but is predicted to be less than about 0.1% of the membrane, even though tremendous increases in ionic conduction and molecular transport take place. This is in reasonable agreement with experimental findings. According to present understanding, the minimum pore size is $r_{min} \approx 1$ nm, which means that the small ions that comprise physiologic saline can be conducted. For larger or more charged species, however, the available fractional aqueous area, $F_{w,s}$, is expected to decrease. This is a consequence of a heterogeneous pore population. With increasing molecular size and/or charge, fewer and fewer pores should participate, and this means that $F_{w,s}$ should decrease as the size and charge of "s" increase.

21. Molecular Transport Owing to Reversible Electroporation

Tremendously increased molecular transport *(33,34)* is probably the most important result of electroporation for biological research (Table 1). Although clearly only partially understood, much of the evidence to date supports the view that electrophoretic transport through pores is the major mechanism for transport of charged molecules *(20,24,35,36)*.

Table 1
Candidate Mechanisms for Molecular Transport Through Pores (20)[a]

Mechanism	Molecular basis
Drift	Velocity in response to a local physical (e.g., electrical) field
Diffusion	Microscopic random walk
Convection	Fluid flow carrying dissolved molecules

[a]The dynamic pore population of electroporation is expected to provide aqueous pathways for molecular transport. Water-soluble molecules should be transported through the pores that are large enough to accommodate them, but with some hindrance. Although not yet well established, electrical drift may be the primary mechanism for charged molecules (20–35).

One surprising observation is the molecular transport caused by a single exponential pulse can exhibit a plateau, i.e., transport becomes independent of field pulse magnitude, even though the net molecular transport results in uptake that is far below the equilibrium value $N_s = V_{cell}c_{ext}$ (37–40). Here N_s is the number of molecules taken up by a single cell, V_{cell} is the cell volume, and c_{ext} is the extracellular concentration in a large volume of pulsing solution.

A plateauing of uptake that is independent of equilibrium uptake ($\overline{n}_s = V_{cell}c_{s,ext}$) may be a fundamental attribute of electroporation. Initial results from a transient aqueous pore model show that the transmembrane voltage achieves an almost constant value for much of the time during an exponential pulse. If the local driving force is therefore almost constant, the transport of small charged molecules through the pores may account for an approximate plateau (24). Transport of larger molecules may require deformation of the pores, but the approximate constancy of $U(t)$ should still occur, since the electrical behavior is dominated by the many smaller pores. These partial successes of a transient aqueous pore theory are encouraging, but a full understanding of electroporative molecular transport is still to be achieved.

22. Terminology and Concepts: Breakdown and Electropermeabilization

Based on the success of the transient aqueous pore models in providing reasonably good quantitative descriptions of several key features of electroporation, the existence of pores should be regarded as an attractive hypothesis (Table 2). With this in mind, two widely used terms, "breakdown" and "electropermeabilization," should be re-examined. First, "breakdown" in the sense of classic dielectric breakdown is mis-

Table 2
Successes of the Transient Aqueous Pore Model[a]

Behavior	Pore theory accomplishment
Stochastic nature of rupture	Explained by diffusive escape of very large pores (10)
Rupture voltage, U_c	Average value reasonably predicted (10,25,64)
Reversible electrical breakdown	Transition from rupture to REB correctly predicted (21)
Fractional aqueous area	$F_{w,ions} \leq 10^{-3}$ predicted; membrane conductance agrees (22)
Small change in capacitance	Predicted to be <2% for reversible electroporation (22)
Plateau in charged molecule transport	Approximate plateau predicted for exponential pulses (24)

[a]Successful predictions of the transient aqueous pore model for electroporation at the present time. These more specific descriptions are not accounted for simply by an increased permeability or an ionizing type of dielectric breakdown. The initial, combined theoretical and experimental studies convincingly showed that irreversible breakdown ("rupture") was not the result of a deterministic mechanism, such as compression of the entire membrane, but could instead be quantitatively accounted for by transient aqueous pores (10). Recent observations of charged molecule uptake by cells that exhibits a plateau, but is far below the equilibrium value cannot readily be accounted for by any simple, long-lasting membrane permeability increase, but is predicted by the transient aqueous pore model.

leading. After all, the maximum energy available to a monovalent ion or molecule for $U \approx 0.5-1$ V is only about one-half to 1 ev. This is too small to ionize most molecules, and therefore cannot lead to conventional avalanche breakdown in which ion pairs are formed (41). Instead, a better term would be "high conductance state," since it is the rapid membrane rearrangement to form conducting aqueous pathways that discharges the membrane under biochemically mild conditions (42). Second, in the case of electropermeabilization, "permeabilization" implies only that a state of increased permeability has been obtained. This phenomenological term is directly relevant only to transport. It does not lead to the concept of a stochastic membrane destruction, the idea of "reversible electrical breakdown" as a protective process in the transition from rupture to REB, or the plateau in molecular transport for small charged molecules. Thus, although electroporation clearly causes an increase in permeability, electroporation is much more, and the abovementioned additional features cannot be explained solely by an increase in permeability.

23. Membrane Recovery

Recovery of the membrane after pulsing is clearly essential to achieving reversible behavior. Presently, however, relatively little is known about the kinetics of membrane recovery after the membrane has been discharged by REB. Some studies have used "delayed addition" of molecules to determine the integrity of cell membranes at different times after pulsing. Such experiments suggest that a subpopulation of cells occurs that has delayed membrane recovery, as these cells are able to take up molecules after the pulse. In addition to "natural recovery" of cell membranes, the introduction of certain surfactants has been found to accelerate membrane recovery, or at least re-establishment of the barrier function of the membrane *(43)*. Accelerated membrane recovery may have implications for medical therapies for electrical shock injury, and may also help us to understand the mechanism by which membranes recover.

24. Cell Stress and Viability

Complete cell viability, not just membrane recovery, is usually important to biological applications of electroporation, but in the case of electroporation, determination of cell death following electroporation is nontrivial. After all, by definition, electroporation alters the permeability of the membrane. This means that membrane-based short-term tests (vital stains, membrane exclusion probes) are therefore not necessarily valid *(29)*. If, however, the cells in question can be cultured, assays based on clonal growth should provide the most stringent test, and this can be carried out relatively rapidly if microcolony (2–8 cells) formation is assessed *(44)*. This was done using microencapsulated cells. The cells are initially incorporated into agarose gel microdrops (GMDs), electrically pulsed to cause electroporation, cultured while in the microscopic (e.g., 40–100 µm diameter) GMDs, and then analyzed by flow cytometry so that the subpopulation of viable cells can be determined *(45,46)*.

Cellular stress caused by electroporation may also lead to cell death without irreversible electroporation itself having occurred. According to our present understanding of electroporation itself, both reversible and irreversible electroporation result in transient openings (pores) of the membrane. These pores are often large enough that molecular transport is expected to be relatively nonspecific. As already noted, for irreversible electroporation, it is plausible that a portion of the cell membrane behaves much like a small planar membrane, and therefore can undergo rupture. In the case of reversible electroporation, significant molecular transport between the intra- and extracellular volumes may lead to a significant chemical imbalance. If this imbalance is too large, recovery may not occur, with cell death being the result. Here it is hypothesized that the volumetric ratio:

$$R_{vol} \equiv (V_{extracellular}/V_{intracellular}) \tag{9}$$

may correlate with cell death or survival *(47)*. According to this hypothesis, for a given cell type and extracellular medium composition, $R_{vol} \gg$ 1 (typical of in vitro conditions, such as cell suspensions and anchorage-dependent cell culture) should favor cell death, whereas the other extreme $R_{vol} \ll 1$ (typical of in vivo tissue conditions) should favor cell survival. If correct, for the same degree of electroporation, significantly less damage may occur in tissue than in body fluids or under most in vitro conditions.

25. Tissue Electroporation

Tissue electroporation is a relatively new extension of single-cell electroporation under in vitro conditions, and is of interest because of possible medical applications, such as cancer tumor therapy *(48–50)*, transdermal drug delivery *(51,52)*, noninvasive transdermal chemical sensing *(4)*, and localized gene therapy *(53,54)*. It is also of interest because of its role in electrical injury *(43,55,56)*. The interest in tissue electroporation is growing rapidly, and may lead to many new medical applications. The basic concept is that application of electric field pulses to tissue generally results in a localized, large electric field developing across the lipid-based barriers within the tissue. This can result in the creation of new aqueous pathways across the barrier, just where they are needed in order to achieve local drug delivery. Relevant barriers are not only the single bilayer membranes of cells, but one or more tissue monolayers in which cells are connected by tight junctions (essentially two bilayers in series per monolayer), and the stratum corneum of the skin, which can be regarded very approximately as about 100 bilayer membranes in series. In such cases, it is envisioned that electroporation is to be used with living human subjects. With this in mind, it is significant that several studies support the view that electroporation conditions can be found that result in negligible damage, both in isolated cells *(57–59)* and in intact tissue in vivo *(60,61)*. Increased use of electroporation for drug delivery implies that a much better mechanistic understanding of electroporation will be needed to secure both scientific and regulatory acceptance.

26. Summary

The basic features of electrical and mechanical behavior of electroporated cell membranes are reasonably well established experimentally. Overall, the electrical and mechanical features of electroporation are consistent with a transient aqueous pore hypothesis, and several features, such as membrane rupture and reversible electrical breakdown, are reasonably well described quantitatively. This gives confidence that "electroporation" is an attractive hypothesis, and that the appearance of temporary pores owing to the simultaneous contributions of thermal fluctuations ("kT energy") and an elevated transmembrane voltage ("electric field energy") is the microscopic basis of electroporation.

Acknowledgments

I thank J. Zahn, T. E. Vaughan, M. A. Wang, R. M. Prausnitz, R. O. Potts, U. Pliquett, J. Lin, R. Langer, L. Hui, E. A. Gift, S. A. Freeman, Y. Chizmadzhev, and V. G. Bose for many stimulating and critical discussions. This work supported by NIH Grant GM34077, Army Research Office Grant No. DAAL03-90-G-0218, NIH Grant ES06010, and a computer equipment grant from Stadwerke Düsseldorf, Düsseldorf, Germany.

References

1. Neumann, E., Sowers, A., and Jordan, C. (eds.) (1989) *Electroporation and Electrofusion in Cell Biology.* Plenum, New York.
2. Tsong, T. Y. (1991) Electroporation of cell membranes. *Biophys. J.* **60,** 297–306.
3. Chang, D. C., Chassy, B. M., Saunders, J. A., and Sowers, A. E. (eds.) (1992) *Guide to Electroporation and Electrofusion.* Academic.
4. Weaver, J. C. (1993) Electroporation: a general phenomenon for manipulating cells and tissue. *J. Cell. Biochem.* **51,** 426–435.
5. Orlowski, S. and Mir, L. M. (1993) Cell electropermeabilization: a new tool for biochemical and pharmacological studies. *Biochim. Biophys. Acta* **1154,** 51–63.
6. Weaver, J. C. (1994) Electroporation in cells and tissues: a biophysical phenomenon due to electromagnetic fields. *Radio Sci.* (in press).
7. Weaver, J. C. and Chizmadzhev, Y. A. Electroporation, in *CRC Handbook of Biological Effects of Electromagnetic Fields,* 2nd ed. (Polk, C. and Postow, E., eds.), CRC, Boca Raton (submitted).
8. Parsegian, V. A. (1969) Energy of an ion crossing a low dielectric membrane: solutions to four relevant electrostatic problems. *Nature* **221,** 844–846.
9. Zahn, M. (1979) *Electromagnetic Field Theory: A Problems Solving Approach,* Wiley, New York.
10. Abidor, I. G., Arakelyan, V. B., Chernomordik, L. V., Chizmadzhev, Yu. A., Pastushenko, V. F., and Tarasevich, M. R. (1979) Electric breakdown of bilayer

membranes: I. The main experimental facts and their qualitative discussion. *Bioelectrochem. Bioenerg.* **6**, 37–52.

11. Pastushenko, V. F., Chizmadzhev, Yu. A., and Arakelyan, V. B. (1979) Electric breakdown of bilayer membranes: II. Calculation of the membrane lifetime in the steady-state diffusion approximation. *Bioelectrochem. Bioenerg.* **6**, 53–62.

12. Chizmadzhev, Yu. A., Arakelyan, V. B., and Pastushenko, V. F. (1979) Electric breakdown of bilayer membranes: III. Analysis of possible mechanisms of defect origin. *Bioelectrochem. Bioenerg.* **6**, 63–70.

13. Pastushenko, V. F., Chizmadzhev, Yu. A., and Arakelyan, V. B. (1979) Electric breakdown of bilayer membranes: IV. Consideration of the kinetic stage in the case of the single-defect membrane. *Bioelectrochem. Bioenerg.* **6**, 71–79.

14. Arakelyan, V. B., Chizmadzhev, Yu. A., and Pastushenko, V. F. (1979) Electric breakdown of bilayer membranes: V. Consideration of the kinetic stage in the case of the membrane containing an arbitrary number of defects. *Bioelectrochem. Bioenerg.* **6**, 81–87.

15. Pastushenko, V. F., Arakelyan, V. B., and Chizmadzhev, Yu. A. (1979) Electric breakdown of bilayer membranes: VI. A stochastic theory taking into account the processes of defect formation and death: membrane lifetime distribution function. *Bioelectrochem. Bioenerg.* **6**, 89–95.

16. Pastushenko, V. F., Arakelyan, V. B., and Chizmadzhev, Yu. A. (1979) Electric breakdown of bilayer membranes: VII. A stochastic theory taking into account the processes of defect formation and death: statistical properties. *Bioelectrochem. Bioenerg.* **6**, 97–104.

17. Litster, J. D. (1975) Stability of lipid bilayers and red blood cell membranes. *Phys. Lett.* **53A**, 193,194.

18. Taupin, C., Dvolaitzky, M., and Sauterey, C. (1975) Osmotic pressure induced pores in phospholipid vesicles. *Biochemistry* **14**, 4771–4775.

19. Powell, K. T., Derrick, E. G., and Weaver, J. C. (1986) A quantitative theory of reversible electrical breakdown. *Bioelectrochem. Bioelectroenerg.* **15**, 243–255.

20. Weaver, J. C. and Barnett, A. (1992) Progress towards a theoretical model of electroporation mechanism: membrane electrical behavior and molecular transport, in *Guide to Electroporation and Electrofusion* (Chang, D. C., Chassy, B. M., Saunders, J. A., and Sowers, A. E., eds.), Academic.

21. Barnett, A. and Weaver, J. C. (1991) Electroporation: a unified, quantitative theory of reversible electrical breakdown and rupture. *Bioelectrochem. Bioenerg.* **25**, 163–182.

22. Freeman, S. A., Wang, M. A., and Weaver, J. C. (1994) Theory of electroporation for a planar bilayer membrane: predictions of the fractional aqueous area, change in capacitance and pore-pore separation. *Biophysical J.* **67**, 42–56.

23. Renkin, E. M. (1954) Filtration, diffusion and molecular sieving through porous cellulose membranes. *J. Gen. Physiol.* **38**, 225–243.

24. Wang, M. A., Freeman, S. A., Bose, V. G., Dyer, S., and Weaver, J. C. (1993) Theoretical modelling of electroporation: electrical behavior and molecular transport, in *Electricity and Magnetism in Biology and Medicine* (Blank, M., ed.), San Francisco, pp. 138–140.

25. Weaver, J. C. and Mintzer, R. A. (1981) Decreased bilayer stability due to transmembrane potentials. *Phys. Lett.* **86A,** 57–59.

26. Benz, R., Beckers, F., and Zimmermann, U. (1979) Reversible electrical breakdown of lipid bilayer membranes: a charge-pulse relaxation study. *J. Membrane Biol.* **48,** 181–204.

27. Pastushenko, V. F. and Chizmadzhev, Yu. A. (1982) Stabilization of conducting pores in BLM by electric current. *Gen. Physiol. Biophys.* **1,** 43–52.

28. Sugar, I. P. and Neumann, E. (1984) Stochastic model for electric field-induced membrane pores: electroporation. *Biophys. Chemistry* **19,** 211–225.

29. Weaver, J. C., Harrison, G. I., Bliss, J. G., Mourant, J. R., and Powell, K. T. (1988) Electroporation: high frequency of occurrence of the transient high permeability state in red blood cells and intact yeast. *FEBS Lett.* **229,** 30–34.

30. Tsoneva, I., Tomov, T., Panova, I., and Strahilov, D. (1990) Effective production by electrofusion of hybridomas secreting monodonal antibodies against Hc-antigen of *Salmonella. Bioelectrochem. Bioenerg.* **24,** 41–49.

31. Weaver, J. C. (1993) Electroporation: a dramatic, nonthermal electric field phenomenon, in *Electricity and Magnetism in Biology and Medicine* (Blank, M., ed.), San Francisco, pp. 95–100.

32. Chernomordik, L. V., Sukharev, S. I., Abidor, I. G., and Chizmadzhev, Yu. A. (1982) The study of the BLM reversible electrical breakdown mechanism in the presence of UO_2^{2+}. *Bioelectrochem. Bioenerg.* **9,** 149–155.

33. Neumann, E. and Rosenheck, K. (1972) Permeability changes induced by electric impulses in vesicular membranes. *J. Membrane Biol.* **10,** 279–290.

34. Kinosita, K. Jr. and Tsong, T. Y. (1978) Survival of sucrose-loaded erythrocytes in circulation. *Nature* **272,** 258–260.

35. Klenchin, V. A., Sukharev, S. I., Serov, S. M., Chernomordik, L. V., and Chizmadzhev, Yu. A. (1991) Electrically induced DNA uptake by cells is a fast process involving DNA electrophoresis. *Biophys. J.* **60,** 804–811.

36. Sukharev, S. I., Klenchin, V. A., Serov, S. M., Chernomordik, L. V., and Chizmadzhev, Y. A. (1992) Electroporation and electrophoretic DNA transfer into cells. *Biophys. J.* **63,** 1320–1327.

37. Prausnitz, M. R., Lau, B. S., Milano, C. D., Conner, S., Langer, R., and Weaver, J. C. (1993) A quantitative study of electroporation showing a plateau in net molecular transport. *Biophys. J.* **65,** 414–422.

38. Prausnitz, M. R., Milano, C. D., Gimm, J. A., Langer, R., and Weaver, J. C. (1994) Quantitative study of molecular transport due to electroporation: uptake of bovine serum albumin by human red blood cell ghosts. *Biophys. J.* **66,** 1522–1530.

39. Gift, E. A. and Weaver, J. C. (1995) Observation of extremely heterogeneous electroporative uptake which changes with electric field pulse amplitude in *Saccharomyces cerevisiae. Biochim. Biophys. Acta* **1234(1),** 52–62.

40. Hui, L., Gift, E. A., and Weaver, J. C. Uptake of Bovine Serum Albumin by Yeast due to Electroporation: Existence of a Plateau as Pulse Amplitude is Increased (in preparation).

41. Lillie (1958) Glass, in *Handbook of Physics* (Condon, E. U. and Odishaw, H., eds.), McGraw-Hill, New York, pp. 8–83, 8–107.

42. Neumann, E., Sprafke, A., Boldt, E., and Wolf, H. (1992) Biophysical digression on membrane electroporation, in *Guide to Electroporation and Electrofusion* (Chang, D. C., Chassy, B. M., Saunders, J. A., and Sowers, A. E., eds.), Academic.

43. Lee, R. C., River, L. P., Pan, F.-S., Ji, L., and Wollmann, R. L. (1992) Surfactant induced sealing of electropermeabilized skeletal muscle membranes *in vivo*. *Proc. Natl. Acad. Sci. USA* **89,** 4524–4528.

44. Gift, E. A. and Weaver, J. C. (1993) Cell survival following electroporation: quantitative assessment using large numbers of microcolonies, in *Electricity and Magnetism in Biology and Medicine* (Blank, M., ed.), San Francisco, pp. 147–150.

45. Weaver, J. C., Bliss, J. G., Powell, K. T., Harrison, G. I., and Williams, G. B. (1991) Rapid clonal growth measurements at the single-cell level: gel microdroplets and flow cytometry. *Bio/Technology* **9,** 873–877.

46. Weaver, J. C., Bliss, J. G., Harrison, G. I., Powell, K. T., and Williams, G. B. (1991) Microdrop technology: a general method for separating cells by function and composition. *Methods* **2,** 234–247.

47. Weaver, J. C. (1994) Molecular basis for cell membrane electroporation. *Ann. NY Acad. Sci.* **720,** 141–152.

48. Okino, M. and Mohri, H. (1987) Effects of a high-voltage electrical impulse and an anticancer drug on *in vivo* growing tumors. *Jpn. J. Cancer Res.* **78,** 1319–1321.

49. Mir, L. M., Orlowski, S., Belehradek, J., Jr., and Paoletti, C. (1991) *In vivo* potentiation of the bleomycin cytotoxicity by local electric pulses. *Eur. J. Cancer* **27,** 68–72.

50. Dev, S. B. and Hofmann, G. A. (1994) Electrochemotherapy—a novel method of cancer treatment. *Cancer Treatment Rev.* **20,** 105–115.

51. Prausnitz, M. R., Bose, V. G., Langer, R. S., and Weaver, J. C. (1992) Transdermal drug delivery by electroporation. Abstract, Proc. Intern. Symp. Control. Rel. Bioact. Mater. 19, Controlled Release Society, July 26–29, Orlando, FL, pp. 232,233.

52. Prausnitz, M. R., Bose, V. G., Langer, R., and Weaver, J. C. (1993) Electroporation of mammalian skin: a mechanism to enhance transdermal drug delivery. *Proc. Natl. Acad. Sci. USA* **90,** 10,504–10,508.

53. Titomirov, A. V., Sukharev, S., and Kistoanova, E. (1991) In vivo electroporation and stable transformation of skin cells of newborn mice by plasmid DNA. *Biochim. Biophys. Acta* **1088,** 131–134.

54. Sukharev, S. I., Titomirov, A V., and Klenchin, V. A. (1994) Electrically-induced DNA transfer into cells. Electrotransfection in vivo, in *Gene Therapeutics* (Wolff, J. A., ed.), Birkhäuser, Boston, pp. 210–232.

55. Gaylor, D. C., Prakah-Asante, K., and Lee, R. C. (1988) Significance of cell size and tissue structure in electrical Trauma. *J. Theor. Biol.* **133,** 223–237.

56. Bhatt, D. L., Gaylor, D. C., and Lee, R. C. (1990) Rhabdomyolysis due to pulsed electric fields. *Plast. Reconstr. Surg.* **86,** 1–11.

57. Hughes, K. and Crawford, N. (1989) Reversible electropermeabilisation of human and rat blood platelets: evaluation of morphological and functional integrity "in vitro" and "in vivo." *Biochim. Biophys. Acta* **981,** 277–287.

58. Mouneimne, Y., Tosi, P.-F., Barhoumi, R., and Nicolau, C. (1991) *Biochim. Biophys. Acta* **1066,** 83–89.

59. Zeira, M., Tosi, P.-F., Mouneimne, Y., Lazarte, J., Sneed, L., Volsky, D. J., and Nicolau, C. (1991) *Proc. Natl. Acad. Sci. USA* **88,** 4409–4413.

60. Belehradek, M., Domenge, C., Orlowski, S., Belehradek, J., Jr., and Mir, L. M. (1993) *Cancer* **72,** 3694–3700.

61. Riviele, J. E., Monterio-Riviere, N. A., Rogers, R. A., Bommannan, D., Tamada, J. A., and Potts, R. O. Pulsatile Transdermal Delivery of LHRH Using Electroporation: Drug Delivery and Skin Toxicology (submitted).

62. Potts, R. O. and Francoeur, M. L. (1990) Lipid biophysics of water loss through the skin. *Proc. Natl. Acad. Sci. USA* **87,** 3871–3873.

63. Bach, D. and Miller, I. R. (1980) Glyceryl monooleate black lipid membranes obtained from squalene solutions. *Biophys. J.* **29,** 183–188.

64. Sugar, I. P. (1981) The effects of external fields on the structure of lipid bilayers. *J. Physiol. Paris* **77,** 1035–1042.

CHAPTER 2

Instrumentation

Gunter A. Hofmann

1. Introduction

The techniques of electroporation and electrofusion require that cells be subjected to brief pulses of electric fields of the appropriate amplitude, duration, and wave form. In this chapter, the term electro cell manipulation (ECM) shall describe both techniques. ECM is a quite universal technique that can be applied to eggs, sperm, platelets, mammalian cells, plant protoplasts, plant pollen, liposomes, bacteria, fungi, and yeast—generally to any vesicle surrounded by a membrane. The term "cells" will be used representatively for any of the vesicles to be manipulated unless specific requirements dictate otherwise.

Electroporation is characterized by the presence of one membrane in proximity to molecules that are to be released or incorporated. One or several pulses of the appropriate field strength, pulse length, and wave shape will initiate this process.

Electrofusion is characterized by two membranes in close contact that can be joined by the application of a pulsed electric field. The close contact can be achieved by mechanical means (centrifuge), chemical means (PEG), biochemical means (avidin-biotin [1]), or by electrical means (dielectrophoresis [2]). Only the electric method is discussed as it relates to ECM instrumentation.

The intent of this chapter is to provide the researcher with a basic understanding of the hardware components and electrical parameters of ECM systems to allow intelligent, economical choices about the best instrumentation for a specific application and to understand its limita-

From: *Methods in Molecular Biology, Vol. 47: Electroporation Protocols for Microorganisms*
Edited by: J. A. Nickoloff Humana Press Inc., Totowa, NJ

tions. Commercial instruments have been available for more than 10 years; the commercial ECM technology has matured and become more costeffective. Rarely is it economical to build one's own instrument. Although articles occasionally appear on how to build an instrument for a few hundred dollars, the plans are generally of poor design, and the cost estimates often do not take into account the researcher's time for electronic development. Furthermore, today's commercial instruments often incorporate measuring circuits for important parameters, which are difficult to develop. The difficulty is that in one housing, there are voltages of many kilovolts and currents of hundreds of Amperes (A) flowing next to low signal/control voltages, typically between 5 and 20 V. Sophisticated design is needed to prevent crosstalk or electromagnetic interference between these different circuits. Thus, it is usually not costeffective to build an instrument unless specific parameters are needed that are not available commercially. Another very important issue is safety. Voltages and currents generated in efficient ECM generators are large enough to induce cardiac arrest. Generators need to be constructed to be safe and foolproof against accidental wrong settings. They must also deliver the pulse to the chamber in such a way that the operator will not, under any circumstances, come in contact with parts carrying high voltage.

A database of over 2500 publications in the field of electroporation and electrofusion is maintained and updated continuously by BTX (San Diego, CA) as a service to the research community. Any researcher may inquire about the BTX Electronic Genetics® Database and request a database search.

2. Components of an ECM System and Important Parameters

Generally, ECM systems consist of a generator providing the electric signals and a chamber in which the cells are subject to the electric fields created by the voltage pulse from the generator. A third optional component is a monitoring system, either built into the generator or connected in line between the generator and the chamber, which measures the electrical parameters as the pulse passes through the system. Each component is discussed in Sections 4.–6. In this section, we discuss the relationship between the electrical parameters, which the ECM system provides, and the parameters that the cells experience.

The biophysical process of electropermeabilization is caused by the electrical environment of the cell in a medium. The main parameter, which describes this environment, is the electric field strength E, measured in V/cm. Though the presence of the cell itself modifies the field in close proximity, knowledge of the average field strength at the location of the cells is sufficient for the purpose of ECM experiments. The electric field is generally created by the application of a potential difference (voltage) between metallic electrodes immersed in the medium containing the cells. For the simple electrode geometry of parallel plates located at a distance d (cm), the electric field is calculated from the applied voltage V as:

$$E = V/d \text{ (V/cm)} \tag{1}$$

Practical values of E used in ECM range from a few hundred V/cm for mammalian cells to many kV/cm for bacteria.

The electric field in the medium gives rise to currents depending on the medium specific resistivity r, which is measured in $\Omega \cdot$ cm. The specific resistivity ranges from a low of about 100 $\Omega \cdot$ cm for saline solutions to many k$\Omega \cdot$ cm for nonionic solutions, such as mannitol. The resulting current density j is:

$$j = E/r \text{ (A/cm}^2) \tag{2}$$

The current produced results in heating of the medium. Saline solutions with a low value of r experience severe heating effects as compared to nonionic solutions for the same electric field and pulse length.

The temperature rise ΔT (°C) can influence the permeabilization mechanism, or lead to excessive heating and evaporation of the medium. It can be calculated for different pulse wave shapes:

Square pulse: $\Delta T = E^2 t/4.2 \, r$, where t is the pulse length in s.
Exponential pulse: $\Delta T = E^2 \tau/8.4 \, r$, where τ is the 1/e time constant (s) (*see* Section 4.2.1.).

Having defined the parameters at the location of the cells, we can relate them to the electrical parameters at the chamber electrodes:

Electrode voltage: $V = E \cdot d$ (V) (plane parallel electrodes)
Chamber current: $J = j \cdot F$ (A), where F is the electrode area, cm^2
Chamber resistance: $R = f(r) \, \Omega$, where $f(r)$ is a function of the chamber geometry.
For plane parallel electrodes, $R = r \, d/F$.

The voltage at the chamber electrodes is not necessarily equal to the voltage that is indicated or even measured in the generator. This relationship is discussed in Section 4.1. The range of the electric field for optimum yield is quite narrow. Deviation of 5–10% from the optimum value can lead to a drop of an order of magnitude in yield. The pulse length is a less sensitive parameter. It is desirable to assure that the optimum electric field is established in the chamber.

3. Volume Requirements

Small volumes of 100 μL to a few milliliters can be treated in a batch mode: fill the chamber, electroporate, or fuse, and empty the chamber. Larger volumes (many milliliters to 1 L) require chambers that might not be available and a high output power level, which generators typically cannot deliver. A good solution to this problem is the use of a flow-through system in which the generator periodically pulses in synchronism with a pump, so that every volume element of cell/transformant mixture is exposed to the desired electric fields and number of pulses as it passes through the chamber. This method requires flowthrough chambers and generators that can pulse automatically, either at a fixed or adjustable repetition rate. Such generators and chambers are available (*see* Tables 1 and 6). For fusion, a continuous flow is not desirable, because fused cells need to be undisturbed for a period of time to round off and complete the fusion process. In this case, a pulsating (stop and go) flowthrough system would be appropriate.

4. Generators

The relationship between the electrical parameters in a generator and the parameters actually delivered to the chambers is important because substantial differences can exist. Following this discussion, different types of generators are described. Table 1 presents a survey of commercially available generator types.

4.1. Actual Voltage Delivered to the Chamber

The momentary power the generators are required to deliver to chambers can far exceed the electrical power available from laboratory outlets. To overcome this limitation, electrical energy is stored in capacitors by charging them slowly at low power to a preset voltage and then discharging them at high power level into the chamber. The voltage V_0 to which the capacitors will be charged can be set and is typically indicated

Table 1
Survey of Electroporation and Electrofusion Generators

Manufacturer	No PS	With PS one pulse length	With PS multiple pulse lengths	Square wave	Electro cell fusion	Stand-alone monitor
		Electroporation				
		Exponential discharge wave form				
IBI (New Haven, CT)			X			
Invitrogen (San Diego, CA)	X					
Bio-Rad (Richmond, CA)		X	X			
BRL (Grand Island, NY)		X	X			
BTX	X	X^a	X^a	X^a	X	X

Abbreviation: PS, power supply.
^{a}R, optional version available with repetitive pulsing for flowthrough applications.

at the front panel of the generators. The actual voltage delivered to the chamber can be substantially lower than what is normally assumed to be the generator output voltage. This effect is caused by the internal resistance of the generator (typically around 1 Ω), which absorbs part of the charging voltage during discharge and is more pronounced in larger chambers (several milliliters) and low resistivity medium. Some generators are also designed with a relatively high internal resistance, which is undesirable, to protect the output switch against high currents. These generators can exhibit a drastic drop in actual voltage delivered to the chamber under certain circumstances. If the internal resistance R_i and the chamber resistance R_c are known, the actual voltage V on the chamber can be calculated as:

$$V = V_0 \cdot R_c/(R_c + R_i) \qquad (3)$$

4.2. Generators for Electroporation

The two types of generators commonly encountered differ by the wave shape of their output: exponential decay wave form or square pulses. Though both can in principle be used for electroporation, it appears that bacteria are transformed more efficiently by exponential wave forms (with some exceptions [3]), whereas some mammalian cell types (4) and plant protoplasts (5) show generally superior transformation results with square waves.

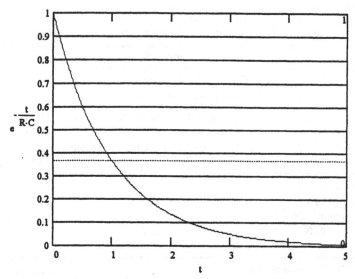

Fig. 1. Exponential decay wave form, representative of the complete discharge of a capacitor into a resistor.

4.2.1. Exponential Wave Form Generators

The voltage of a capacitor C (capacity measured in Farad or, more conveniently, in microfarad) discharging into a resistor R (Ω) follows an exponential decay law (Fig. 1):

$$V = V_o \cdot exp\ (-t/RC) \tag{4}$$

The pulse length of such a discharge wave form is commonly characterized by the "$1/e$ time constant." This is the time required for the initial voltage to decay to $1/e \approx 1/3$ of the initial value ($e = 2.718\ ...$is the basis of natural logarithms). This time constant can be conveniently calculated from the product of R and C, where C is the storage capacitor in the generator and R is the total resistance into which the capacitor discharges, which can have several components. Figure 2 shows a general circuit diagram of an exponential decay generator.

The power supply slowly charges the capacitor to the desired voltage and does not play a role during the discharge. The internal resistance R_i of the capacitor is on the order of 0.5–1 Ω for electrolytic capacitors and, in normal operation, is much smaller than any other resistance in the circuit and can therefore be neglected. The resistor R_L is installed in some instruments to limit the current in the circuit, especially in case of an arc in the chamber, which would result in high currents because the chamber

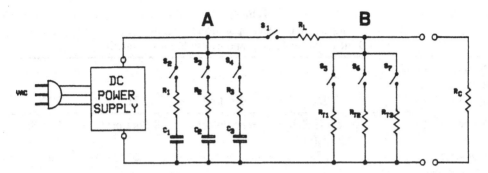

Fig. 2. General circuit diagram of an exponential decay wave form generator. $C_{1,2,3}$ are the energy storage capacitors, which have an internal resistance $R_{1,2,3}$. They can be added to the circuit by switches $S_{2,3,4}$ in order to vary the total capacitance. Closing switch S_1 allows the charged capacitors to discharge to the output and into the chamber, represented by the resistor R_c. R_L is a discharge current-limiting resistor, which is needed in some designs. R_{T1}, R_{T2}, and R_{T3} are timing resistors, which can be added to the circuit by switches $S_{5,6,7}$. If the output voltage is measured at (**A**), instead of (**B**), incorrect readings of the actual voltage on the chamber will result.

resistance drops to very low values during an arc. The size of this resistor is determined by the maximum current capability of the switch. As a result of the presence of R_L, the voltage at the chamber is reduced by the voltage drop across R_L, which can be substantial. Furthermore, some instruments measure the peak discharge voltage at the point A instead of directly across the chamber at point B, resulting in incorrect readings. Use of instruments that do not have a built-in current-limiting resistor provides advantages. One needs to be aware that without a current-limiting resistor, arcs appear more violent because of higher current flow. However, if the instrument and chamber stand are designed correctly, this should be of no consequence. It should be noted here that arcing in the chamber occurs mostly at high field strengths (above 10 kV/cm) and is a statistical effect.

The resistance R_t is a timing resistor that, typically, can be selected to adjust the pulse length. Maintaining a low value, relative to the chamber resistance R_c, serves the function of determining pulse length. Often, the size of the capacitance can be changed by connecting one or more capacitors in parallel. Since the time constant is determined by the product of resistance and capacitance, either variable can be used to adjust it. Keeping the resistance as low as possible, well below the chamber resis-

Table 2
Comparison of Electroporation Generators
with Built-in Power Supply and Multiple Pulse Length

Manufacturer of electroporation system	IBI	Bio-Rad	BRL	BTX
Model number	geneZAPPER™ 450/2500	Gene Pulser®	Cell-Porator™	ECM® 600
# of Unit components required	2	3	2	1
Voltage range	50–2500	50–2500	0–2500	50–2500
# of Pulse length settings	34	8	8	126
Maximum field strength (kV/cm)	12.5	25	16.6	40
Maximum current, A	130	120	40	>1000
Monitoring	Partial	Partial	Partial	Full
Actual voltage	No	No	Yes	Yes
Actual pulse length	Yes	Yes	No	Yes
Safety design	Operator shock proof	Safety interlock	Safety interlock	Operator shock, arc, and short circuit proof
Data base	No	No	No	Yes
Warranty	1 yr	1 yr	1 yr	2 yr

tance, is generally desirable. Sometimes the chamber resistance is too low for the timing resistors to be effective. In this case, the chamber resistance itself will determine the pulse length, which then can be adjusted only by varying the capacitance.

For the characterization of the pulse into the chamber, only two parameters need to be known: the peak voltage and the $1/e$ pulse length. It is convenient to use a generator with a built-in measuring circuit that measures the pulse parameters at the output of the instrument (Point B in Fig. 2). Table 2 shows a comparison of the main features of commercially available exponential discharge generators with built-in power supply, multiple pulse length capability, and at least some monitoring.

If only a limited number of applications are planned, such as *E. coli* transformation, a generator with a fixed pulse length will be sufficient. This simplifies the generator design and reduces costs. To reduce costs

Table 3
Exponential Decay Generator Options and Costs

Fixed time constant t	Fixed t	Variable t
No power supply (PS) ~$1000	With PS $1000–2000	With PS and monitoring $4000–5000

even further, it is also possible to use an external electrophoresis power supply and eliminate a built-in supply. Table 3 shows generators available with increasing flexibility and cost options. The maximum voltage is typically 2500 V. The time constant for fixed pulse length is typically 5 ms.

4.2.2. Square-Wave Generators

Square-wave pulses appear to have advantages for certain applications, such as transfection of mammalian cell lines and plant protoplasts, though no generalization can be made. Each cell line needs to be individually investigated to determine whether use of square-wave pulses would be advantageous. In general, square waves do not appear to result in higher transformation yields for bacteria, although there are some protocols that give good results (*6*; Xing Xin, Texas Heart Institute, personal communication). Square waves are used almost exclusively for in vivo applications of electroporation, such as electrochemotherapy, where drugs are electroporated into tumor cells. These generators are more difficult to build because the square-wave pulse is produced by a partial discharge of a large capacitor, which requires the interruption of high currents against high voltages. In the past, their costs were higher than exponential discharge generators, and the range of parameters was more limited. However, recent advances in solid-state switching technology have lowered costs. A square-wave generator is now available that can deliver up to 3000 V into a 20-Ω load at costs comparable to exponential discharge units (*see* Table 1).

4.3. Generators for Electrofusion

If nonelectrical means of cell–cell contact are used, any electroporation generator can also be used for electrofusion. If it is desirable to induce cell–cell contact by dielectrophoresis, the generator needs to produce an alternating wave form (ac) over a longer period of time, typically seconds, before the fusion pulse is applied.

The optimal frequency appears to be around 1 MHz (*7*). Above and below this frequency, the viability of mammalian cells, at least, appears

Table 4
Mouse Egg Fusion with Different Wave Forms and Chambers

AC wave form and chamber type	# of Eggs	% Fusion	% Developed	Stability of development
Nonsinusoidal, wire chamber	20	100	75	Stable
Sinusoidal, rectangular bar chamber	24	87.5	52.4	Less stable

to suffer. Nonionic fusion media are desirable to reduce the generation of heat and turbulence. A pure sinusoidal wave form is not necessary and, possibly, not even advantageous. It is, however, important that there is no net dc component in the wave form. Higher harmonics in the wave form appear to produce better fusion results. Table 4 compares results obtained with different wave forms and chambers *(8)*.

Commercial fusion generators are available (Table 1) that allow the sequential application of ac wave forms and fusion pulses, which are generally of the square-wave type.

4.4. Generators with Other Wave Forms

Researchers have experimented with wave forms other than exponential and square. It is apparent that for some applications, special wave forms have certain advantages. Bursts of radio frequency electric fields (a few 100 kHz) appear to be more benign to cells and might be advantageous when fusing cells of widely different sizes *(9,10)*. However, such generators are not presently available commercially, and are difficult and expensive to build with high-power levels.

5. Chambers

There are many choices in chambers for ECM. In general, chambers need to create the required field strength from the voltage delivered to the electrodes by the generator; they need to contain the appropriate volume, and need to be sterilized or sterilizable, easily filled, emptied, and if reused, easily cleaned. Table 5 gives the main trade-off parameters in the selection of chambers. The following describes only the more frequently used chambers in the field.

Table 5
Chamber Trade-off Parameters

Small volume	Large volume
Disposable	Reusable
Homogeneous field	Inhomogeneous field
Visualization of cells	Cells obscured
Batch process	Flowthrough
Aluminum electrodes	Stainless steel or noble material
Presterilized	Sterilizable
Small gap	Large gap

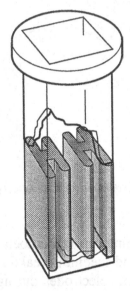

Fig. 3. Disposable electroporation cuvet with molded-in aluminum electrodes.

5.1. Small-Volume Chambers

Disposable, presterilized cuvets with molded-in aluminum electrodes (Fig. 3) are most frequently used for electroporation. They are available with different gap sizes, typically 1 mm (for bacteria), and 2 and 4 mm (for mammalian cells and plant protoplasts). The electric field in these cuvets is quite homogeneous. Some workers clean and reuse cuvets to reduce costs.

Reusable, parallel plate electrode assemblies (Fig. 4) that fit into spectrophotometer cuvets are available. They are also available in different gap sizes.

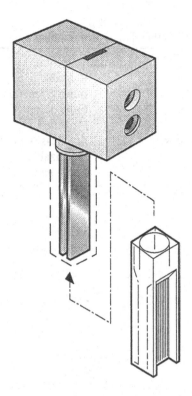

Fig. 4. Reusable parallel plate electrode assembly to fit into spectrophotometer cuvets.

Note that chemical cleaning or even autoclaving might not remove
cell debris, transformant, and medium breakdown products that might
have been deposited onto the electrodes during a pulse. Only a good
mechanical cleaning will remove the debris.

Electrodes on microslides are used to visualize the fusion process
under a microscope. Parallel wires (Fig. 5), separated by 1 mm or less,
produce divergent fields that favor dielectrophoretic pearl chain forma-
tions of cells. For small gaps (<1 mm), a meander-type electrode configuration (Fig. 6) allows visualization of the fusion process. Electrodes with
square bars (Fig. 7), which provide a more homogeneous electric field, can
also be mounted on microslides for visualization of embryo manipulation.

5.2. Large-Volume Chambers

Intermediate-size chambers with a volume of a few milliliters can be
built with parallel bars (Fig. 8). The electrodes can be flat to create

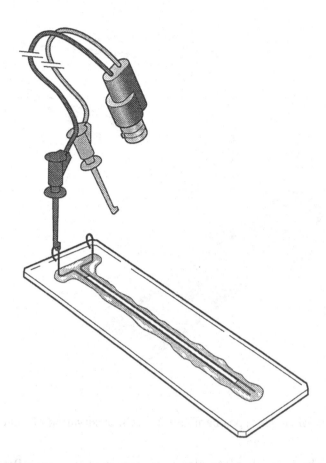

Fig. 5. Parallel wire electrodes mounted on a microslide for visual observation of the electrofusion process. These electrodes create an inhomogeneous electric field, which is preferable for dielectrophoresis.

homogeneous fields or they can have grooves to create divergent fields for fusion. A convenient implementation of a large-volume chamber with a volume up to 50 mL is an array of parallel plate electrodes fitted into a plastic Petri dish (Fig. 9). The gap between the electrodes can be 2 mm for mammalian cells or 10 mm for embryo and fish egg electroporation. Generally, such large volumes need a high resistivity medium because the chamber resistance with saline solution, such as PBS, would be very low. Partial filling of the chamber will reduce the resistance proportionally. As an example, 10 mL of PBS in a 10-cm diameter Petri dish with 2-mm spaced electrodes resulted in a resistance of 0.4 Ω. Some generators can generate sufficient voltage to transform mammalian cells even

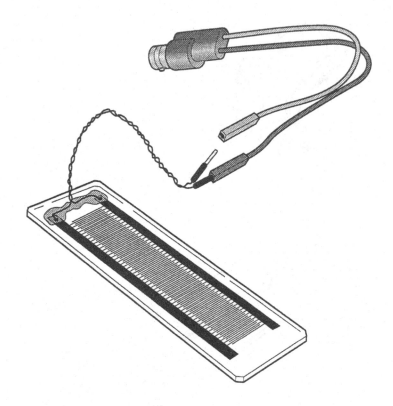

Fig. 6. Meander-type chamber for visual observation of fusion.

with PBS. The parallel plate electrode configuration in a Petri dish is also very useful for electroporation of adherent cells, if the electrodes are situated so they touch the Petri dish bottom. Instead of parallel plates, an array of concentric electrodes can also be used to create a large-volume ECM chamber in a Petri dish *(11)*.

5.3. Small-Volume Flowthrough Chambers for the ECM of Large Volumes

If it is required to transform large volumes (above 50 mL), it is economical to pulse the generator repetitively in synchrony with a pump that pushes the medium with the cells and transformants through a relatively small chamber. The repetition rate and pumping speed can be arranged so that every volume element receives one or, if desired, multiple pulses. Care needs to be taken in the design of the flowthrough chamber to minimize dead volume. Repetitive pulse generators are com-

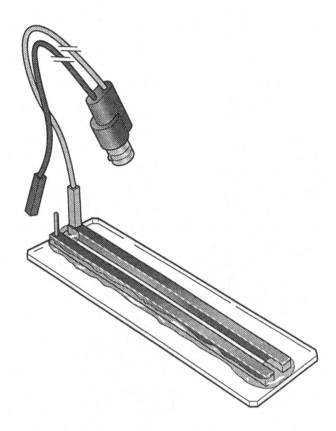

Fig. 7. Rectangular electrodes mounted on a microslide for visual observation, generating homogeneous electric fields.

mercially available for exponential decay, as well as for square-wave form output.

5.4. Chamber Material

Despite an oxide layer present on the surface, aluminum (Al) electrodes appear to give satisfactory results in disposable chambers. The commercially available presterilized cuvets use embedded Al electrodes. Stainless steel (SS) is used more often for reusable chambers. SS can be mechanically cleaned more easily. Gold plating is an option for SS as well as Al, but it appears that the increase in yield does not justify the additional costs. Comparative electrofusion experiments of embryos in either SS or gold-plated chambers did not show a substantial difference in fusion yield. Over 100 fusion experiments were performed using gold-plated electrodes, with a

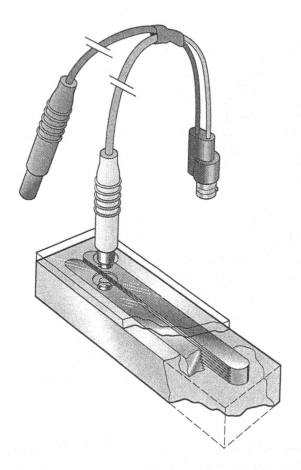

Fig. 8. Intermediate-volume chamber with parallel bar electrodes. The electrodes can be flat to create homogeneous fields or have grooves to create inhomogeneous fields.

fusion yield of >90%; with care, similar results can be achieved with SS electrodes (James M. Robl, U. of Massachusetts, personal communication). A comparison of plant protoplast fusion yields using a large-volume parallel plate chamber made of SS or a gold-plated concentric ring chamber (both for Petri dishes) showed a consistently higher yield for the gold-plated chamber *(10)*. Table 6 shows a survey of commercially available chambers.

6. Measuring ECM Parameters

By following an established protocol, it is generally not necessary to measure the ECM parameters, especially if the same types of generator

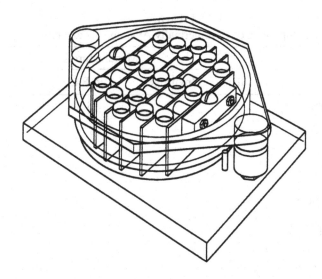

Fig. 9. Petri dish electrodes for large-volume electroporation of mammalian cells in suspension or adherence, and electroporation of fish eggs.

and chamber are used (different generators might give different output voltages for the same charging voltage setting). However, when pursuing new applications with an instrument that does not have built-in monitoring, it is desirable to measure the voltage and pulse actually delivered to the chamber to allow accurate reporting and reproducible performance. A commercial instrument is available to monitor ECM parameters specifically, display them, and print them out (Table 1). A measuring system can be assembled consisting of a digital oscilloscope (bandwidth should be 100 MHz), and a high-voltage probe attenuating the voltage signal 1:1000, with a voltage range up to 3 kV. Commercial generators and chambers are typically constructed so that at no place in the circuit is the high-voltage potential easily accessible. Therefore, adapters need to be placed in line between the generator and the chamber so that a voltage probe can be connected. Before any measurements are performed, the grounding situation must be understood and verified with the manufacturer. Sometimes neither of the two outputs of the generator is at the ground potential of the oscilloscope (which is normally tied to the power line ground), depending on the design of the discharge circuit. It is still possible to perform measurements in this case by disconnecting the oscilloscope from the power line earth/ground, either by inserting an iso-

Table 6
Comparison of Commercially Available Chambers

Chamber types	Manufacturers				
	IBI	Invitrogen	Bio-Rad	BRL	BTX
Cuvets: disposable	3	3	3	3	3
Cuvets: reusable, homogeneous field	3				2
Cuvets: reusable, divergent field	1				1
Microslides homogeneous field					2
Microslides divergent field					2
Meander					1
Flat electrode divergent field					1
Petri dish electrodes					2
96-Well plate electrode					1
Flowthrough					2
Sandwich					2
In vivo electrodes					3

[a]The numbers indicate available variations, typically in the gap size.

lation transformer between the line and the oscilloscope or by disconnecting the oscilloscope ground lead to the power line with an insulating plug. These plugs can be recognized by regular three prongs on one side and only two receptacles on the other side with the earth/ground wire separate, which should not be connected. During the pulse, the chassis of the oscilloscope will attain a potential difference to the laboratory ground and should not be touched. These kinds of measurements obviously are hazardous, and should be performed with extreme care and only by trained personnel. If it is desirable to measure both the voltage output and the current, a convenient, contactless way is to route one lead to the chamber through a current transformer. Several manufacturers provide these elements (e.g., Pearson Electronics Inc. [Palo Alto, CA], Current Transformer Model Nr. 411). The most important specifications to verify the usefulness of a current transformer are the peak current capability (e.g., 5000 A for the 411) and the limit of the product of current × pulse length (e.g., 0.2 A · s for the 411) to avoid saturation of the current transformer before the pulse has passed completely. Measuring current and voltage allows one to determine the chamber resistance as a function of time from Ohm's law, $R = V/I$. Through the geometry of the chamber, the specific resistivity of the cell/medium suspension as a function of time can then be determined, which might be of interest for biophysical investigations of the ECM process, because lysis of cells results in an increase of the medium conductivity.

Acknowledgments

I want to thank my colleagues for helpful comments, and especially Linda Hull for editing the manuscript.

References

1. Tsong, T. Y. and Tomita, M. (1993) Selective B lymphocyte-myeloma cell fusion. *Methods in Enzymol.* **220**, 238–246.
2. Pohl, H. A. (1978) *Dielectrophoresis.* Cambridge University Press, London.
3. Meilhoc, E., Masson, J.-M., and Teissie, J. (1990) High efficiency transformation of intact yeast cells by electric field pulses. *Biotechnology* **8(3)**, 223–227.
4. Takahashi, M., Furukawa, T., Saito, H., Aoki, A., Koike, T., Moriyama, Y., Shibata, A. (1991) Gene transfer into human leukemia cell lines by electroporation: experiments with exponentially decaying and square wave pulse. *Leukemia Res.* **15(6)**, 507–513.
5. Saunders, J., Rhodes, S. C., and Kaper, J. (1989) Effects of electroporation profiles on the incorporation of viral RNA into tobacco protoplasts. *Biotechniques* **7(10)**, 1124–1131.
6. Xie, T. and Tsong, T. (1992) Study of mechanisms of electric field-induced DNA transfection III, Electric parameters and other conditions for effective transfection. *Biophys. J.* **63**, 28–34.
7. Hofmann, G. H. (1989) Cells in electric fields—physical and practical electronic aspects of electro cell fusion and electroporation, in *Electroporation and Electrofusion in Cell Biology* (Neumann, E., Sowers, A., and Jordan, C., eds.), Plenum, New York, pp. 389–407.
8. Nagata, K. and Imai, H. (1992) The difference of electro fusion rate for the pronuclear transplantation of mouse eggs between three different electro generators. The 7th Eastern Japan Animal Nuclear Transplantation Research Conference.
9. Chang, D. C. (1989) Cell fusion and cell poration by pulsed radio-frequency electric fields, in *Electroporation and Electrofusion in Cell Biology* (Neumann, E., Sowers, A., and Jordan, C., eds.), Plenum, New York, pp. 215–227.
10. Tekle, E., Astumian, R. D., and Chock, P. B. (1991) Electroporation by using bipolar oscillating electric field: an improved method for DNA transfection of NIH 3T3 cells. *PNAS* **88**, 4230–4234.
11. Motumura, T., Akihama, T., Hidaka, T., and Omura, M. (1993) Conditions of protoplast isolation and electrical fusion among citrus and its wild relatives, in *Techniques on Gene Diagnosis and Breeding in Fruit Trees* (Hayashi, T., et al., eds.), FTRS, Japan, pp. 153–164.

CHAPTER 3

Direct Plasmid Transfer Between Bacterial Species and Electrocuring

Helen L. Withers

1. Introduction

A common limitation in any molecular biological study is the introduction of DNA, whether bacteriophage or plasmid, into a recipient strain. Electroporation has a number of advantages over other more traditional methods for the introduction of DNA into cells, the most significant being the amount of DNA required is reduced, and in some procedures, prepared plasmid DNA is completely unnecessary. Since electroporation utilizes a physical rather than biological mechanism of DNA transfer, the technique has the power to overcome many of the barriers that are normally found when using more traditional techniques. For example, electroporation can be utilized for the direct movement of nonconjugative plasmids across species barriers, making it a powerful tool in molecular biology.

Electrotransfer or electroduction allows the movement of plasmid DNA from a donor bacteria into a genetically identifiable recipient. The recipient may be of the same species, giving rise to intraspecies transfer *(1)*, or a completely different species or genera giving interspecies transfer *(2–4)*. The ability to transfer plasmid DNA directly between two different strains has many advantages, such as time reduction for analytic procedures, ability to harvest plasmid DNA from bacterial strains that are resistant to normal cell-lysis procedures, increase plasmid DNA yield, and decrease the risk of contamination of slow-growing bacteria. For example, the transfer of plasmid DNA directly from *E. coli* to caulobacters bypasses the need for plasmid purification, conjugation pro-

From: *Methods in Molecular Biology, Vol. 47: Electroporation Protocols for Microorganisms*
Edited by: J. A. Nickoloff Humana Press Inc., Totowa, NJ

cedures, and the modification of *E. coli* to act as a donor strain *(4)*. Similarly, the ability to transfer plasmid DNA directly from *Mycobacterium spp.* to *E. coli* obviates the contamination risks of long-term mycobacterial cultures. Relative ease of DNA extraction from *E. coli* compared to *Mycobacterium* is an added benefit *(2)*.

In addition to the ability to introduce DNA either directly between cells or via a DNA preparation stage, electroporation has been used to cure *E. coli* of plasmid DNA *(5)*. Electrocuring and direct transfer procedures of prokaryotes involve two steps: preparation of cells and the electroporation procedure. Both of these steps can affect the overall efficiency and success of DNA transfer or curing.

2. Materials

1. LB: 1.0% (w/v) bacto-tryptone, 0.5% (w/v) bacto-yeast extract, and 0.5% (w/v) NaCl (*see* Note 1) *(6)*.
2. 2xYT: 1.6% (w/v) bacto-tryptone, 1.0% (w/v) bacto-yeast extract, and 0.5% (w/v) NaCl (*see* Note 1).
3. Antibiotics and other nutritional supplements as required (these will vary according to specific bacterial host requirements) (*see* Note 2).
4. SOC recovery medium: 2% (w/v) bacto-tryptone, 0.5% (w/v) bacto-yeast extract, 10 mM NaCl, 2.5 mM MgCl$_2$, 10 mM MgSO$_4$, and 20 mM glucose.
5. GY dilution fluid: 10 mM Tris-HCl, pH 7.4, 10 mM MgSO$_4$, and 0.01% (w/v) gelatin.

3. Methods

3.1. Preparation of Electrocompetent Cells

3.1.1. Escherichia coli

E. coli is one of the easiest bacteria to culture in the laboratory. Growth can be obtained on the simplest defined media through to the most complex preprepared media and under temperatures that range from 15–42°C. However, the rate of growth and the final cell density achieved depend on the mutation load that the organism is carrying. It is therefore important to optimize the media and growth conditions for each specific *E. coli* strain to be used. It is also important to ensure the use of a freshly grown culture, since subsequent recovery of electrotransformants will depend on the viability of the initial culture.

1. Prepare an overnight broth culture of the recipient cells by inoculating 2 mL of LB or 2xYT with a colony from an agar plate (this culture can be grown either with shaking or stationary, at 37°C).

2. Add 500 µL of the overnight culture to 500 mL of broth (LB or 2xYT), which after further overnight incubation with shaking at 37°C, provides sufficient cells to carry out 25 separate electroporations (allow 20 mL broth/electroporation). These volumes can be scaled up or down as necessary for individual requirements. Under the above conditions, cell densities of between 1×10^8 and 1×10^{11} cells/mL are routinely obtained (*see* Note 3).
3. Cells should be chilled on ice for 15 min prior to centrifugation (10 min, 2860*g*) (*see* Note 4).
4. Resuspend the resulting pellet of cells in 10 mL of prechilled sterile distilled water. This washing step should be carried out at least three times.
5. Resuspend the pellet in 1 mL of prechilled sterile distilled water, and transfer to a 1.5-mL tube to harvest the cells before resuspending in a final working volume of 40 µL of prechilled sterile distilled water/20 mL of starting culture (*7*). This final volume is somewhat dependent on the amount of cell growth that has been achieved. However, the minimum volume required for covering the bottom of a 0.2-cm electroporation cuvet is 40 µL.
6. Incubate the cells on ice for at least half an hour before performing the electroporation step (*see* Note 5).

3.1.2. Other Bacterial Species

Each different bacterial culture to be used for any electroporation procedure should be grown under optimal conditions for that specific organism (*see* Note 1). The preparation of electrocompetent cells from other bacterial species should be carried out as for *E. coli* regardless of whether they are to be recipient or donor (*see* Section 3.1.1.).

It has been reported that *Mycobacterium spp.*, which is to be used as a donor strain, can be prepared simply by collecting a dozen colonies from a fresh agar plate and resuspending in 10% glycerol (*2*) (*see* Note 6).

3.2. General Method for Electroporation

1. Transfer 40 µL of cells, either freshly prepared or thawed from frozen stocks (*see* Note 5), to a prechilled 0.2-cm electroporation cuvet. If necessary, the cells should be gently "knocked" down to the bottom of the cuvet, so that the mixture coats the entire base of the cuvet (*see* Note 7). Ideally, the cell solution should be of such a consistency that it will adhere as a drop to the side of the cuvet prior to displacement to the base of the cuvet. Leave on ice for a minute.
2. Dry the outside of the cuvet containing the cells thoroughly, and place into the cuvet holder (*see* Note 8).

3. Pulse the cells once using the following electroporation parameters: 2.4 kV, 25 µF, and 200 Ω. A time constant of between 4.5 and 4.7 ms is usually achieved with these conditions.
4. Add 1 mL of SOC immediately to the cells, and mix well.
5. The cells can either be left in the cuvet for a stationary incubation period, or can be transferred to a 10-mL sterile tube for incubation with shaking. In both cases, the incubation period should be not <1 h at 37°C (*see* Note 9).
6. Plate 100-µL aliquots of serial dilutions in GY on appropriate selective media (*see* Notes 2 and 10).
7. A summary of this basic method is shown in Fig. 1. Control samples can be taken at the steps indicated in Fig. 1. These should be plated onto nonselective media to provide a check on cell viability throughout the procedure (*see* Note 3).

3.3. Curing E. coli *of a Plasmid*

Heery et al. *(5)* showed that it is possible to use electroporation to produce plasmid-free *E. coli* strains.

1. An overnight culture of *E. coli*-containing plasmid should be grown in either 10 mL of LB or 2xYT containing the appropriate antibiotic. The cells are pelleted and washed in sterile distilled water as previously described (Section 3.1.1.). The conditions for electroporation are the same as for the general procedure (Section 3.2.).
2. After incubation with 1 mL of SOC media at 37°C has occurred and dilutions have been prepared, 100 µL of each dilution are plated out onto both selective and nonselective media.
3. Once incubated, a comparison between the selective (plasmid-containing colonies) and the nonselective (nonplasmid and plasmid-containing colonies) plates can be made and the efficiency of curing can then be assessed.
4. Inoculate colonies from the nonselective plates onto both selective and nonselective media to locate and confirm that a plasmid-free isolate has been obtained. Heery et al. *(5)* reported a curing efficiency of 80–90%. However, in our hands, the level of curing is dependent on both the *E. coli* strain being cured and the plasmid to be cured. However, curing of plasmids by electroporation generates higher frequencies of plasmid-free cells than traditional curing methods.

3.4. Direct Transfer of Plasmid DNA from Donor Cell to Recipient
3.4.1. Between Different E. coli *Strains*

The ability to transfer plasmid DNA from one host to another directly removes the need to prepare plasmid DNA, thus saving time. This varia-

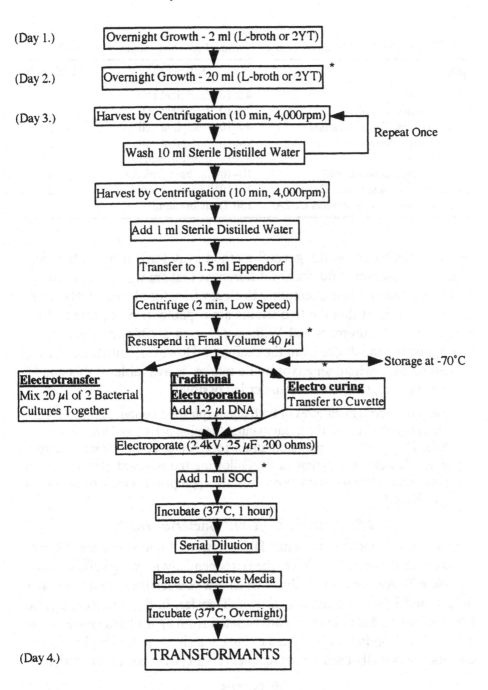

Fig. 1. Flowchart illustrating the steps involved in the electroporation procedure. Asterisk indicates points where control samples can be taken to test viability.

Table 1
A List of Transformation Efficiencies of Some Intra- and Interspecies Transfers

Species	Efficiency	Reference
E. coli-E. coli	4×10^{-5}/recipient cell	*3*
S. typhimurium-E. coli	7×10^{-4}/recipient cell	*3*
S. typhimurium-S. typhimurium	4×10^{-3}/recipient cell	*3*
E. coli-S. typhimurium	2×10^{-6}/recipient cell	*3*
E. coli-E. coli	3×10^3 transformants	*1*
Mycobacterium bovis-E. coli	10–100/*M. bovis* colony	*2*
Mycobacterium smegmatis-E. coli	1×10^4/*M. smegmatis* colony	*2*
E. coli-Caulobacter crescentus CB2A	150 Transformants	*4*

tion is again based on the general method as described in Section 3.2. There are, however, a number of differences that are worth highlighting.

It is important when attempting to electrotransfer plasmid DNA from donor to recipient that the two strains are capable of being selected for independently. Summers and Withers *(1)* used antibiotic selection for distinguishing the donor from the recipient. However, nutritional markers can be used where an auxotrophic marker is available. If this condition can be met, then electrotransfer is possible.

1. Prepare overnight cultures of both donor (plasmid-containing) and recipient (plasmid-free) in the usual manner (Section 3.1.) (*see* Note 3).
2. Mix 20 µL of each prepared strain together prior to transferring to a prechilled electroporation cuvet. Following the standard electroporation procedure, plate serial dilutions on to the appropriate selective agar plates (*see* Note 11).

3.4.2. Between Different Bacterial Hosts

Efficiency of transfer of plasmid DNA between different species of bacteria has been shown to be higher when exponentially growing cells are used (*see* Note 3). An exception to this is when *Mycobacterium* is used as a donor for plasmid DNA transfer into *E. coli*. Transfer, in this instance, can be achieved when donor colonies are isolated from agar plates (*see* Section 3.1.2.). Electroporation is carried out using a single pulse employing the conditions normally used for the recipient strain (*see* Note 11 and Table 1).

4. Notes

1. LB and 2xYT are suitable for culturing most members of the Enterobacteriaceae family. It is important to use healthy bacterial cultures for

electroporation procedures. Therefore, use a medium that has been optimized for each species of bacteria to be used.

2. When selecting for β-lactamase production, use carbenicillin rather than ampicillin. This will prevent the growth of satellite colonies (nontransformed cells) when the cell density is high.

3. Cultures of *E. coli* should be prepared by inoculating at least 20 mL of LB and growing with shaking overnight. Overnight cultures of *E. coli* yield between 10^8 and 10^{11} cells/mL; other bacterial species may yield lower numbers. The final cell density will vary with media and growth conditions, but viability of the cells should remain at a constant level throughout the preparation and electroporation procedures. The cells should be resuspended to a final concentration of approx 10^{10} cells/mL.

4. At all times during preparation and the subsequent electroporation procedure, cells should be maintained on ice, since this will help cell survival, thus increasing the level of transfer achieved. Centrifugation should preferably be carried out in a refrigerated centrifuge. However, a standard bench-top centrifuge is satisfactory as long as the cells are maintained on ice between each harvesting step for at least 10–15 min.

5. Prepared cells can be stored at –70°C. In this case, the final 10 mL wash and subsequent resuspension should be carried out using chilled 10% (v/v) glycerol. The cells should be aliquoted into 40-μL lots, frozen rapidly in a mixture of methanol/dry ice or liquid nitrogen, and stored at –70°C. These cells can be stored up to 6 mo without loss of potential transformability. Cells prepared in sterile distilled water only can also be frozen, but there can be a 10-fold loss of electrocompetence. In both cases, cells should be thawed slowly on ice.

6. Using cells that have been isolated directly from agar plates may cause arcing or a low time constant. If this occurs, try using cells grown in broth and prepared by the method described in Section 3.1.1.

7. Try to avoid trapping air bubbles in the cell mixture once in the electroporation cuvet. The presence of air bubbles can lower the transformation efficiency. This results in a lower than expected time constant.

8. Always ensure that the cuvet is well chilled prior to adding the cell mixture. Dry the outside of the cuvet to prevent arcing.

9. Always incubate the cells for a minimum of 1 h at a temperature permissive for growth, before plating onto selective media. This allows for expression of genes involved in antibiotic resistance. If no colonies appear on selective agar plates, it is worth considering the following action. Allow extra time for the expression phase (up to 3 h if necessary) and also decrease the amount of antibiotic added to agar plates, since there may be some difference in achieved levels of resistance to specific antibiotics with

different bacterial hosts. Finally, try the electroporation again, testing for viability of host and donor strains during the procedure (Fig. 1) (*see* Note 3).

10. The range of dilutions plated depends on a number of factors, including cell density and the time constant achieved at electroporation. This needs to be calibrated to individual experimental requirements, but as a loose rule of thumb, dilutions of 10^{-1}, 10^{-2}, and 10^{-3} will usually provide an appropriate number of transformants per plate.

11. There can be a lot of variation in the transfer efficiency between different bacterial species. Table 1 summarizes some of the transfer efficiencies that have been obtained.

References

1. Summers, D. K. and Withers, H. L. (1990) Electrotransfer: direct transfer of bacterial plasmid DNA by electroporation. *Nucleic Acids Res.* **18,** 2192.

2. Baulard, A., Jourdan, C., Mercenier, A., and Locht, C. (1992) Rapid mycobacterial plasmid analysis by electroduction between *Mycobacterium spp.* and *E. coli. Nucleic Acids Res.* **20,** 4105.

3. Pfau, J. and Youderian, P. (1990) Transferring plasmid DNA between different bacterial species with electroporation. *Nucleic Acids Res.* **18,** 6165.

4. Gilchrist, A. and Smit, J. (1991) Transformation of freshwater and marine Caulobacters by electroporation. *J. Bacteriol.* **173,** 921–925.

5. Heery, D. M., Powell, R., Gannon, F., and Dunican, L. K. (1989) Curing of a plasmid from *E. coli* using high-voltage electroporation. *Nucleic Acids Res.* **17,** 10,131.

6. Kennedy, C. K. (1971) Induction of colicin production by high temperature or inhibition of protein synthesis. *J. Bacteriol.* **108,** 10–19.

7. Heery, D. M. and Dunican, L. K. (1989) Improved efficiency M13 cloning using electroporation. *Nucleic Acids Res.* **17,** 8006.

CHAPTER 4

Transfer of Episomal and Integrated Plasmids from *Saccharomyces cerevisiae* to *Escherichia coli* by Electroporation

Laura Gunn, Jennifer Whelden, and Jac A. Nickoloff

1. Introduction

Yeast shuttle vectors are common tools in molecular and cellular studies of *Saccharomyces cerevisiae*, including studies of gene structure–function relationships, DNA repair, and recombination *(1)*. Shuttle vectors are manipulated easily in vitro, introduced into *Escherichia coli*, from which large quantities of pure plasmid DNA can be prepared, and then "shuttled" to a yeast host cell for analysis. The three essential features of yeast shuttle vectors are sequences that allow replication and selection in *E. coli* (an origin of replication and an antibiotic resistance marker), and a yeast selectable marker (usually a gene encoding an enzyme in metabolic pathway, such as *URA3* or *HIS3; 2,3*). Other elements may be added to confer properties required for particular applications *(4)*. Plasmids may exist in yeast as freely replicating circular molecules (episomes) of one of three types. Derivatives with an autonomously replicating sequence *(ARS)* from the 2-μm circle (a natural yeast plasmid; *5*) exist at high copy number (30–50/cell) and segregate to daughter cells with high efficiency. Plasmids with an *ARS* from a yeast chromosome have medium to high copy numbers, but they are rapidly lost because they segregate to daughter cells with low efficiency *(6)*. Adding a yeast centromere *(CEN)* to an *ARS* plasmid reduces the copy

From: *Methods in Molecular Biology, Vol. 47: Electroporation Protocols for Microorganisms*
Edited by: J. A. Nickoloff Humana Press Inc., Totowa, NJ

number to 1–2/cell, and increases stability (loss rates about 1%/generation). Adding telomere sequences to an *ARS/CEN* plasmid creates a yeast artificial chromosome (YAC), which is a freely replicating, low copy number linear molecule *(7)*.

DNA may also be introduced into yeast as linear DNA fragments integrated into host chromosomes. *ARS* elements are not necessary elements of integrating plasmids, but they may be included; *CEN* elements cannot be included, since integration creates a lethal dicentric chromosome. Linear DNA fragments may be integrated via one or two homologous reciprocal recombination events, producing duplicated or transplaced markers, respectively *(8,9)*. Such manipulations allow precise alterations of specific yeast chromosomal loci.

Because plasmid DNA is easily isolated from *E. coli,* plasmids in yeast are often transferred back to *E. coli* for detailed characterization. For example, molecular cloning may involve the selection of a complementing activity encoded by a DNA fragment from a plasmid library and the subsequent transfer (or "rescue") from yeast to *E. coli* for DNA sequence analysis. Plasmid rescue may also be used to confirm the structure (genotype) of a mutant gene introduced into yeast in order to correlate a genotype with a phenotype. In our laboratory, we frequently rescue plasmids that have undergone a homologous recombination event in yeast for structural analysis of recombinant products (e.g., ref. *10*).

Protocols for several types of electroporation-based plasmid rescue schemes are described in this chapter. Only circular DNA transforms *E. coli* efficiently *(11–13)*. Thus, only episomal plasmids can be rescued without intermediate enzymatic treatments. High copy number circular plasmids can be rescued directly (without steps to purify yeast DNA) by using a dual electric pulse procedure *(14)*. Direct transfer of low copy number plasmids is too inefficient to be generally useful, so a "semidirect" transfer method is used in which plasmid DNA is released from yeast cells by mechanical means, but that does not require steps to purify DNA. Relatively pure yeast DNA is required to rescue integrated plasmids from yeast chromosomes or regions of YACs linked to bacterial vector sequences. Purified yeast DNA is then digested with an appropriate restriction enzyme to release bacterial vector sequences linked to the fragment of interest, treated with T4 DNA ligase under conditions that favor fragment circularization over concatemerization, and electroporated into *E. coli* to effect plasmid rescue.

2. Materials

2.1. Yeast Culture and DNA Preparation

1. Yeast strains: Any strain harboring bacterial vector sequences is suitable (*see* Note 1).
2. YPD: 1% bacto-yeast extract (Difco, Detroit, MI), 2% bacto-peptone (Difco), and 2% dextrose. For solid medium, add 2% agar. Autoclave to sterilize. For this solution and all others, sterilize by autoclaving and store at room temperature unless noted otherwise.
3. Yeast minimal media (Bio 101, Vista, CA): Specific media requirements depend on the genotype of the yeast strain. Special formulations may be created using recipes in ref. *(1)*.
4. Lysis solution: 50 mM Tris-HCl, pH 7.5, 20 mM EDTA, and 1% SDS; prepare fresh from 10X stock solutions.
5. Acid washed glass beads (450 μ; Sigma, St. Louis, MO): Autoclave under excess dH$_2$O.
6. 5M potassium acetate (KAc).
7. 5M NaCl.
8. Polyethylene glycol (PEG; 6000 mol wt; Sigma): 30% solution.
9. TE: 10 mM Tris-HCl, pH 8.0, 0.5 mM EDTA.
10. Siliconized glass beads, 200–300 μm (diameter) (Sigma): Wash with CHCl$_3$ to remove machine oil, siliconize by treating for 5 min with 5% dimethyldichlorosilane:95% CHCl$_3$, wash with H$_2$O, and then 95% ethanol. Repeat this wash sequence five times and autoclave under excess dH$_2$O.
11. Sepharose CL-6B (Pharmacia, Piscataway, NJ): Wash with an equal volume of TE six times, prepare a 60% slurry, and autoclave. Store at 4°C.
12. 27-gage syringe needle.
13. 5X stop mix: 25% glycerol, 0.25% SDS, 5 mM EDTA, 0.125% bromophenol blue, and 0.125% xylene cyanol FF; filter-sterilize.
14. Restriction enzymes and buffers: Used to excise integrated plasmids from yeast chromosomal DNA. The specific enzyme(s) used will depend on which restriction sites flank the integrated plasmid. Restriction enzymes are normally sold with concentrated buffers.
15. 10X ligase buffer: 500 mM Tris-HCl, pH 7.8, 100 mM MgCl$_2$, 100 mM dithiothreitol, 10 mM ATP, and 250 μg/mL bovine serum albumin.
16. T4 DNA ligase (New England Biolabs, Beverly, MA).
17. 3M sodium acetate (NaAc), pH 7.0: sterilize by autoclaving.

2.2. Bacterial Culture and Electroporation

1. Electrocompetent *E. coli* are available from a variety of commercial sources, such as Stratagene (La Jolla, CA) and Gibco-BRL (Bethesda, MD). Suitable strains include DH5α, HB101, XL1-Blue, and SURE cells

(*see* Note 2). Alternatively, electrocompetent cells can be prepared as described in Chapters 5 and 8.

2. Antibiotic stock solutions: ampicillin (100 mg/mL); and kanamycin (40 mg/mL). Store at –20°C.
3. Luria broth (LB) antibiotic plates: 5 g/L yeast extract (Difco), 10 g/L Bacto-tryptone (Difco), and 10 g/L NaCl (Sigma). Adjust to pH 7.0, then add agar (15 g/L), and autoclave. Cool to 45–50°C on a stir plate, add 1 mL/L of an appropriate antibiotic stock solution, mix completely, and pour into Petri dishes. Store at 4°C.
4. SOC: Dissolve 2 g bacto-tryptone (Difco) and 0.5 g yeast extract (Difco) in 100 mL of H_2O. Add 200 µL of 5M NaCl and 250 µL of 1M KCl. Adjust to pH 7.0 and autoclave. Cool to room temperature, and then add 1 mL of 1M $MgCl_2$ and 2 mL of 20% (w/v) glucose. Store at room temperature.

3. Methods

Each of the following procedures requires competent bacterial cells with transformation efficiencies $>2 \times 10^9$/µg of pure supercoiled plasmid DNA (such as pUC19). Because relatively few transformants are obtained in some procedures, it is important that transformation efficiencies are maximized. This is accomplished by thawing frozen cells on ice immediately before use; precooling all tubes, DNA solutions, and electroporation cuvets to 4°C on ice; and adding room temperature SOC immediately following the electric pulse(s). Discussions about optimizing *E. coli* transformation in Chapters 5 and 8 are also relevant to the procedures described below.

3.1. Direct Transfer of High Copy Number Plasmids from Yeast to E. coli

This procedure is based on the report by Marcil and Higgins *(14)*. Although only tested with high copy number plasmids, the reported efficiencies suggested that it may also be effective with low copy number plasmids. However, no transformants were obtained in four attempts to transfer a 7-kbp low copy number *ARS/CEN* plasmid from yeast to *E. coli* (unpublished results). We currently use the procedures described in Sections 3.2. or 3.3. to rescue *ARS/CEN* plasmids.

1. Thaw frozen competent *E. coli* on ice, and transfer 40 µL of cells to a fresh 1.5-mL tube.
2. Using a plastic pipet tip, transfer a yeast colony (grown for 2–3 d at 30°C) to the bacterial cells, and mix well by pipeting.
3. Transfer the yeast/bacterial cell suspension to a cold electroporation cuvet with a 1-mm electrode gap.

4. Pulse once at 750 V, 50 µF, 100 Ω (*see* Note 3). Incubate on ice for 30–60 s.
5. Pulse a second time at 1500 V, 50 µF, 150 Ω (*see* Note 4). Immediately add 1 mL of SOC.
6. Transfer the suspension to a 15-mL tube, and incubate with agitation at 37°C for 1 h.
7. Centrifuge at 3000*g* for 5 min, suspend cells in 100–200 µL of the supernatant, and transfer the entire suspension to a single LB plate containing an appropriate antibiotic (*see* Notes 5 and 6).

3.2. Semidirect Transfer of Low Copy Number Plasmids from Yeast to **E. coli**

1. Streak yeast cells to a minimal medium plate to maintain unstable plasmids, and incubate at 30°C for 2 d (*see* Note 7).
2. Add 0.3 mL of acid-washed glass beads to a 1.5-mL tube. Glass beads are most easily dispensed with a 10- or 25-mL plastic pipet held vertically to avoid adding dH_2O. If excess dH_2O is transferred, remove it with a Pasteur pipet.
3. Using a toothpick, transfer about 2×10^7 cells to the glass beads. This quantity is approximately the number obtained from a streak $2–3 \times 10$ mm (*see* Note 8).
4. Vortex vigorously for 5 min, heat to 100°C for 1 min, and then incubate on ice for at least 2 min.
5. Add 20–40 µL of electrocompetent *E. coli* to the treated yeast suspension, mix briefly with a pipet, and transfer to a chilled electroporation cuvet with a 2-mm electrode gap.
6. Pulse once with 2500 V, 25 µF, 200 Ω, and immediately add 1 mL of SOC.
7. Perform steps 6 and 7 in Section 3.1. (*see* Note 9).

3.3. High-Efficiency Transfer of Low Copy Number Plasmids from Yeast to **E. coli**

Although the procedure outlined in Section 3.2 normally yields sufficient quantities of bacterial transformants for most applications, occasionally a higher transfer efficiency may be required. For example, very large plasmids may transfer with reduced efficiency, or many bacterial transformants may be required, such as when it is necessary to determine the distribution of plasmid types within a population of yeast cells that may not be clonal with respect to plasmid content (e.g., ref. *10*). The following procedure requires the preparation of yeast DNA (steps 1–7; based on ref. *15*), and therefore, is more time-consuming than the direct and semidirect transfer methods. However, it is highly reproducible, and it yields 3- to 30-fold more transformants than the procedure in Section 3.2.

1. Inoculate 2 mL of minimal media in a 15-mL tube with a single yeast colony, and incubate with agitation until the culture reaches stationary growth phase (about 2 d for most strains). Centrifuge at 3000g for 5 min, discard supernatant, suspend cells in 4 mL of YPD, and incubate as above for 1 d (*see* Note 10).

2. Harvest cells by centrifugation as above, and discard the supernatant. Cell pellets may be stored for at least 6 mo at –20°C.

3. Wash cells once with 1 mL dH$_2$O, suspend in 0.5 mL of lysis buffer and pour into a 1.5-mL tube containing 0.3 mL of acid-washed glass beads (as in Section 3.2., step 2).

4. Vortex for 1 min, and then place on ice for 1 min. Repeat vortex/ice cycles five times.

5. Incubate at 65°C for 10 min. Add 200 µL of 5*M* KAc and 150 µL of 5*M* NaCl. Mix well and incubate for 20 min on ice.

6. Centrifuge for 10 min at 13,000g. Transfer 900 µL of the supernatant to a 1.5-mL tube; do not transfer any pellet material. Precipitate DNA by adding 300 µL of 30% PEG, mixing well, and incubating on ice for 20 min.

7. Centrifuge as above. Rinse DNA pellet with cold 80% ethanol. Dry DNA and suspend in 50 µL of TE (*see* Note 11).

8. Mix 1 µL of DNA with 20 µL of electrocompetent *E. coli,* transfer to a cuvet with a 2-mm electrode gap, pulse once with 2500 V, 25 µF, 200 Ω, and immediately add 1 mL of SOC (*see* Note 12).

9. Transfer cell suspension to a 15-mL tube, and incubate at 37°C for 1 h. Then transfer an appropriate amount to an LB antibiotic plate (*see* Note 13).

3.4. Transferring Integrated Plasmids from Yeast to E. coli

1. Inoculate a yeast colony into 4 mL of YPD, and incubate for 1–2 d at 30°C with agitation until culture reaches stationary phase. Integrated plasmids are normally stable, so there is no need to culture such strains in minimal medium.

2. Prepare yeast DNA (Section 3.3., steps 2–7) (*see* Note 14).

3. Purify DNA by CL-6B spin-column chromatography (*16*) as follows. Add 5 µL of 5X stop mix to the 50 µL yeast DNA solution, vortex, and then incubate for 5 min at 65°C. Prepare a column during this incubation by carefully poking a hole in the bottom of a 1.5-mL tube (with or without a cap) using a 27-gage needle, and add 50 µL of siliconized glass beads. Mix the Sepharose CL-6B slurry, and add 10 times the volume of the DNA solution to be purified (i.e., 550 µL). Place the column on a second 1.5-mL receiving tube. Centrifuge at 1500g for 2 min in a horizontal rotor. Replace the receiving tube with a new tube, and add DNA evenly to the column. Centrifuge as above, and discard the column tube. The purified DNA will

elute into the receiving tube without a significant change in volume or concentration (*see* Note 15).

4. Digest 1–2 µg of purified yeast DNA (approx 10 µL) with an appropriate restriction enzyme in a total volume of 20 µL to excise the plasmid from the chromosomal DNA.

5. Purify digested DNA through CL-6B (step 3). Dilute the sample to 44 µL. Add 5 µL of 10X ligase buffer, and 400 U (1 µL) of T4 DNA ligase. Mix by tapping gently, and incubate at 15°C for 4–24 h (*see* Note 16).

6. Precipitate DNA by adding 5 µL of 3*M* NaAc and 150 µL of 100% etha- nol. Mix well, incubate at –20°C for 30 min, and centrifuge at 13,000*g* for 10 min. Rinse pellet with cold 80% ethanol, air-dry the DNA pellet for 10– 15 min, and suspend in 5 µL of dH$_2$O. Mix 2 µL of DNA with 20 µL of electrocompetent *E. coli,* and transfer to an electroporation cuvet with a 2-mm electrode gap.

7. Pulse with 2500 V, 25 µF, 200 Ω, and immediately add 1 mL of SOC. Perform steps 6 and 7 in Section 3.1. (*see* Note 17).

4. Notes

1. We employ derivatives of the well-characterized yeast strain YPH250 *(17)*. Note, however, that laboratory yeast strains often have bacterial vector sequences integrated into chromosomal loci as a result of past chromo- somal manipulations. Such sequences are not problematic when rescuing episomal elements, since they will transform *E. coli* efficiently only if they are released from chromosomal DNA by digestion with a restriction enzyme and circularized with DNA ligase (Section 3.4.). However, when rescuing integrated plasmids, these extra plasmid sequences will often be res- cued as efficiently as desired plasmids (depending on the sizes of the res- cued plasmids; *see* Note 17), and additional screening steps may be required to identify desired plasmids. Alternatively, shuttle vectors may be constructed with a different antibiotic resistance marker (such as kanamycin) than that present in the host strain (normally ampicillin). In this example, *E. coli* transformants selected on plates with kanamycin will carry only the desired shuttle vector.

2. We use *E. coli* strains DH5α and HB101 most often for plasmid rescue. The safest strategy is to transfer into the same *E. coli* strain from which the shuttle vector was originally isolated.

3. Pulsing with 750 V, 50 µF, and 100 Ω should produce a time constant of about 5 ms. These settings are optimal for electroporation of DNA into yeast *(18)* and are therefore expected to be near optimal for electroporation of DNA out of yeast. If cuvets with 2-mm electrode gaps are used, pulse at 1500 V, 25 µF, 200 Ω, which also produces a 5-ms time constant.

4. The second pulse is optimal for transformation of *E. coli* and should be adjusted for individual strains. For example, we use cuvets with 2-mm electrode gaps and deliver the second pulse at 2500 V, 25 μF, 200 Ω.

5. Yeast does not grow on LB plates and will not interfere with growth of bacterial transformants. However, if too many bacterial cells are present on a plate, "pools" of dead cells will form and prevent growth of transformants. This problem is common if a large volume of *E. coli* cells is electroporated or if electrocompetent cells are highly concentrated.

6. A high copy number 2-μm circle-based plasmid (30–50 copies/cell) yielded up to 1000 transformants using the direct transfer procedure *(14)*.

7. Selection pressure should be applied continuously for markers carried by unstable *ARS* plasmids, *ARS/CEN* plasmids, and YACs by maintaining strains on appropriate minimal media. This will improve transfer efficiency by ensuring that a high proportion of the cells carry the unstable element.

8. Similar transfer efficiencies were obtained when $1–3 \times 10^7$ cells were used; arcing is common when $>4 \times 10^7$ cells were used.

9. We tested many alternative procedures to release plasmid DNA from yeast cells that were both simple and fast to perform. In each of the following procedures, we used a Bio-Rad (Richmond, CA) Gene Pulser set at 25 μF, 200 Ω, and *E. coli* cells with a transformation efficiency of 6×10^9 transformants/μg of pUC19. None of the following procedures were successful, or they yielded only one or a few transformants *(19)*.

   a. We performed dual electric pulse procedures, similar to those described in ref. *(14)*, using an initial pulse of 1.0, 1.5, or 2.0 kV and a second pulse of 2.5 kV, both 1- and 2-mm electrode gaps, and cells from a single 2-d-old colony or a streak (approx 3×20 mm). Cells from streaks were either not treated or treated before being added to 20 or 40 μL of electrocompetent *E. coli*. Yeast cell treatments included washing (suspension in 1 mL dH$_2$O, centrifugation at 3000g for 5 min, and resuspension in 10 μL of dH$_2$O), grinding (by vortexing for 5 min with glass beads, with or without a subsequent centrifugation as above to pellet cell debris), and digesting cell walls (by washing cells in 100 μL of 0.9M sorbitol, centrifuging as above, resuspending in 10 μL of 0.9M sorbitol, 1 mg/mL zymolyase 100T [Kirin, Japan], and incubating for 1 h at 37°C). Zymolyase digestions were either performed alone or they were followed by grinding with glass beads.

   b. We performed single electric pulse procedures with voltages of 0.5–2.5 kV using cuvets with 1- or 2-mm electrode gaps and 20 or 40 μL of electrocompetent *E. coli*. In these experiments, yeast cells were from a streak and were treated by grinding with glass beads.

c. We performed single electric pulse procedures with a voltage of 2.5 kV using cuvets with 2-mm electrode gaps, and 20 or 40 µL of *E. coli*. Yeast cells from a streak were subjected to one of ten treatments:

   i. Heating (100°C for 1 min).
   ii. Freezing (on dry ice for 1 min).
   iii. Freezing and then heating.
   iv. Grinding in the presence of 5% Triton X-100 (a nonionic detergent; Sigma).
   v. Heating and then grinding.
   vi. Heating with 5% Triton X-100 and then grinding.
   vii. Grinding with 5% Triton X-100 and then heating.
   viii. Grinding, then freezing, and then thawing (65°C for 2 min).
   ix. Heating, then freezing, and then thawing.
   x. Grinding, then freezing, and then heating.

d. Finally, we performed dual electric pulse procedures in which yeast cells from a streak were treated either by heating, grinding, or grinding and then heating. Treated yeast cells were subjected to a pulse of 1.5 kV in cuvets with 2-mm electrode gaps in the absence of *E. coli*, then 20 µL of *E. coli* cells were added, and a second pulse of 2.5 kV was applied. Interestingly, plasmid transfer was inefficient when heating preceded grinding, whereas the most efficient and reproducible procedure simply reverses these steps (Section 3.2., step 4), which typically yielded 5–20 transformants from a yeast strain carrying a 7.0-kbp plasmid.

10. By growing cells initially to stationary phase in minimal medium, a large number of cells are obtained, nearly all of which have unstable plasmid elements. Relatively few cells lose plasmids during the short burst of growth after transfer to YPD, but cells reach a higher density in this rich medium compared to minimal medium. Thus, this two-stage growth scheme maximizes both the total cell yield and the fraction of cells retaining unstable elements. Although similar results are expected if cells are grown to stationary phase in larger volumes of minimal medium, this requires longer incubation periods. If the initial inoculum is larger (i.e., from a streak or patch), only 1 d of growth in minimal medium is required to reach stationary phase.

11. Resuspension of yeast DNA can be accelerated by pipeting with a 200-µL tip. Make sure DNA is completely in solution before proceeding with electroporation. This DNA is often not pure enough for restriction enzyme digestions (*see* Note 14).

12. If arcing occurs, use a smaller volume of DNA, a larger volume of competent *E. coli* cells, or deionize the DNA solution by passing it through a Sepharose CL-6B spin column (Section 3.4., step 3). Higher transfer efficiencies are obtained by using 2 µL of DNA and 40 µL of competent *E. coli*.

13. Using *E. coli* with a transformation efficiency of $5 \times 10^9/\mu g$ of pUC19, the high efficiency transfer procedure typically yields 30–300 transformants. Therefore, if only one or a few transformants are required, only a fraction of the transformed cells need to be plated. In this case, there is no need to concentrate the cells prior to plating (Section 3.1., step 7), since 100–200 μL of the cell suspension will normally contain 3–60 transformants.

14. Restriction enzyme digestion of yeast DNA prepared by this method is not reproducible without further purification by CL-6B spin-column chromatography (5 min) or by a second ethanol precipitation (30–60 min).

15. The dyes should be retained by the spin column, indicating that small molecules (salts, SDS) were also retained. If dye elutes with the DNA, rerun the sample in a new column. The volume of CL-6B used is normally 10-fold larger than the sample size, but it can be increased slightly until dyes are completely retained in the column. However, DNA yield may be reduced if a too large an excess of CL-6B is used. The dyes also serve to identify columns to which DNA has been added. Samples must be applied evenly to the column. Otherwise they pass down the side of the tube and not through the column material. Fifty columns can be prepared and run in 30 min using repeating pipets.

16. Ligations are performed at low DNA concentrations (<50 $\mu g/mL$) to favor monomolecular reactions (recircularization) instead of concatemerization reactions. The DNA is subsequently concentrated by ethanol precipitation, since it is not concentrated enough in these dilute ligation reactions to transform *E. coli* efficiently.

17. When plasmids integrated into yeast chromosomes are transferred to *E. coli* having a transformation efficiency of $5 \times 10^9/\mu g$ of pUC19, 20–1000 transformants are usually obtained. If no transformants are obtained, digest more DNA and increase the volume of the ligation proportionally (to maintain DNA concentration at or below 50 $\mu g/mL$), but do not increase the volume in which precipitated DNA is suspended, such that a greater amount of DNA is added to cells. Yields may also be improved by using more T4 DNA ligase, increasing the incubation time of the ligation reaction, or by using cuvets with narrower electrode gaps (which increases the strength of the electric field, but also increases the chance of arcing). Transformation efficiency is also affected by plasmid size. Since the size of transferred plasmids depends on the positions of the restriction sites used to excise bacterial vector sequences from chromosomal DNA, it is possible to transfer plasmids of different sizes and with different sequences linked to vector sequences from a single strain simply by digesting with different enzymes.

Acknowledgments

This work was supported by grant CA55302 to JAN from the National Cancer Institute of the National Institutes of Health.

References

1. Sherman, F. (1991) Getting started with yeast. *Methods Enzymol.* **194,** 3–21.
2. Rose, M., Grisafi, P., and Botstein, D. (1984) Structure and function of the yeast *URA3* gene: expression in *Escherichia coli. Gene* **29,** 113–124.
3. Struhl, K. and Davis, R. W. (1977) Production of a functional eukaryotic enzyme in Escherichia coli: cloning and expression of the yeast structural gene for imidazole-glycerolphosphate dehydratase (*his3*). *Proc. Natl. Acad. Sci. USA* **74,** 5255–5259.
4. Schneider, J. C. and Guarente, L. (1991) Vectors for expression of cloned genes in yeast: regulation, overproduction, and underproduction. *Methods Enzymol.* **194,** 373–388.
5. Broach, J. R. and Volkert, F. C. (1991) Circular DNA plasmids of yeasts, in *The Molecular and Cellular Biology of the Yeast Saccharomyces cerevisiae* (Broach, J. R., Pringle, J. R., and Jones, E. W., eds.), Cold Spring Harbor Laboratory, Cold Spring Harbor, NY, pp. 297–331.
6. Fitzgerald-Hayes, M., Clarke, L., and Carbon, J. (1982) Nucleotide sequence comparisons and functional analysis of yeast centromere DNAs. *Cell* **29,** 235–244.
7. Burke, D. T. and Olson, M. V. (1991) Preparation of clone libraries in yeast artificial chromosome vectors. *Methods Enzymol.* **194,** 251–270.
8. Scherer, S. and Davis, R. W. (1979) Replacement of chromosome segments with altered DNA sequences constructed in vitro. *Proc. Natl. Acad. Sci. USA* **76,** 4951–4955.
9. Rothstein, R. J. (1983) One-step gene disruption in yeast. *Methods Enzymol.* **101,** 202–211.
10. Sweetser, D. B., Hough, H., Whelden, J. F., Arbuckle, M. A., and Nickoloff, J. A. (1994) Fine-resolution mapping of spontaneous and double-strand break-induced gene conversion tracts in *Saccharomyces cerevisiae* reveals reversible mitotic conversion polarity. *Mol. Cell. Biol.* **14,** 3863–3875.
11. Conley, E. C. and Saunders, J. R. (1984) Recombination-dependent recircularization of linearized pBR322 plasmid DNA following transformation of Escherichia coli. *Mol. Gen. Genet.* **194,** 211–218.
12. Deng, W. P. and Nickoloff, J. A. (1992) Site-directed mutagenesis of virtually any plasmid by eliminating a unique site. *Anal. Biochem.* **200,** 81–88.
13. Ray, F. A. and Nickoloff, J. A. (1992) Site-specific mutagenesis of almost any plasmid using a PCR-based version of unique-site elimination. *Biotechniques* **13,** 342–346.
14. Marcil, R. and Higgins, D. R. (1992) Direct transfer of plasmid DNA from yeast to *E. coli* by electroporation. *Nucleic Acids Res.* **20,** 917.
15. Fujimura, H. and Sakuma, Y. (1993) Simplified isolation of chromosomal and plasmid DNA from yeasts. *Biotechniques* **14,** 538–539.

16. Nickoloff, J. A. (1994) Sepharose spin column chromatography: a fast, nontoxic replacement for phenol:chloroform extraction/ethanol precipitation. *Mol. Biotechnol.* **1**, 105–108.

17. Sikorski, R. S. and Hieter, P. (1989) A system of shuttle vectors and yeast host strains designed for efficient manipulation of DNA in *Saccharomyces cerevisiae.* *Genetics* **122**, 19–27.

18. Becker, D. M. and Guarente, L. (1991) High-efficiency transformation of yeast by electroporation. *Methods Enzymol.* **194**, 182–187.

19. Gunn, L. and Nickoloff, J. A. (1994) Rapid transfer of low copy number episomal plasmids from *Saccharomyces cerevisiae* to *Escherichia coli* by electroporation. *Mol. Biotechnol.*, in press.

CHAPTER 5

Production of cDNA Libraries by Electroporation

Christian E. Gruber

1. Introduction

In the late 1970s and early 1980s, several laboratories *(1–4)* reported that an applied electric field could induce transient pores in eukaryotic cell membranes. In addition to small molecules, such as sucrose and dyes, DNA could be introduced into these electropermeabilized cells *(5–7)*. More recently, the electroporation process has been used to transform bacterial cells with plasmid DNA efficiently *(8,9)*.

The gram-negative bacteria, especially *Escherichia coli,* are generally more efficiently electrotransformed than the gram-positive bacteria *(10,11)*. A high field strength (12.5–16.7 kV/cm) applied to some strains of electrocompetent *E. coli* yields transformation efficiencies (>1 × 10^{10} transformants/µg plasmid DNA) *(8,9,12)* that are 10–20 times the level obtained with the most productive chemically competent cells. Additionally, up to 80% of all electrocompetent *E. coli* cells are capable of taking up DNA *(9)*. Nontransforming DNA competes with vector DNA in both chemical transformation and electrotransformation reactions *(13)*. However, owing to the high frequency of electrotransformation, competition from the nontransforming DNA byproduct (i.e., nonvector DNA) of DNA cloning is minimized.

An efficient method that produces complete cDNA libraries from small amounts of starting material (i.e., mRNA) is a major goal of cDNA cloning technology. Traditionally, λ vectors yield a higher cDNA clon-

From: *Methods in Molecular Biology, Vol. 47: Electroporation Protocols for Microorganisms*
Edited by: J. A. Nickoloff Humana Press Inc., Totowa, NJ

ing efficiency *(14)* than plasmid vectors *(15)* largely because of the high-efficiency in vitro packaging reactions. Nevertheless, plasmid vectors are attractive because they require less subcloning of the cDNA insert than λ vectors, and plasmids generally accommodate larger cDNA inserts than λ vectors *(16)*. Now, with the high cDNA cloning efficiencies attainable with electroporation, plasmid cloning rivals λ cloning and may become the preferred method for the production of large, complex cDNA libraries.

2. Materials

2.1. cDNA Synthesis and Cloning Kits/Plasmid Vectors

Various cDNA synthesis and cloning kits are commercially available (*see* product catalogs of Gibco/BRL [Gaithersburg, MD], Life Technologies [Gaithersburg, MD], Stratagene [La Jolla, CA], and so forth). Prokaryotic or eukaryotic plasmid vectors can be purchased from these same companies or from the American Type Culture Collection (Rockville, MD).

2.2. Eletrocompetent Cells

Gram-negative bacteria are made electrocompetent more easily than the gram-positive species *(10,11)*. Although electrocompetent *Agrobacterium* can be purchased from Gibco/BRL, *Escherichia coli* is the major electrocompetent bacteria sold commercially. Table 1 lists some strains of *E. coli* suitable for cDNA cloning. Certain strains (i.e., DH10B) will not restrict cDNA that contains methylated cytosine and adenine residues. Several strains (i.e., DH12S) may be used for the production of single-stranded DNA for subtractive cDNA libraries *(17,18)*. Additionally, many strains are *rec*A1 (the SURE cells are *rec*B, *uvr*C, *umu*C, and *sbc*C), which increases the stability of the cDNA inserts. Although these strains were genetically engineered for the production and maintenance of cDNA libraries, this does not preclude the use of other bacteria for this purpose. Most species of gram-negative and gram-positive bacteria can be made electrocompetent.

2.3. Solutions

1. SOB medium (without magnesium): 20 g/L bacto-tryptone, 5 g/L bacto-yeast extract, 0.584 g/L NaCl, and 0.186 g/L KCl. Adjust pH to 7.0 with NaOH, and autoclave. Store at 4°C for up to 3–6 mo.
2. 2*M* glucose: 36.04 g/100 mL glucose. Filter-sterilize and store at 4°C for up to 3–6 mo.

Table 1

Commercially Available Electrocompetent *Escherichia coli* Cells

	DH5α[b,c]	DH10B[c]	DH10B/p3[c]	MC1061/p3[b]	DH12S[c]	JS5[a]	XL1-Blue MRF[d]	XL1-Blue[d]	SURE[d]	ABLE[d]
cDNA library construction	X	X	X	X	X	X	X	X	X	X
Cloning methylated cDNA	X	X	X		X	X	X	X	X	X
Single-stranded DNA rescue (F')					X	X	X	X	X	
Blue/white color selection	X	X	X		X	X	X	X	X	
Lack of homologous recombination	X	X	X		X	X	X	X	X	

[a]Bio-Rad Laboratories (Hercules, CA).
[b]ClonTech Laboratories, Inc. (Palo Alto, CA).
[c]Gibco/BRL Life Technologies, Inc.
[d]Stratagene.

3. $2M$ Mg^{2+} stock: 20.33 g/100 mL $MgCl_2 \cdot 6H_2O$ and 24.65 g/100 mL $MgSO_4 \cdot 7H_2O$. Autoclave and store at 4°C for 6 mo to 1 yr.
4. SOC medium: Add 98 ml of SOB medium to 1 mL of sterile $2M$ glucose and 1 mL of sterile $2M$ Mg^{2+} stock. Mix and store at 4°C for 3–6 mo.
5. WB solution: Add 100 mL of redistilled glycerol to 900 mL of distilled water. Mix and filter-sterilize. Store at 4°C for 3–6 mo.
6. 1X TE buffer: 10 mM Tris-HCl, pH 7.5, and 1 mM EDTA. Autoclave and store at 4°C for 6 mo to 1 yr.
7. 5X T4 DNA ligase buffer: 250 mM Tris-HCl, pH 7.6, 50 mM $MgCl_2$, 5 mM ATP, 5 mM DTT, and 25% (w/v) PEG 8000. Sterilize and store at −20°C for up to 1 yr.
8. Stabilization solution: Add 40 mL of redistilled glycerol to 60 mL of SOC medium. Mix and filter-sterilize. Store at 4°C for 3–6 mo.

2.4. Equipment

1. Electroporation pulse generator: An electroporation apparatus capable of generating a pulse length of 4 ms and a field strength of 16,600 V/cm.
2. Sterile electroporation chambers: Disposable chambers or cuvets containing two electrodes separated by an average distance of 0.1–0.15 cm.
3. Bacterial incubator: Capable of maintaining a temperature of 37°C and able to shake 500 mL to 2.8 L flasks at a setting of 275 rpm.

3. Methods

3.1. Preparation of Electrocompetent E. coli Cells

As shown in Table 1, many strains of *E. coli* can be purchased as frozen electrocompetent cells. Usually, these strains will be sufficient for the production of plasmid-based cDNA libraries. However, if other gram-negative bacterial strains are deemed more appropriate, the protocol described below, for the production of electrocompetent *E. coli*, may be used as a starting point.

Gram-positive bacteria tend to be much more difficult to make electrocompetent than the gram-negative bacteria. In fact, the highest reported efficiencies of transformation (10^7 transformants/µg DNA) *(10)* are far below the values obtained with *E. coli*. Because of these limitations, gram-positive bacteria are infrequently, if ever, used for the production of cDNA libraries. Nevertheless, procedures for generating electrocompetent gram-positive bacteria can be found in the literature *(10,11,19–21)*.

1. Pick a single colony from a freshly streaked agar plate and inoculate 50 mL of SOB medium (without magnesium) in a 500-mL flask. Incubate overnight at 37°C with shaking at 275 rpm.

2. On the next day, dilute 0.5 mL of this overnight culture into 500 mL of SOB medium (without magnesium) in a 2.8-L flask. Grow cells for 2–3 h at 37°C with shaking at 275 rpm until an optical density$_{550}$ (OD$_{550}$) of 0.8 is reached. It is important to harvest the cells in the early or middle phases of their logarithmic growth. Cells taken in the late logarithmic or stationary phases will have decreased electrocompetence *(8,22)*.
3. Harvest the cells by transfer into chilled centrifuge bottles, followed by centrifugation at 2600g for 10 min at 4°C. Carefully remove the supernatant.
4. The cell pellet must be thoroughly washed to remove all salts carried over from the growth medium. If the conductivity of the cell preparation is not reduced, arcing will occur during the electroporation, resulting in sample loss *(9,20)*. Resuspend the cell pellet in 500 mL of sterile ice-cold WB solution. Centrifuge the cell suspension, as in step 3, for 15 min, and carefully remove the supernatant.
5. Resuspend the cell pellet, as in step 4, centrifuge for 15 min, and carefully remove the supernatant.
6. Resuspend the cell pellet in the WB that remains in the centrifuge bottle, measure the volume, and determine the OD$_{550}$. Adjust the final volume (usually 2 mL) of the cell suspension to an OD$_{550}$ of 200–300 U/mL. Cells can be used immediately or can be frozen in 0.2-mL aliquots in microcentrifuge tubes using a dry ice/ethanol bath. Store frozen cells at –70°C. These cells should maintain their electrocompetence for up to 1 yr.

3.2. cDNA Synthesis and Ligation

Owing to the abundant number of commercially available cDNA synthesis and cloning kits, the generation of double-stranded cDNA from mRNA is now fairly straightforward. The preferred cDNA kits are those that employ a reverse transcriptase lacking the ribonuclease H activity to generate first-strand cDNA *(22,23)*. This enzyme produces higher yields of first-strand cDNA and greater full-length cDNA synthesis than other reverse transcriptases *(24,25)*.

Before synthesizing double-stranded cDNA, the type of plasmid cloning vector should be considered. Depending on the downstream application, cDNA can be cloned into the appropriate multifunctional prokaryotic or eukaryotic plasmid vectors (*see* ref. *16* for review). For the most part, these plasmid vectors are commercially available.

To optimize the cDNA cloning potential or colony-forming units, various ligation reactions containing different cDNA-to-vector-mass ratios should be examined. For more information, *see* Notes 2 and 3.

1. Add the appropriate amount of cDNA and plasmid vector to a sterile 1.5-mL microcentrifuge tube at room temperature. Add 4 µL of 5X T4 DNA

ligase buffer and enough autoclaved water to bring the volume to 19 µL. Add 1 µL (1 U) of T4 DNA ligase, and mix by pipeting.

2. Incubate this reaction for 3 h at room temperature.

3. To prevent arcing problems, the conductivity of the DNA sample must be minimized. Ligation reactions may be diluted *(26)*, drop-dialyzed *(27,28)*, phenol-chloroform-extracted, and microfiltered *(29)* or ethanol-precipitated *(30)*. Although all of these protocols are effective, the ethanol precipitation is the preferred method, since it yields the most concentrated cDNA sample. Add 5 µL (1 µg/µL) of yeast tRNA or 10 µg of glycogen and 12.5 µL of 7.5M NH_4OAc to the ligation reaction. Add 70 µL of absolute ethanol (–20°C). Vortex the mixture thoroughly and immediately centrifuge at room temperature for 20 min at 14,000g.

4. Remove the supernatant carefully, and overlay the pellet with 0.5 mL of 70% ethanol (–20°C). Centrifuge for 2 min at 14,000g, and remove the supernatant. Dry the ligated cDNA at 37°C for 10 min to evaporate residual ethanol.

5. Add 5 µL of TE to the dried pellet, vortex, and collect the contents of the tube by brief centrifugation. The DNA solution is now ready for electroporation.

3.3. Electroporation

1. Label all necessary 15-mL tubes (e.g., Falcon 2059) and autoclaved microcentrifuge tubes (one each per transformation). Add 1 mL of SOC medium to each Falcon tube, and set at room temperature. Dispense 1 µL of DNA solution to each microcentrifuge tube, and place on ice. Always determine the transformation efficiency of the electrocompetent cells by including one control tube (use 1 µL = 10 pg of plasmid DNA) for every set of electroporation reactions. This information can be used to troubleshoot problems with the cDNA cloning.

2. Place the required number of sterile electroporation chambers (1/transformation) and frozen electrocompetent cells on ice.

3. After the cells are thawed (approx 10 min on ice), mix gently and transfer the appropriate amount of cells (e.g., 24 µL/transformation using the Gibco/BRL Cell-Porator Electroporation System and 40 µL/transformation using the Bio-Rad Gene Pulser) to a microcentrifuge tube containing DNA. Using a micropipet, pipet this DNA–cell slurry gently to mix and place into an electroporation chamber between the electrode bosses.

4. Pulse this DNA–cell mixture once at 4°C using the optimal conditions for the appropriate cell strain (a pulse length of approx 4 ms and a field strength of 16,600 V/cm are used for DH10B; *see* Notes 4–6 for more information). By chilling the chambers and DNA on ice and conducting the electroporation at 4°C, arcing across the chamber is minimized, and the efficiency of transformation is enhanced *(22,31)*.

5. Following the electric pulse, SOC medium must be added immediately to the DNA–cell mixture. Without SOC medium, the transformation efficiency may be reduced threefold in the first minute after electroporation *(9)*. Using a sterile Pasteur pipet (15 cm), remove a portion (about 500 μL) of the SOC medium from the designated Falcon tube, and gently rinse the electroshocked cells into the bottom of the chamber. Pipet the diluted cells into the Falcon tube containing the remainder of the SOC. Repeat steps 3–5 using a new electroporation chamber each time until all DNA samples are electroporated. Do not refreeze the cells or subsequent transformations will deliver lower efficiencies.

6. Incubate the electroporated cells for 1 h at 37°C with shaking before plating on selective plates.

3.4. Plating and Storage

1. Label the required number of autoclaved microcentrifuge tubes (equal to the number of electroporated samples), add 1 mL of SOC to each of these tubes, and recap. After the 1-h incubation, remove 2 μL from each electroporated sample (from the control sample remove 20 μL), and add to the appropriately marked tube. Gently mix these diluted cells.

2. To the balance of the transformation within the Falcon tubes, add an equal volume (approx 1 mL) of a sterile stabilization solution. Mix and place at –70°C. The titer of these frozen transformed cells will remain unchanged for up to 1 yr.

3. Place 100 μL of SOC into the middle of each 100-mm agar plate containing the appropriate antibiotic (2 × number of transformations). From each diluted transformation, transfer 5 and 50 μL to two separate agar plates, and immediately spread this mixture uniformly over the surface of the plates. As indicated in Table 1, certain strains of *E. coli* containing the ø80d/*lac*ZΔM15 marker provide α-complementation of the β-galactosidase gene from pUC19 or similar vectors, and therefore, can be used for blue/white color selection. Therefore, if X-gal (final concentration = 0.01%) is included in the agar, background colonies (vector containing no cDNA inserts) will be blue and recombinant colonies (vector containing cDNA inserts) will be white. To perform blue/white color selection with vectors that contain the *lac*I repressor gene, isopropylthio-β-galactoside (50 μL of 0.1*M* IPTG) must first be spread on the X-gal plates.

4. After spreading the diluted transformations, incubate the agar plates overnight at 37°C. The following day, count the number of viable colonies and determine the transformation efficiency (recombinants/μg vector DNA) of the control, the cloning efficiency (recombinants/μg cDNA), and the size of the cDNA library (total recombinants/electroporation).

The transformation efficiency can be determined by the following equation:

$$\text{Recombinants/µg} = (\text{recombinants on agar plate/ng vector DNA used in transformation}) \times (1 \times 10^3 \text{ ng/µg} \times \text{dilution factor}) \quad (1)$$

For example, if 10 ng of vector DNA yields 25 colonies when 5 µL from 1 mL of a 1:500 dilution is plated, then:

$$\text{Recombinants/µg} = (25 \text{ recombinants/10 ng}) \times (1 \times 10^3 \text{ ng/µg}) \times (2 \times 10^2) \times (5 \times 10^2) = 2.5 \times 10^8 \quad (2)$$

The cloning efficiency can be determined by the same equation, except the denominator would be ng of cDNA used in the transformation. If 2 ng of cDNA yield 25 colonies, then the recombinants/µg cDNA = 1.25×10^9.

The size of the cDNA library can be determined by using a different equation:

$$(\text{Total recombinants/ligation}) = \text{recombinants on agar plate} \times \text{dilution factor} \times 5 \ (\# \text{ of electroporations/ligation}) \quad (3)$$

Therefore, in the example above, if 5 µL of a 1:500 dilution yields 25 colonies then:

$$(\text{Total recombinants/ligation}) = 1.25 \times 10^7 \quad (4)$$

Generally, cDNA libraries are constructed in small plasmid DNA vectors (2–6 kbp). The transformation efficiencies of these vector–cDNA ligations range from 1.0×10^8 to 1.0×10^9 recombinants/µg of vector DNA. Cloning efficiencies can vary between 5×10^8 and 5×10^9 recombinants/µg of cDNA. Finally, the size of the library is usually between 5×10^6 and 5×10^7 recombinants/ligation. Electrocompetent bacteria that yield $<1 \times 10^{10}$ transformants/µg control plasmid DNA and the use of larger plasmid vectors (>6 kbp) in cDNA cloning may result in lower efficiency values than stated above (*see* Note 1).

4. Notes

1. Large plasmid size may affect the transformation efficiency of electrocompetent bacteria. Although this effect has not been thoroughly studied, every *E. coli* strain tested to date *(12,32–34)* shows a reduced transformation efficiency (transformants/µg plasmid DNA) with increasing plasmid size (2.6–200 kbp). Therefore, to enhance the representation of large cDNA

inserts within the cDNA library, an effort should be made to minimize the size of the plasmid cloning vector (i.e., <6 kbp).

2. To avoid the introduction of multiple plasmid–cDNA inserts into individual bacterial cells, do not saturate their transformation capacity *(35)*. The saturation capacity (no additional transformants after a given DNA concentration) of electrocompetent *E. coli* cells is about 10-fold higher than for high-efficiency chemically competent cells. Nevertheless, care should be taken to add no more than 10 ng of ligated vector DNA (for plasmids <6 kbp)/electroporation of commercially prepared cells (saturation capacity of 10–100 ng of pUC19) *(12,33)*. To determine the saturation capacity of other electrocompetent bacteria, electroporate different amounts of pUC19 ranging from 0.01 to 100 ng/reaction.

3. One major benefit of electroporation is the ability to create a large, complex cDNA library of >1 × 10^7 total recombinants from one ligation *(36)*. However, achieving this level of recombination requires the prior determination of the cDNA saturation level. As mentioned in Note 2, 10 ng of vector DNA/electroporation will be adequate in most cases. Based on this, 50 ng of vector DNA should be used/ligation (one electroporation requires one-fifth of the ligation). Various amounts of cDNA should be ligated to this quantity of vector DNA and one-fifth of these ligation reactions electroporated into the appropriate bacterial cells. Figures 1 and 2 show results from an optimization procedure *(36)*. In this example, the cDNA cloning efficiencies are >1 × 10^9 and remain essentially unchanged with cDNA concentrations from 0.1 to 3 ng/ligation. In Fig. 1, the decreased efficiency from 3 to 20 ng of cDNA reflects the approaching saturation of the available vector DNA. These higher cDNA concentrations yield >1 × 10^7 total recombinants from one ligation reaction (Fig. 2).

4. The electroporation of bacterial cells requires very high field strengths. Field strength is defined as the voltage applied to the cell suspension divided by the distance between the electrodes. The optimum field strength tends to be species- and strain-specific, and for *E. coli* is 12.5–16.7 kV/cm (DH10B and DH12S = 16.6 kV/cm) *(35)*. Generally, the *E. coli* electroporation conditions can be applied to most of the gram-negative bacteria with minor adjustments (i.e., 1–2 kV/cm increments).

5. The other parameter that affects the transformation efficiency is pulse length. Pulse length can be controlled by changing the capacitance or the resistance. For many bacteria, a pulse length of 2–6 ms is appropriate (DH10B and DH12S = 4 ms). To some extent, the pulse length and field strength counterbalance one another *(9)*. Therefore, increasing the pulse length can marginally compensate for lower field strength.

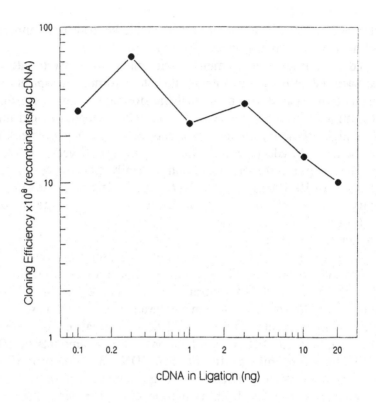

Fig. 1. Cloning efficiency vs the amount of cDNA in the ligation reaction. Different amounts of *Eco*RI-adapted HeLa cDNA (0.1, 0.3, 1, 3, 10, or 20 ng) were ligated to 50 ng of *Eco*RI-digested, dephosphorylated pSPORT 1 vector DNA in a 20-µL reaction for 3 h at room temperature. The ligation reaction mixtures were ethanol precipitated, and the DNA pellets were washed with 70% ethanol and dissolved in 5 µL of TE buffer. One microliter of the cDNA ligation was mixed with 24 µL of ELECTROMAX DH10B™ cells (Gibco/ BRL Life Technologies). After electroporation and a 1-h incubation at 37°C in 1 mL of SOC medium, appropriate dilutions of the cells were grown on LB plates containing 100 µg/mL ampicillin. Reprinted from ref. *36* with permission.

6. Another indicator for optimal field strength and pulse-length conditions is the percentage of cell survivors. For *E. coli*, a cell survival rate of 50–80% will yield the highest transformation efficiency *(35)*. Excessive field strength or pulse length leads to irreversible cell membrane damage and <50% cell survival, as evidenced by a slightly foamy cell suspension after

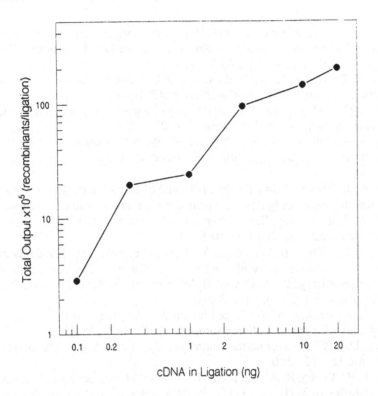

Fig. 2. Total recombinant output vs the amount of cDNA in the ligation reaction. Reprinted from ref. *36* with permission.

electroporation. Attempts to compensate for lower field strength by repeated pulses of the cell suspension will reduce the cell viability and decrease the transformation efficiency *(31)*.

References

1. Kinosita, K., Jr. and Tsong, T. Y. (1977) Hemolysis of human erythrocytes by a transient electric field. *Proc. Natl. Acad. Sci. USA* **74,** 1923–1927.
2. Baker, P. F. and Knight, D. E. (1978) A high voltage technique for gaining rapid access to the interior of secretory cells. *J. Physiol.* **284,** 30.
3. Gauger, B. and Bentrup, F. W. (1979) A study of dielectric membrane breakdown in the *Fucus* egg. *J. Membrane Biol.* **48,** 249–264.
4. Zimmermann, U., Vienken, J., and Pilwat, G. (1980) Development of drug carrier systems: electric field induced effects in cell membranes. *J. Electroanal. Chem.* **116,** 553–574.
5. Wong, T. K. and Neumann, E. (1982) Electric field mediated gene transfer. *Biochem. Biophys. Res. Commun.* **107,** 584–587.

6. Potter, H., Weir, L., and Leder, P. (1984) Enhancer-dependent expression of human k immunoglobulin genes introduced into mouse pre-B lymphocytes by electroporation. *Proc. Natl. Acad. Sci. USA* **81,** 7161–7165.

7. Fromm, M. L., Taylor, P., and Walbot, V. (1986) Stable transformation of maize after gene transfer by electroporation. *Nature* **319,** 791–793.

8. Calvin, N. M. and Hanawalt, P. C. (1988) High-efficiency transformation of bacterial cells by electroporation. *J. Bacteriol.* **170,** 2796–2801.

9. Dower, W. J., Miller, J. F., and Ragsdale, C. W. (1988) High efficiency transformation of *E. coli* by high voltage electroporation. *Nucleic Acids Res.* **16,** 6127–6145.

10. Dower, W. J., Chassy, B. M., Trevors, J. T., and Blaschek, H. P. (1992) Protocols for the transformation of bacteria by electroporation, in *Guide to Electroporation and Electrofusion* (Chang, D. C., Chassy, B. M., Saunders, J. A., and Sowers, A. E., eds.), Academic, San Diego, pp. 485–499.

11. Trevors, J. T., Chassy, B. M., Dower, W. J., and Blaschek, H. P. (1992) Electrotransformation of bacteria by plasmid DNA, in *Guide to Electroporation and Electrofusion* (Chang, D. C., Chassy, B. M., Saunders, J. A., and Sowers, A. E., eds.), Academic, San Diego, pp. 265–290.

12. Smith, M., Jessee, J., Landers, T., and Jordan, J. (1990) High efficiency bacterial electroporation: 1×10^{10} *E. coli* transformants/µg. *Focus* **12,** 38–40.

13. Hanahan, D. (1983) Studies on transformation of *Escherichia coli* with plasmids. *J. Mol. Biol.* **166,** 557–580.

14. Huynh, T. V., Young, R. A., and Davis, R. W. (1985) Constructing and screening cDNA libraries in λgt10 and λgt11, in *DNA Cloning: A Practical Approach* (Glover, D. M., ed.) IRL, Oxford, pp. 49–78.

15. Hanahan, D. (1985) Techniques for transformation of *E. coli,* in *DNA Cloning: A Practical Approach* (Glover, D. M., ed.) IRL, Oxford, pp. 109–135.

16. Sambrook, J., Fritsch, E. F., and Maniatis, T. (1989) *Molecular Cloning: A Laboratory Manual,* 2nd ed. Cold Spring Harbor Laboratory, Cold Spring Harbor, NY.

17. Rubenstein, J. L., Bruce, A. J., Ciaranello, R. D., Denney, D., Porteus, M. H., and Usdin, T. B. (1990) Subtractive hybridization system using single-stranded phagemids with directional inserts. *Nucleic Acids Res.* **18,** 4833–4842.

18. Li, W.-B., Gruber, C. E., Lin, J.-J., Lim, R., D'Alessio, J. M., and Jessee, J. A. (1994) The isolation of differentially expressed genes in fibroblast growth factor stimulated BC3H1 cells by subtractive hybridization. *BioTechniques* **16,** 722–729.

19. Chassy, B. M. and Flickinger, J. L. (1987) Transformation of *Lactobacillus casei* by electroporation. *FEMS Microbiol. Lett.* **44,** 173–177.

20. Fiedler, S. and Wirth, R. (1988) Transformation of bacteria with plasmid DNA by electroporation. *Anal. Biochem.* **170,** 38–44.

21. Dunny, G. M., Lee, L. N., and LeBlanc, D. J. (1991) Improved electroporation and cloning vector system for gram-positive bacteria. *Appl. Environ. Microbiol.* **57,** 1194–1201.

22. Dower, W. J. (1990) Electroporation of bacteria: a general approach to genetic transformation, in *Genetic Engineering—Principles and Methods,* vol. 12 (Setlow, J. K., ed.) Plenum, New York, pp. 275–296.

23. Kotewicz, M. L., Sampson, C. M., D'Alessio, J. M., and Gerard, G. F. (1988) Isolation of cloned Moloney murine leukemia virus reverse transcriptase lacking ribonuclease H activity. *Nucleic Acids Res.* **16,** 265–277.

24. Gerard, G. F., D'Alessio, J. M., and Kotewicz, M. L. (1989) cDNA synthesis by cloned Moloney murine leukemia virus reverse transcriptase lacking RNase H activity. *Focus* **11,** 66–69.

25. D'Alessio, J. M., Gruber, C. E., Cain, C., and Noon, M. C. (1990) Construction of directional cDNA libraries using the SUPERSCRIPT Plasmid System. *Focus* **12,** 47–50.

26. Willson, T. A. and Gough, N. M. (1988) High voltage *E. coli* electrotransformation with DNA following ligation. *Nucleic Acids Res.* **16,** 11,820.

27. Heery, D. M. and Dunican, L. K. (1989) Improved efficiency M13 cloning using electroporation. *Nucleic Acids Res.* **17,** 8006.

28. Jacobs, M., Wnendt, S., and Stahl, U. (1990) High-efficiency electrotransformation of *Escherichia coli* with DNA from ligation mixtures. *Nucleic Acids Res.* **18,** 1653.

29. Kobori, M. and Nojima, H. (1993) A simple treatment of DNA in a ligation mixture prior to electroporation improves transformation frequency. *Nucleic Acids Res.* **21,** 2782.

30. Zabarovsky, E. R. and Winberg, G. (1990) High efficiency electroporation of ligated DNA into bacteria. *Nucleic Acids Res.* **18,** 5912.

31. Taketo, A. (1988) DNA transfection of *Escherichia coli* by electroporation. *Biochim. Biophys. Acta* **949,** 318–324.

32. Leonardo, E. D. and Sedivy, J. M. (1990) A new vector for cloning large eukaryotic DNA segments in *Escherichia coli*. *BioTechnology* **8,** 841–844.

33. Lin, J.-J., Jessee, J., and Bloom, F. (1992) DH12S: a new electrocompetent *E. coli* strain for the production of highly purified single-stranded DNA using phagemid vectors. *Focus* **14,** 98–101.

34. Siguret, V., Ribba, A.-S., Cherel, G., Meyer, D., and Pietu, G. (1994) Effect of plasmid size on transformation efficiency by electroporation of *Escherichia coli* DH5α. *BioTechniques* **16,** 422–426.

35. Hanahan, D., Jessee, J., and Bloom, F. (1991) Plasmid transformation of Escherichia coli and other bacteria, in *Methods in Enzymology,* vol. 204, Academic, San Diego, CA, pp. 63–113.

36. Gruber, C. E. (1992) High-efficiency cDNA cloning: a comparison of electroporation and in vitro packaging. *BioTechniques* **12,** 804–808.

CHAPTER 6

Electroporation of RNA into *Saccharomyces cerevisiae*

Daniel R. Gallie

1. Introduction

The cytoplasmic regulation of gene expression has received increased attention in recent years. In order to assess directly the in vivo impact of regulatory elements on the translational efficiency and stability of an mRNA in higher eukaryotes, a variety of methods that deliver RNA directly to the cytoplasm have been developed to avoid any potential complications associated with transcription, pre-mRNA processing, or nucleocytoplasmic transport. Studies in yeast, however, have been largely limited to DNA-based constructs, either as episomes or as genomically integrated genes. Moreover, certain types of mRNAs can not be generated in vivo, e.g., uncapped messages or poly(A)⁻ mRNAs. In vitro translation lysates have been developed for yeast *(1–3)*. However, lysates derived from higher eukaryotes do not reflect the full cytoplasmic regulation observed in vivo. For studies focusing on post-transcriptional regulation in yeast, an RNA-based delivery system provides an in vivo approach to the analysis of posttranscriptional regulatory mechanisms. Electroporation has been used as a DNA delivery method for a wide range of prokaryotes *(4)* and eukaryotes *(5)*. Electroporation of intact yeast has already proven useful for the introduction of DNA *(6)*, protein *(7)*, and small molecules *(8,9)*. The procedure for electroporation of yeast detailed below was originally described by Everett and Gallie *(10)*.

In theory, any RNA, whether it is mRNA, tRNA, rRNA, or ribozymes, can be introduced into yeast using electroporation. The RNA can be

From: *Methods in Molecular Biology, Vol. 47: Electroporation Protocols for Microorganisms*
Edited by: J. A. Nickoloff Humana Press Inc., Totowa, NJ

purified from a biological source or synthesized in vitro. We commonly synthesize mRNA in vitro from reporter genes, such as luciferase *(luc)* from firefly or β-glucuronidase from *E. coli,* that are placed under the control of a bacterial phage promoter, e.g., SP6, T7, or T3 *(11)*. A poly(A)$_{50}$ tract, introduced downstream of the reporter gene coding region, allows the production of polyadenylated mRNA with a uniform poly(A) tail length of 50 adenylate residues *(12)* *(see* Note 1).

Since RNA delivery to intact yeast cells is extremely inefficient, it is necessary to remove the cell wall partially. The cell wall does not have to be completely removed, because only portions of the cell membrane need to be exposed for RNA uptake. Depending on the degree of expression required, the extent of cell-wall removal can be limited in order to permit the cells to recover more quickly. The degree of wall digestion must be determined empirically owing to variation between strains and the needs of the experiment *(see* Note 2). These partially spheroplasted cells do require osmotic support, suggesting that the integrity of the cell wall is compromised.

2. Materials

2.1. Yeast Strains and Media

Strain CRY1 *MATa can1-100 ade2-1 his3-11,15 leu2-3,112 trp1-1 ura3-1*, a derivative of W303a, was obtained from R. Fuller and C. Brenner (Department of Biochemistry, Stanford University School of Medicine, Stanford University, CA, 94305) *(see* Note 3). *Saccharomyces cerevisiae* is grown on standard YPD medium.

2.2. Solutions and Reagents for RNA Synthesis

1. 10X T7 buffer: 400 m*M* Tris-HCl, pH 7.5, 60 MgCl$_2$, and 20 m*M* spermidine. Use autoclaved H$_2$O for this and all following solutions.
2. 2.5 mg/mL BSA (RNase-free, DNase-free).
3. 5 m*M* each of ATP, GTP, CTP, and UTP, all at pH 7.5.
4. 100 m*M* dithiothreitol (DTT).
5. 10 m*M* m^{7}GpppG.
6. RNasin ribonuclease inhibitor (Promega, Madison, WI).
7. T7 RNA polymerase (New England Biolabs, Beverly, MA).
8. 100 m*M* EDTA, pH 8.0.
9. Phenol:chloroform (1:1).
10. Ether.
11. 7*M* Ammonium acetate.

12. Isopropanol.
13. 80% ethanol.

2.3. Solutions and Reagents for RNA Electroporation

1. YPD medium: 1% yeast extract, 2% bacto-peptone, and 2% glucose.
2. YPD-sorbitol medium: YPD medium supplemented with $1M$ sorbitol.
3. Buffer A: 50 mM Tris, pH 7.5, 1 mM MgCl$_2$, 30 mM dithiothreitol, 15 mM β-mercaptoethanol, and $1M$ sorbitol.
4. 1 mg/mL Zymolyase (100T, ICN, Costa Mesa, CA) dissolved in buffer A and centrifuged briefly at 12,000g.
5. Sterile $1M$ sorbitol.

2.4. Solutions and Reagents for Luciferase Assays

Protocol without Coenzyme A:

1. Luciferase assay buffer: 25 mM Tricine, pH 7.5, 15 mM MgCl$_2$, 7 mM β-mercaptoethanol and 2 mM ATP.
2. 0.5 mM luciferin in luciferase assay buffer.

Protocol with Coenzyme A:

1. Cell-culture lysis reagent (CCLR, Promega): 100 mM Tricine, pH 7.8, 2 mM dithiothreitol, 2 mM 1,2-diaminocyclohexane-N,N,N',N'-tetraacetic acid, 10% glycerol, and 1% Triton X-100.
2. Luciferase assay reagent (LAR, Promega): 20 mM Tricine, pH 7.8, 1.07 mM (MgCO$_3$)$_4$Mg(OH)$_2$ · 5H$_2$O (magnesium carbonate), 2.67 mM MgSO$_4$, 0.1 mM EDTA, 33.3 mM DTT, 270 μM coenzyme A, 470 μM luciferin, and 500 μM ATP.
3. 0.5 mM luciferin in LAR buffer.

3. Methods
3.1. RNA Synthesis

1. The linearized DNA should be at a final concentration of 0.5 μg/μL.
2. Starting with the DNA in a microfuge tube, add the reagents in order to a final concentration of 40 mM Tris-HCl, pH 7.5, 6 mM MgCl$_2$, 2 mM spermidine, 100 μg/mL BSA, 0.5 mM each of ATP, CTP, and UTP, plus 160 μM GTP, 1 mM m^{7}GpppG, 10 mM dithiothreitol, 0.3 u/μL RNasin ribonuclease inhibitor, and 0.5 u/μL T7 RNA polymerase.
3. Incubate the reaction at 37°C for 3 h.
4. Stop the reaction by adding 1/10 vol of 100 mM EDTA.
5. Extract once with phenol:chloroform (1:1).
6. Transfer the aqueous (top) layer to a fresh microfuge tube.
7. Extract once with 1 vol of ether, and dispose of the ether (top) layer.

8. Add 1/10 vol 7M ammonium acetate and 1 vol isopropanol, and place at –20°C for at least 1 h to precipitate the RNA.
9. Pellet the RNA by centrifugation at 10,000g at 4°C for 15 min.
10. Wash the RNA pellet five times with 80% ethanol.
11. Resuspend in 10–100 µL water (1–10 µL will be used for the electroporation), keeping the microfuge tube on ice for 30 min to allow the RNA to dissolve completely (*see* Note 4).

3.2. Spheroplast Conditions

1. Harvest cells from 15 mL of midlog culture (*see* Note 5) by centrifugation at 1000g for 10 min, and resuspend in 5 mL of sterile buffer A.
2. Add 1 mL of 1 mg/mL Zymolyase to the cells. Place the cells in a Petri dish, and shake gently (30 rpm) at 30°C for 30 min.
3. Harvest the spheroplasts by centrifugation at 100g for 5 min, and wash twice in 5 mL of buffer A, taking care to resuspend the spheroplasts by gentle agitation.
4. Following the second wash, resuspend the spheroplasts by gentle agitation in 1 mL of buffer A. Then add cells to 10 mL of YPD-sorbitol medium. Allow the spheroplasts to recover by gentle shaking at 30°C for 90 min (*see* Note 6).
5. Harvest the spheroplasts by centrifugation at 100g for 5 min, and wash twice with 5 mL of sterile 1M sorbitol in order to remove any salts that will impair efficient electroporation (*see* Note 7).
6. Resuspend the spheroplasts in sterile 1M sorbitol by gentle agitation to yield a final concentration of 1×10^8/mL.

3.3. Electroporation Conditions

1. Aliquot 180 µL of Zymolase-treated yeast into microfuge tubes, and place on ice (*see* Note 8).
2. The mRNA to be delivered (typically 1 µg) is aliquoted in a second microfuge tube and kept on ice (*see* Note 9).
3. Mix the RNA with the yeast immediately before electroporation by transferring the yeast to the tube containing the RNA and then into a standard plastic microcuvet (*see* Note 10).
4. Electroporate immediately at 800 V, 21 µF, 1000 Ω using electrodes with a 0.2-cm gap (*see* Note 11).
5. Add 0.5 mL YPD-sorbitol medium immediately following electroporation to the electroporation cuvet, and if using removable electrodes, rinse the electrodes in the cuvet containing the spheroplast–YPD–sorbitol mixture to remove any adhering spheroplasts (*see* Note 12).
6. Transfer the spheroplasts to a test tube, and incubate at 30°C for several hours (*see* Note 13).

Luciferase activity can be detected within minutes following delivery, suggesting that the introduced mRNA is immediately recruited for translation. Translation of *luc* mRNA virtually ceases by 100 min after mRNA delivery, a result of the complete turnover of the mRNA. For this reason, the yeast is allowed to incubate for at least this period of time following RNA delivery. This permits the preparation, electroporation, and the expression analysis of constructs in yeast to be carried out within a single day. More stable mRNAs will require that the cells be allowed a longer incubation period to translate the introduced mRNA fully.

3.4. Luciferase Assays

Although any RNA can be delivered to yeast using electroporation (*see* Note 14), we often use luciferase as a reporter mRNA because of the ease of the enzyme assay. However, luciferase is subject to feedback inhibition and Coenzyme A prevents this feedback inhibition. There are two basic assay buffers for luciferase. The first contains no Coenzyme A and, therefore, is less expensive. The second contains Coenzyme A and is more than an order of magnitude more sensitive *(13)*. In yeast, the level of expression from delivered RNA is usually so high that the more sensitive buffer is not usually required *(10,14)*.

Assay without Coenzyme A:

1. Harvest the spheroplasts by centrifugation at 100*g* for 5 min following incubation, and resuspend in 0.5 mL of luciferase assay buffer.
2. Sonicate for 5 s on medium power using a microtip. Pellet the cell debris by centrifugation at 10,000*g* for 5 min at 4°C.
3. Add an aliquot of the cell extract to luciferase assay buffer to a total volume of 100 µL. Measure luciferase activity using 0.5 m*M* luciferin dissolved in luciferase assay buffer and a luminometer, such as a Monolight 2010 (Analytical Luminescence Laboratory, San Diego, CA) (*see* Note 15).

Assay with Coenzyme A:

1. Harvest the spheroplasts following the incubation, and resuspend in 0.5 mL of CCLR buffer.
2. Sonicate for 5 s on medium power using a microtip, and pellet the cell debris as described above.
3. Add an aliquot of the cell extract to LAR buffer to a total volume of 100 µL. Measure luciferase activity using 0.5 m*M* luciferin dissolved in LAR buffer and a luminometer.

4. Notes

1. The template plasmids are linearized with *DraI*, which cuts at a site immediately downstream of the $poly(A)_{50}$ tract. The concentration of the template DNAs is quantitated spectrophotometrically following linearization. Capping is carried out concomitant with transcription by including m^7GpppG in the transcription reaction.

2. We have found that a 15-min digestion with Zymolyase is sufficient to result in 60% of the level of expression observed for cells digested for 90 min.

3. Data from only one strain of *S. cerevisiae* were included in this protocol. However, we have examined several strains of *S. cerevisiae,* and all have been amenable to RNA electroporation.

4. Under these transcription conditions, >95% of the mRNA is capped. Each RNA construct is synthesized in triplicate in separate transcription reactions, so that any variability in RNA yield is reflected in the expression data. The integrity and relative quantity of RNA can be determined by formaldehyde-agarose gel electrophoresis as described *(11)*. Quantitation of RNA yields can also be determined by trace radiolabeling or spectrophotometrically. Virtually all mRNAs of higher and lower eukaryotes contain a cap and poly(A) tail. Both of these mRNA elements are essential for efficient translation. The cap serves as the binding site for eukaryotic initiation factor (eIF)-4F, which is an important regulatory point in the initiation of translation *(15,16)*. The poly(A) tail serves as the binding site for the poly(A) binding (PAB) protein and plays an essential role in translational initiation *(17)*. We have demonstrated that the cap and poly(A) tail are functionally codependent in higher eukaryotes and, to a lesser extent, in *S. cerevisiae* as well *(14)*. It is therefore important to synthesize mRNA as capped and polyadenylated RNA in order to achieve a high level of expression in vivo.

5. Cell age can influence translational activity and, as a consequence, the level of expression from the introduced mRNA. We found that late-log-phase cells electroporated with luc-A_{50} mRNA are only 52% as active (1.2×10^7 light U/mg protein) as midlog cells (2.3×10^7 light U/mg protein) and stationary-phase cells are just 3.6% as active (8.5×10^5 light U/mg protein) as midlog phase cells. Osmoticum is important for spheroplast survival. Spheroplasts prepared at either 0.25, 0.5, 0.75, or $1M$ sorbitol were electroporated in sorbitol of equivalent concentration. Table 1 demonstrates that luciferase expression increases with the increase in osmoticum to $1M$ sorbitol. Significant cell lysis owing to low osmoticum is observed only at $0.25M$ sorbitol.

6. Spheroplasting of yeast is known to disrupt active translation, and therefore, spheroplasts are generally allowed a recovery period. Yeast electro-

Table 1
The Effect of Osmoticum
on Electroporation Efficiency

Sorbitol concentration, M	Luciferase activity, light U, $\times 10^6$/mg protein
0.25	0.028
0.50	3.7
0.75	11.5
1.0	32.7

porated immediately following spheroplasting is 11-fold less active in translation of *luc* mRNA (2.5×10^6 U/mg protein) than following a 90-min recovery in YPD-sorbitol medium (2.8×10^7 U/mg protein) for an equivalent number of cells. However, a prolonged recovery will reduce RNA delivery. Expression from spheroplasts allowed a 90-min recovery period before electroporation is 11-fold higher (3.2×10^7 light U/mg protein) than from spheroplasts allowed a 20-h recovery period (2.9×10^6 light U/mg protein). The lower expression in the yeast allowed to recover for 20 h may be the result of cell-wall regeneration, a process that would impede efficient RNA uptake.

7. The presence of salts can significantly alter electroporation efficiency. Very low salt conditions are required in the electroporation of prokaryotes *(4)*. In contrast, electroporation of higher eukaryotes uses lower voltages, and salts are commonly present in the electroporation buffer *(18,19)*. Yeast are similar to bacteria in this respect. The presence of residual salts will reduce electroporation efficiency. It is important, therefore, to wash the yeast thoroughly with 1M sorbitol after the recovery from spheroplasting.

8. Expression increases with cell concentration over a range from 1×10^6/mL to 5×10^8/mL *(10)*. The volume of the sample has a small influence on the level of expression. Expression decreases when the volume is >150 µL. This could be a result of the increase in volume or a decrease in reporter mRNA concentration. We have chosen a volume of 200 µL because of the greater reproducibility between samples.

9. The dose response of yeast electroporated with mRNA is linear over a wide range of concentrations. We tested *luc*-A_{50} mRNA over a range of 0.5–15 µg and found that the dose response is linear throughout most of this range with only a slight decrease in linearity at the higher levels of mRNA (Fig. 1). We typically use 1 µg mRNA, which falls well within the linear range and thereby does not overwhelm the translational machinery. Carrier RNA had no beneficial effect on mRNA uptake. No effect on

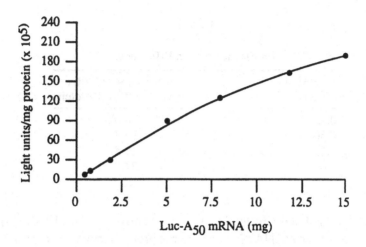

Fig. 1. Dose–response analysis of RNA electroporation in spheroplasts. From 0.5–15 µg *luc*-A_{50} mRNA were electroporated into 1×10^8 yeast in a constant volume.

luciferase expression was observed when up to 100 µg carrier RNA (*E. coli* tRNA) were used. As a result, carrier RNA is not included in our electroporations.

10. The presence of RNase activity in a cell preparation can have a significant impact on the efficiency of delivery during RNA electroporation as well as on reproducibility. We have not detected significant RNase activity in our washed spheroplast preparations. Other yeast strains may have different levels of extracellular RNases. To test a spheroplast preparation for extracellular RNase activity, we mix *luc* mRNA with the spheroplasts and allow the cell–*luc* mRNA mixture to incubate at room temperature for varying amounts of time before electroporating. A reduction in the resulting yield of luciferase activity is suggestive of RNase activity. Normally, only 5 s are required to mix the yeast with the mRNA, transfer to the cuvet, and electroporate. Therefore, RNase activity, if it were to be present in a strain, should not present a problem.

11. The electroporation conditions described here are those for the GeneZapper (IBI, New Haven, CT). RNA delivery using a different electroporation apparatus may require empirical optimization for the voltage, capacitance, and resistance. Starting with 21 µF of capacitance, spheroplasts should be electroporated with 1–10 µg mRNA over a range of voltages. An example of the influence of voltage on the resulting level of expression is shown in Fig. 2. Luciferase expression increases to 750–1000 V, after which expres-

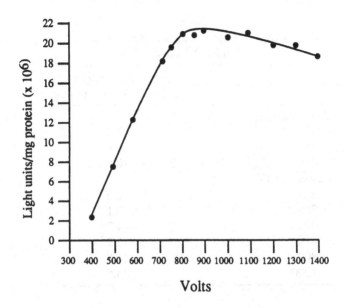

Fig. 2. Voltage as a parameter in yeast electroporation efficiency. One microgram *luc*-A$_{50}$ mRNA was used for electroporation at each voltage tested. Luciferase activity was used a measure of electroporation efficiency.

sion levels decrease. Once the optimal voltage has been determined, a range of resistances should be tested. The effect that resistance can have on RNA delivery is shown in Table 2. Maximum expression is observed at the highest resistance employed (1000 Ω).

12. The electroporation of each mRNA construct is carried out in triplicate, and each cell extract is assayed in duplicate. We have determined the error associated with RNA electroporation in yeast by performing ten replicate electroporations of yeast with 1 µg of *luc*-A$_{50}$ mRNA. The average and standard error was calculated to be $1.8 \times 10^7 \pm 5.1 \times 10^5$.

13. The length of the incubation will depend on the half-life of the RNA used and the nature of the experiment. For *luc* mRNA, we incubate for 3 h to ensure that all the *luc* mRNA has been degraded.

14. The *Escherichia coli* β-glucuronidase *(uidA)* gene has gained widespread use in plant studies because of the low endogenous β-glucuronidase activity in most plant species. We have shown that β-glucuronidase mRNA is a useful reporter gene in yeast *(10)*. We detect no endogenous β-glucuronidase activity in *S. cerevisiae* under the assay conditions for the *E. coli* β-glucuronidase enzyme. Moreover, the translation of β-glucuronidase mRNA constructs is enhanced to a greater extent by the

Table 2
Effect of Capacitance and Resistance
on RNA Delivery

	Luciferase activity, light U, $\times 10^6$/mg protein
Capacitance, µF	
7	8.4
14	16.8
21	20.2
Resistance, Ω	
100	2.5
200	9.1
400	14.7
600	15.2
800	18.0
1000	23.5

addition of a poly(A) tail than are *luc* constructs *(12)*, and therefore, it represents a more sensitive reporter gene for measuring the impact of the poly(A) tail on expression.

15. Reporter gene expression can be normalized to protein. Determine protein concentration by the method described by Bradford *(20)*.

References

1. Altmann, M., Sonenberg, N., and Trachsel, H. (1989) Translation in *Saccharomyces cerevisiae:* initiation factor 4E-dependent cell-free system. *Mol. Cell. Biol.* **9,** 4467–4472.
2. Blum, S., Mueller, M., Schmid, R., Linder, P., and Traschsel, H. (1989) Translation in *Saccharomyces cerevisiae:* initiation factor 4A-dependent cell-free system. *Proc. Natl. Acad. Sci. USA* **86,** 6043–6046.
3. Gasior, E., Herrera, F., Sadnik, I., McLaughlin, C. S., and Moldave, K. (1979) The preparation and characterization of a cell-free system from *Saccharomyces cerevisiae* that translates natural messenger ribonucleic acid. *J. Biol. Chem.* **254,** 3965–3969.
4. Miller, J. F., Dower, W. J., and Tompkins, L. S. (1988) High-voltage electroporation of bacteria: genetic transformation of *Campylobacter jejuni* with plasmid DNA. *Proc. Natl. Acad. Sci. USA* **85,** 856–860.
5. Potter, H. (1988) Electroporation in biology: methods, applications, and instrumentation. *Anal. Biochem.* **174,** 361–373.
6. Becker, D. M. and Guarente, L. (1991) High-efficiency transformation of yeast by electroporation. *Methods Enzymol.* **194,** 182–187.
7. Uno, I., Fukami, K., Kato, H., Takenawa, T., and Ishikawa, T. (1988) Essential role for phosphatidylinositol 4,5-bisphosphate in yeast cell proliferation. *Nature* **333,** 188–190.

8. Bartoletti, D. C., Harrison, G. I., and Weaver, J. C. (1989) The number of molecules taken up by electroporated cells: quantitative determination. *FEBS Lett.* **256,** 4–10.

9. Weaver, J. C., Harrison, G. I., Bliss, J. G., Mourant, J. R., and Powell, K. T. (1988) Electroporation: high frequency of occurrence of a transient high-permeability state in erythrocytes and intact yeast. *FEBS Lett.* **229,** 30–34.

10. Everett, J. G. and Gallie, D. R. (1992) RNA delivery in *Saccharomyces cerevisiae* using electroporation. *Yeast* **8,** 1007–1014.

11. Melton, D. A. Kreig, P. A., Rebagliati, M. R., Maniatis, T., Zinn, K., and Green, M. R. (1984) Efficient *in vitro* synthesis of biologically active RNA and RNA hybridization probes from plasmids containing a bacteriophage SP6 promoter. *Nucleic Acids Res.* **12,** 7035–7056.

12. Gallie, D. R., Feder, J. N., Schimke, R. T., and Walbot, V. (1991) Post-transcriptional regulation in higher eukaryotes: the role of the reporter gene in controlling expression. *Mol. Gen. Genet.* **228,** 258–264.

13. Leuhrsen, K. R., de Wet, J. R., and Walbot, V. (1992) Transient expression analysis in plants using firefly luciferase reporter gene. *Methods Enzymol.* **216,** 397–414.

14. Gallie, D. R. (1991) The cap and poly(A) tail function synergistically to regulate mRNA translational efficiency. *Genes & Dev.* **5,** 2108–2116.

15. Rhoads, R. E. (1988) Cap recognition and the entry of mRNA into the protein synthesis initiation cycle. *TIBS* **13,** 52–56.

16. Sonenberg, N. (1988) Cap-binding proteins of eukaryotic messenger RNA: functions in initiation and control of translation. *Prog. Nucleic Acid Res. Mol. Biol.* **35,** 173–207.

17. Munroe, D. and Jacobson, A. (1990) Tales of poly(A): a review. *Gene* **91,** 151–158.

18. Callis, J., Fromm, M., and Walbot, V. (1987) Expression of mRNA electroporated into plant and animal cells. *Nucleic Acids Res.* **15,** 5823–5831.

19. Fromm, M. E., Callis, J., Taylor, L. P., and Walbot, V. (1987) Electroporation of DNA and RNA into plant protoplasts. *Methods Enzymol.* **153,** 351–366.

20. Bradford, M. M. (1976) A rapid and sensitive method for the quantitation of microgram quantities of protein utilizing the principle of protein-dye binding. *Anal. Biochem.* **72,** 248–254.

CHAPTER 7

Electrofusion of Yeast Protoplasts

Herbert Weber and Hermann Berg

1. Introduction

Electrofusion is an important method in genetics and biotechnology. The first morphological *(1)* and genetical *(2)* evidence for yeast electrofusion was presented at the 5th Bioelectrochemical Symposium held at Weimar in 1979. Protoplasts from auxotrophic yeast strains were fused in a macrochamber (volume > 0.2 mL) by single-capacitor discharge pulses with cell-membrane contact facilitated by PEG *(3–10; see also Note 6)*. Early experiments tested the viability of several yeast strains with field strength E > 5 kV/cm (Table 1). It could be demonstrated further *(11)* that cells from the stationary growth phase are much more resistant to killing by electric fields ≤20 kV/cm than cells from the exponential phase *(see Note 2)*. Knowing the specific threshold of a strain, a lower E value should be applied for the expansion of statistical small hydrophobic pores in the lipid bilayer to larger hydrophilic pores as the first prerequisite for fusion. The driving force for this electroporation process is the enhancement of the transmembrane voltage $\Delta\phi_m$ from its normal value of about 100 to about 1000 mV *(12)*, depending on the protoplast radius r according to the general relation:

$$\Delta\phi_m = 1.5 \, E \, r \cos \theta \qquad (1)$$

where θ = the angle between the field vector and the surface element under consideration.

The second prerequisite for electrofusion of protoplasts is the close contact between regions of two cell membranes *(see Note 7)*, where pores

From: *Methods in Molecular Biology, Vol. 47: Electroporation Protocols for Microorganisms*
Edited by: J. A. Nickoloff Humana Press Inc., Totowa, NJ

Table 1
Effect of Electric Field Strength on Viability
of Different Yeast Species (Averages of Two Experiments)

Yeast species		% Survivors at 20 kV/cm
Saccharomyces cerevisiae	7	11
Saccharomycopsis lipolytica (*Yarrowia lipolytica*)	18	14
Saccharomyces cerevisiae	28	27
Lodderomyces elongisporus	39	44
Saccharomycopsis lipolytica	68	82

are prevalent as $\theta \Rightarrow 0$. During the pulse or afterward in the resealing process, an intermingling of disturbed bilayers occurs (as shown in Fig. 1); subsequently, the cell wall will be regenerated. Some models try to explain the mechanism in more detail *(13,14)*. PEG or other polymers, such as dextran, albumin, and polylysine, at concentrations <1% *(15,16)* increase fusion yield after dielectrophoresis (*see* Note 6). Higher concentrations (e.g., >40%) lead to cell-membrane contact without application of dielectrophoresis. Adsorption and dehydration at the surface increase the hydrophobic interaction forces and cell–cell bridging. Therefore, the energetic barrier for the electrical formation of hydrophilic pores is reduced, leading to a synergistic effect. As a consequence, the level of the electric field can be decreased. The kinetics of the fusion process can be observed microscopically using a microchamber with flat electrodes on a slide *(17)*. Only electrofusion offers this possibility in combination with preceding dielectrophoresis. Electrofusion overcomes natural barriers of hybridization of cells with identical mating types *(18)*, of polyploid strains *(19)*, and of different species *(3,4)*, which normally do not copulate (*see* Note 1). This chapter describes procedures involving a single pulse, and a pulse preceded by low-voltage dielectrophoresis using macro- and microchambers, respectively.

2. Materials
2.1. Equipment for Electrofusion in the Macrochamber

A schematic diagram of the pulse generator is shown in Fig. 2. The pulse of the initial field strength Eo shows an exponential decay because of the discharge of capacitor in the range of 25 nF up to 25 µF, loaded by a high-voltage generator up to 25 kV. A closed Plexiglas™ chamber con-

Fusion

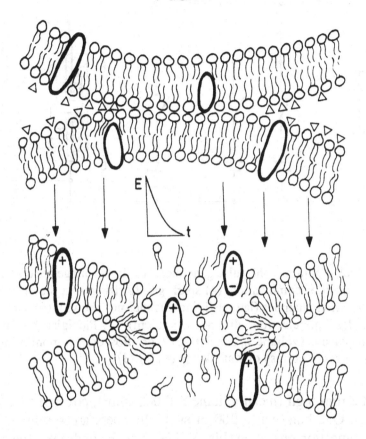

Fig. 1. Scheme of first stages of electrofusion. (**Top**) Before pulse. (**Bottom**) After an exponential pulse. Lipids and polarized proteins are disrupted.

sists of two stainless-steel cylindrical electrodes of variable distance d (Fig. 2). There is also an open macrochamber similar to an optical cuvet with vertical rectangular electrodes.

2.2. Equipment for Electrofusion in the Microchamber

A microchamber consists of thin wires or sputtered alloy electrodes mounted on a glass slide, allowing microscopic observation or video recording of the fusion process. The electrode material consists of Ni-Cr electrodes of about 1.5 μm thickness covered by a 35-nm SiO_2 layer for protection. The microchamber is 10 mm long, 0.05–0.2 mm wide, and

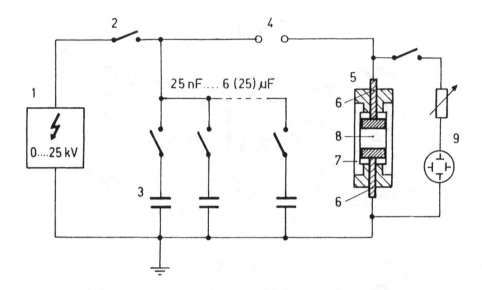

Fig. 2. Block diagram of the pulse generator. (1) High-voltage generator, (2) switch, (3) capacitor combination, (4) spark gap for triggering the discharge, (5) electroporation chamber (lateral filling channel not shown), (6) stainless-steel electrodes (interelectrode distance $d = 1$ cm), (7) Plexiglas™ chamber body, (8) suspension (volume $v_0 \geq 500$ µL), (9) oscilloscope for monitoring the discharge process (Courtesy of Katenkamp et al. *[23]*).

about 0.002 mm high (mean volume > 0.002 mm³), covered by a thin cover glass. One hundred to 200 of such chambers are sputtered in a meander connection on a slide (Fig. 3). Under such conditions, the close contact of cell membranes can be easily attained by dielectrophoresis (1 MHz sine waves; *15–17*).

2.3. Strains

To select and identify hybrids and to distinguish hybrids from parent strains after fusion, it is necessary to use genetically marked strains. Any two strains with different nutritional markers are suitable, since fusion products may be selected on medium lacking nutrients required by both parent strains. In all cases, crossfeeding, back-mutation, and contamination must be excluded.

2.4. Media for Yeast Cultivation

1. YEPD medium: 1% yeast extract, 2% peptone, 2% dextrose, pH 5.5–6.0.
2. Minimal media:

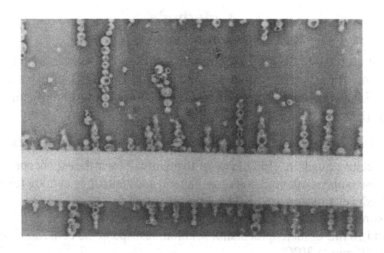

Fig. 3. Pearl chain formation of protoplasts between electrodes in a microchamber produced by dielectrophoresis.

a. MM *(S. cerevisiae)*: 0.2% KH_2PO_4, 0.1% $MgSO_4$, 0.1% $(NH_4)_2SO_4$, 2% dextrose, 200 μg/L thiamin, 2 μg/L biotin.

b. MM *(S. lipolytica)*: 0.5% $NH_4H_2PO_4$, 0.25% KH_2PO_4, 0.1% $MgSO_4 \cdot 7H_2O$, 1.0% glucose, 20 mg/L Ca $(NO_3)_2 \cdot 4H_2O$, 2 mg/L $FeCl_3 \cdot 6H_2O$, 0.5 mg/L H_3BO_4, 0.4 mg/L $MnSO_4 \cdot H_2O$, 0.4 mg/L $ZnSO_4$, 0.1 mg/L $CaCl_2$, 0.1 mg/L $CuSO_4 \cdot 5H_2O$, 0.1 mg/L Kaliumjodid (Merck, Germany), 100 μg/L thiamine, pH 6.5.

2.5. Solutions and Reagents for Protoplasting

1. Preparation medium: 22.44 g/L Na_2HPO_4 and 73.80 g/L $(NH4)_2SO_4$, adjust pH to 6.0 with 1M citric acid.
2. Tris-EDTA solution: 2.42 g Tris-HCl, 7.45 g EDTA in 200 mL dH_2O. Adjust pH to 8.0 with 1M KOH.
3. Mercaptoethanol solution: 2 μL of β-mercaptoethanol, 50 mL Tris-EDTA solution.
4. Tris-HCl solution: 4.85 g Tris-HCl in 200 mL dH_2O, pH 9.2.
5. L-Cystein solution: 0.175 g of L-cystein, in 10 mL of Tris-HCl solution.
6. Lyophilized snail enzyme (Helicase) or another lytic enzyme having β-glucuronidase activity (Glusulase, Novozym, or Sulfatase); for 1 g yeast, dissolve 160 mg snail enzyme in 15 mL of preparation medium.
7. Polyethylene glycol (PEG) solution: 40% PEG (mol wt 6000), 0.01M $CaCl_2$ and 1.4M sorbitol.

3. Methods

3.1. Maintenance or Strains

Yeast strains (*see* Note 1) are maintained on YEPD slants in a refrigerator and for experiments cultivated in liquid YEPD on a reciprocal shaker. If strains were obtained from slants, they should be precultivated several times in liquid YEPD.

3.2. Isolation and Stabilization of Protoplasts

1. Precultivate cells in YEPD. Harvest 10 mL of culture in the exponential growth phase (approx 10^8 cells/mL) (*see* Note 3) and centrifuge at 500g for 5 min.
2. Wash cells twice with dH_2O.
3. Estimate the weight of yeast mass within the tube.
4. Add 10 mL of mercaptoethanol solution, resuspend the cells, and incubate for 10 min at 30°C.
5. Centrifuge at 500g for 5 min, wash twice in 18 mL of preparation medium, and discard the supernatant.
6. Resuspend cells in 15 mL of preparation medium containing 160 mg lyophilized snail enzyme/1 g of yeast cells. Incubate for 30 min at 28°C with gentle shaking.
7. Monitor microscopically the formation of protoplasts. Cells become spherical and, under a phase-contrast microscope, protoplasts appear bright. Confirm protoplasting by adding a drop of dH_2O to the edge of the coverslip: Protoplasts will burst.
8. Centrifuge protoplast suspension for 5 min at 500g, and discard the supernatant.
9. Resuspend in 10 mL of 1.4M sorbitol (*see* Note 4) by stirring with a glass rod to separate protoplasts from yeast cells and cell debris.
10. Estimate the percentage of protoplasts by counting an aliquot of cells in a counting chamber before and after addition of dH_2O.
11. For electrofusion experiments, use protoplasts immediately.
12. Mix fresh protoplast suspensions of both fusion partners in equal proportions at a concentration of about 10^7–10^8 protoplasts/mL.
13. Centrifuge at 500g for 5 min, and discard the supernatant.
14. Resuspend protoplasts in 2 mL of PEG solution (or other volume corresponding to the capacity of the discharge chamber).

3.3. Electrofusion in a Macrochamber

1. Sterilize the chamber with ethanol and/or UV light.
2. Add 0.2 mL or more of the protoplast suspension to the closed chamber through the lateral opening, and seal with a plug. Fill open macrochambers as optical cuvets. Avoid introducing air bubbles between electrodes.

3. Cool chambers to 4°C prior to applying the pulse to decrease the resealing rate.
4. Pulse once using these conditions: 3–15 kV/cm, 2–10 µF, ≥ 1 kΩ, time constant 3–20 ms.
5. Plate cells on stabilized selective media (*see* Note 11).

3.4. Electrofusion in a Microchamber

In contrast to the macrochamber, the close contact of protoplast membranes in the microchamber is achieved by dielectrophoresis. The dipole induction of cells causes their movement along electric field lines leading to "pearl chains" of 2–20 protoplasts touching one or both electrodes (Fig. 3). The application of rectangular electric pulses by a commercial generator connected with an amplifier leads to protoplast fusion. The advantages of this technique are the required small volumes of protoplast suspension, low voltages (about 20V) between electrodes, no additional fusogens, or low concentrations of fusogens (<1%) and the ability to document the fusion process by video recording.

1. Pipet <0.2 µL of protoplast suspension onto the slide under a cover glass.
2. Cool the slide with two Peltier elements to 10°C for a more effective fusion process.
3. Dielectrophorese for 30 s with ≥ 300 V/cm, 1 MHz frequency.
4. Pulse 1–5 times with a square wave at 1–5 kV/cm, with time constant of 1–200 ms at 0.5–2 s intervals.

3.5. Regeneration and Selection of Fusion Products

Successful application of protoplast fusion is dependent on subsequent culture and regeneration (reversion) of protoplasts to the typical vegetative yeast state (*see* Note 5). Macrochamber fusion product detection requires that products are able to form reproductive cells leading to colonies and that they can be clearly differentiated from parent strains or revertants (*see* Note 11). The most common method is the regeneration on solid medium. A great advantage of most yeast protoplasts is their capability to regenerate their cell wall even in the presence of PEG. Therefore, samples of the fusion mixture containing PEG can be plated directly on the surface of solid agar plates. After a few days, small colonies will appear (Fig. 4). Alternatively, the fusion mixture may be added to molten agar cooled to 40–45°C in prewarmed Petri dishes. Regeneration is also possible in YEPD medium without agar containing 35% PEG.

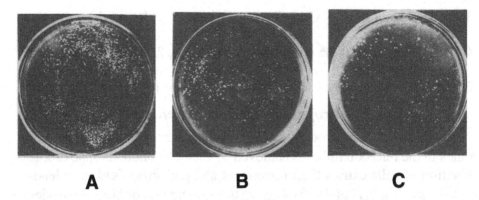

Fig. 4. Formation of prototrophic colonies from fused protoplasts of *Saccharomycopsis (Yarrowia) lipolytica* on minimal medium after electric field pulses in presence of PEG. **(A)** 10 kV/cm, **(B)** 15 kV/cm, **(C)** 20 kV/cm.

1. Prepare Petri dishes with osmotically stabilized YEPD-agar medium.
2. Make serial dilutions of protoplasts.
3. Count the protoplasts in serial dilutions using a counting chamber.
4. Plate serial dilutions, and incubate at 30°C for 2–3 d (*see* Notes).
5. Compare the number of colonies on plates with the number of protoplasts in serial dilutions to determine regeneration frequency.
6. Select fusion products by replicating cells to minimal medium.

4. Notes

1. Electrofusion has been applied successfully to the following yeast species: *Lodderomyces elongisporus (3,4)*, *Saccharomyces cerevisiae (5,18–22)*, *Saccharomyces diastaticus (20,21)*, and *Saccharomycopsis (Yarrowia) lipolytica (3,4)*.
2. Different yeast species and strains vary dramatically in their sensitivity to electric field pulse treatment. This is also true for cells in different growth phases *(11)*. For intraspecific fusion, the optimal field strength is about 3–5 kV/cm after PEG treatment in a macrochamber *(3–5)* and 10–12 kV/cm after dielectrophoresis agglutination, even when two successive pulses in a microchamber are used *(18–20)*. For optimal interspecific fusion using PEG, a field strength of about 10 kV/cm may be necessary.
3. The most efficient protoplasting is obtained when cells are harvested after 14–18 h of cultivation. This is necessary to obtain a high number of viable protoplasts liberated within a short time. Liberation of protoplasts should be finished not later than 30 min after addition of lytic enzymes. This is important for an effective fusion process as well as for regeneration of protoplasts after fusion.

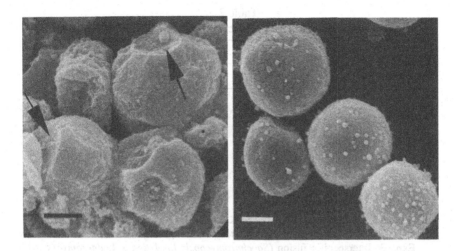

Fig. 5. Scanning electron micrograph effect of PEG treatment on morphology of protoplasts of *Saccharomycopsis (Yarrowia) lipolytica*. (1) Protoplasts in stabilizing sorbitol medium before treatment, (2) protoplasts after addition of PEG. **Note:** depressions (arrow) from adjacent protoplasts. Bar represents 1 μm.

    If insufficient protoplasts form, alternative pretreatment may be necessary, using other enzymes, prolonged treatment with β-mercaptoethanol, or 15 min of pretreatment with L-cystein solution before enzyme treatment. Dithiothreitol, thioglycolic acid, or other sulfhydryl reagents may also be useful before treatment.

4. Cells devoid of their wall are osmotically fragile, and the maintenance of protoplasts is dependent on an external osmotic support. To osmotically stabilize protoplasts, all media are supplemented with 1.4*M* sorbitol; other sugars, salts, and sugar alcohols may also be used.

5. Regeneration of protoplasts may be done not only in or on agar media, but also in 12% gelatin containing 15% yeast extract and 2% glucose, plus 0.1–0.3% agar as a hardening agent. It has been shown that electric field treatment does not significantly influence protoplast regeneration rates. Therefore, the observed deformation of protoplasts by high concentrations of PEG (*see* Fig. 5) or field pulses *(4)* is reversible and without consequence for regeneration.

6. There are two important factors to control in electrofusion experiments. First, protoplasts should be used immediately after liberation. Second, to achieve a very intimate contact between membranes of protoplasts, PEG or another polymer, such as dextran or polylysine, is used as a chemical agglutination agent. Control experiments suggest that the presence of PEG

Table 2
Intraspecies and Intergeneric Electrofusion of Yeast Protoplasts

Electric field strength, kV/cm	No. of fusion products		
	Expt. 1	Expt. 2	Expt. 3
0	28	8	8
1.25	—	7	—
2.5	—	47	—
3.75	—	218	—
5.0	553	62	—
10.0	8	—	662
15.0	1	—	160
20.0	—	0	66

Expts. 1 and 2: Intraspecific fusion *(Saccharomycopsis lipolytica)*
Expt. 3: Interspecific fusion *(Saccharomycopsis lipolytica* × *Lodderomyces elongisporus).*

greatly enhances the fusion rate by electric field pulses. Experiments without PEG yielded no fusion products. The same negative result was obtained when PEG was added after the field pulse treatment. This suggests that the necessary close contact between protoplasts is facilitated by PEG, whereas the electric field pulse causes membrane labilization and poration. Intimate contact can also be established by sedimentation of the protoplasts in an appropriate centrifugation chamber.

7. Electron microscopy demonstrated that after treatment of protoplasts with PEG, deep depressions are seen, resulting from protoplast contact (Fig. 5). In this situation, the electric field pulse greatly stimulates membrane breakdown and cell fusion.

8. The yield of yeast protoplast fusion depends on several factors, including yeast species, the method used to produce protoplasts, the viability of protoplasts, and electric field parameters. Generally, electrofusion frequencies vary from 10^{-4} to 10^{-2}, and are one or two orders of magnitude higher than for chemical fusion. In several cases, treatment with pronase *(22)* or addition of albumin *(20)* to protoplast suspensions yielded a higher fusion frequency. The fusion yield in the microchamber can be enhanced by more than 50% in the presence of PEG, dextran, or polylysine at concentrations 100 times lower than for classical chemical fusion (>1% polymer).

9. The maximal field strength of exponential pulses should be <10 kV/cm (Table 2) for both intraspecific fusion of protoplasts from *Saccharomycopsis (Yarrowia) lipolytica* and intergeneric fusion of strains of *Sm. (Yarrowia) lipolytica* and *Lodderomyces elongisporus.*

10. Postpulse handling after the electric field pulse seems to be important. Therefore, electrodes should not be removed from the chamber until after a 30-min postpulse incubation at 28°C. At this time, samples can be taken out for selection of fusion products. This incubation time is necessary to complete the fusion process *(3–5,20)*.

11. Different types of genetic markers are effective for selecting fusion products. No significant differences were observed using nutritional markers or respiration deficiencies. Genetic analysis (haploidization, marker analysis, genetic stability) allows one to determine whether the prototrophic colonies are true hybrid fusion products. Genetic analysis showed that hybrid genomes often consist of the genome of one of the parental strains with a few genes or chromosomes from the other fusion partner *(19)*. Cytological analysis allows estimation of the number of nuclei per cell.

References

1. Senda, M., Takeda, J., Abe, S., and Nakamura, T. (1979) Induction of cell fusion of plant protoplasts by electrical stimulation. *Plant & Cell Physiol.* **20,** 1441–1443.
2. Jacob, H.-E., Förster, W., Weber, H., and Berg, H. (1979) *5th Symposium on Bioelectrochemistry,* Weimar, Poster Abstract F7.
3. Weber, H., Förster, W., Jacob, H.-E., and Berg, H. (1981) Enhancement of yeast protoplast fusion by electric field effects, in *Current Developments of Yeast Research* (Stewart, G. G. and Russell, I. I., eds.), Pergamon, Toronto, pp. 219–224.
4. Weber, H., Förster, W., Jacob, H.-E., and Berg, H. (1981) Microbiological implication of electric field effects. III. Stimulation of yeast protoplast fusion by electric field pulses. *Z. Allg. Mikrobiol.* **21,** 555–562.
5. Weber, H., Förster, W., Berg, H., and Jacob, H.-E. (1981) Parasexual hybridization of yeasts by electric field stimulated fusion of protoplasts. *Curr. Genet.* **4,** 165,166.
6. Jacob, H.-E., Förster, W., and Berg, H. (1981) Microbiological implication of electric field effects. II. Inactivation of yeast cells and repair of their cell envelope. *Z. Allg. Mikrobiol.* **21,** 295–233.
7. Berg, H. , Augsten, K., Bauer, E., Förster, W., Jacob, H.-E., Kurischko, A., Mühlig, P., and Weber, H. (1983) Cell fusion by electrical field pulses. *Studia Biophysica* **94,** 93–96.
8. Shivarova, N., Grigorova, R., Förster, W., Jacob, H.-E., and Berg, H. (1983) Fusion of *Bacillus thuringiensis* protoplasts by high electric field pulses. *Bioelectrochem. Bioenerg.* **11,** 181–185.
9. Berg, H. (1985) Electromagnetic field effects on cell membranes and cell metabolism, in *Electrochemistry in Research: A Development* (Kalvoda, R. and Parsons, eds.), Plenum, New York, pp. 227–241.
10. Berg, H., Förster, W., and Jacob, H.-E. (1987) Electric field effects on biological membranes: electroincorporation and electrofusion, in *Bioelectrochemistry II* (Milazzo, G. and Blank, M., eds.), Plenum, New York, pp. 135–166.
11. Jacob, H.-E., Förster, W., and Berg, H. (1979) Effects of electric discharge on biological matter. *Proc. 19th Hungarian Ann. Meet. Biochem.* Budapest, 69–71.

12. Tomov, T. and Toneva, I. (1992) Influence of electric pulses on the transmembrane potential in yeast. *Bioelectrochem. Bioenerg.* **27**, 33–36.
13. Neumann, E. (1992) Membrane electroporation and direct gene transfer. *Bioelectrochem. Bioenerg.* **28**, 247–267.
14. Dimitroff, D. (1993) Kinetic mechanisms of membrane fusion mediated by electric fields. *Bioelectrochem. Bioenerg.* **32**, 99–124.
15. Zhang, L., Fiedler, V., and Berg, H. (1991) Modification of barley protoplasts by membrane active agents. *Bioelectrochem. Bioenerg.* **26**, 87–96.
16. Zhang, L. and Berg, H. (1992) Electrofusion yield modified by membrane-active substances, in *Charge and Field Effects in Biosystems—3* (Allen, M., Cleary, S., Sowers, A., and Shillady, D., eds.), Birkhäuser, Boston, pp. 497–502.
17. Katenkamp, U., Jacob, H.-E., Kerns, G., and Dalchow, E. (1989) Hybridization of *Trichoderma resii* protoplasts by electrofusion. *Bioelectrochem. Bioenerg.* **22**, 57–67.
18. Halfmann, H. J., Emeis, C. C., and Zimmermann, U. (1983) Electrofusion of haploid *Saccharomyces* yeast cells of identical mating type. *Arch. Microbiol.* **134**, 1–4.
19. Halfmann, H. J., Emeis, C. C., and Zimmermann, U. (1983) Electrofusion and genetic analysis of fusion products of haploid and polyploid *Saccharomyces* yeast cells. *FEMS Microbiol. Lett.* **20**, 13–16.
20. Schnettler, R. and Zimmermann, U. (1985) Influence of the composition of the fusion medium on the yield of electrofused yeast hybrids. *FEMS Microbiol. Lett.* **27**, 195–198.
21. Förster, E. and Emeis, C. C. (1985) Quantitative studies on the viability of yeast protoplasts following dielectrophoresis. *FEMS Microbiol. Lett.* **26**, 65–69.
22. Halfmann, H. J., Röcken, W., Emeis, C. C., and Zimmermann, U. (1982) Transfer of mitochondrial function into a cytoplasmic respiratory-deficient mutant of *Saccharomyces cerevisiae. Jpn. J. Gen.* **53**, 41–49.
23. Katenkamp, U., Atrat, P., and Müller, E. (1991) Influence of the physiological state on the electric field mediated transformation efficiency of intact mycobacterial cells. *Bioelectrochem. Bioenerg.* **25**, 285–294.

Chapter 8

Escherichia coli Electrotransformation

Elizabeth M. Miller and Jac A. Nickoloff

1. Introduction

Electroporation is now being used to transfer into cells a variety of macromolecules, including DNA, RNA, protein, fluorescent dyes, and some chemotherapeutic agents. With electrotransformation, as many as 80% of the cells receive exogenous DNA *(1)*. Chemical and natural methods of DNA transfer into bacteria are also effective for transforming bacteria, but they are more cumbersome and time-consuming, and generally less efficient than electrotransformation.

In this chapter, methods are discussed for introducing exogenous DNA into *E. coli* reproducibly at high efficiencies using electroporation. These procedures are effective with *E. coli* strains commonly used in the laboratory, such as DH5α and HB101, and yield similar efficiencies with a mismatch repair defective strain, BMH 11-71 mut S *(2)*. Among the variables that influence transformation efficiency, the most important are the growth phase of the cell culture at the time cells are harvested and the temperature during cell preparation. Harvested cells are washed several times in cold (4°C), low-ionic-strength buffer, or water to reduce the conductivity of the cell suspension. During these washes, they are concentrated to 10^9–10^{10} cells/mL. Transformation is achieved simply by adding DNA to cells and applying a high-voltage exponential decay pulse. Transformants are normally selected on agar plates containing an antibiotic, although for certain applications, selection is applied to transformed cells *en masse* in liquid culture *(3,4)*.

From: *Methods in Molecular Biology, Vol. 47: Electroporation Protocols for Microorganisms*
Edited by: J. A. Nickoloff Humana Press Inc., Totowa, NJ

2. Materials

1. *E. coli* strains are available from a variety of commercial sources (e.g., Stratagene, La Jolla, CA; Gibco-BRL, Bethesda, MD). The most commonly used strains include DH5α, HB101, the JM series (e.g., JM101, JM103, JM109), BB4, XL1-Blue, and the SOLR and SURE cells (*see* Note 1).
2. Luria broth (LB): 5 g/L yeast extract (Difco, Detroit, MI), 10 g/L bacto-tryptone (Difco) and 10 g/L NaCl (Sigma, St. Louis, MO). Adjust to pH 7.0 and autoclave. Store at room temperature. Add 15 g/L agar to make solid medium.
3. Freezing medium: 10% (v/v) glycerol. Autoclave and store at 4°C.
4. SOC: Dissolve 2 g bacto-tryptone (Difco) and 0.5 g yeast extract (Difco) in 100 mL of H_2O. Add 200 μL of $5M$ NaCl and 250 μL of $1M$ KCl. Adjust to pH 7.0 and autoclave. Cool to room temperature. Then add 1 mL of $1M$ $MgCl_2$ and 2 mL of 20% (w/v) glucose. Store at room temperature.
5. Antibiotic stock solutions (for selection of transformants): ampicillin (100 mg/mL) and kanamycin (40 mg/mL). Store at –20°C.
6. Selective plates: Autoclave LB agar, cool to 45–50°C while stirring, add 1 mL of appropriate antibiotic stock solution, stir for 5 min to ensure homogeneity, and pour into Petri dishes (*see* Note 2).
7. X-GAL/IPTG plates: 2% X-GAL (w/v) in *N,N*,-dimethylformamide; store in the dark at –20°C for up to 6 mo. Isopropylthio-β-D-galactoside (IPTG): 100 mM in H_2O. Store at –20°C indefinitely. Immediately before use, spread 50 μL of IPTG and 40 μL of X-GAL on an LB agar plate containing an appropriate antibiotic (*see* Note 3).

3. Methods

3.1. Preparation of Competent Cells

The following procedure will produce 1.5–2.0 mL of competent cells from 1 L of cultured cells, sufficient for 40–100 transformations.

1. Inoculate 50 mL of LB with a single colony. Incubate at 37°C overnight with agitation.
2. Prewarm 500 mL of LB in each of two 1-L flasks to 37°C.
3. Transfer 25 mL of the overnight culture into each flask. Incubate at 37°C with agitation until the cultures reach an A_{600} of 0.4 (*see* Note 4).
4. Transfer cultures to sterile ice-cold 500-mL centrifuge bottles. Cap and chill on ice for 20 min (*see* Note 5).
5. Centrifuge at 3000g for 10 min at 4°C. Discard medium and resuspend each cell pellet in 250 mL ice-cold sterile H_2O. Centrifuge at 3000g for 10 min at 4°C (*see* Note 6).
6. Resuspend cell pellets in 40 mL of ice-cold water, and transfer to 50-mL tubes. Centrifuge at 3000g for 7 min at 4°C.

7. Resuspend cell pellets in 20 mL of ice-cold 10% glycerol, and centrifuge at 3000*g* for 5 min at 4°C. Aspirate as much of the supernatant as possible, and then pool the pellets in 1.5–2 mL of ice-cold 10% glycerol (*see* Notes 7 and 8).
8. Aliquot convenient volumes of the cell suspension into 1.5-mL tubes (assume 20–40 μL/electrotransformation). Cells may be used immediately or flash-frozen in an ethanol/dry ice bath and stored at –80°C. There is no significant loss of competence for at least 6 mo.

3.2. Electroporation and Selection of Transformed Cells

1. Precool DNA samples and cuvets (0.1- or 0.2-cm electrode gap) on ice. Thaw the competent cells on ice (*see* Note 9).
2. Transfer 1–2 μL of a DNA solution (in TE or other low-ionic-strength buffer; *see* Notes 10–14) to a 1.5-mL tube, and chill on ice. Add 40 μL of cells to the tube, mix the cells and DNA by pipeting up and down once, and then transfer the mixture to a chilled electroporation cuvet. Cap the cuvet, and then gently shake or tap it to ensure that the cell/DNA mixture is at the bottom of the cuvet and that it forms a complete bridge between the two electrodes (*see* Note 15).
3. Remove any moisture from the outside of the cuvet. Then place it in the electroporation device, and deliver an exponential decay pulse of 2.5 kV, 25-μF capacitance, and 200-Ω resistance. These settings should produce a time constant of 4.5–4.8 ms (*see* Note 16).
4. Immediately add 1 mL of SOC (at room temperature) to the cuvet, and then transfer the entire mixture to a sterile 15-mL screw-cap tube. Incubate for 1 h at 37°C with gentle shaking (*see* Note 17).
5. Plate appropriate volumes on LB agar plates containing an appropriate antibiotic (*see* Note 18) and X-GAL/IPTG if blue-white color screening is possible (*see* ref. 5). Transformants will arise after an overnight incubation at 37°C.

3.3. Determining Transformation Efficiency

It is useful to determine the level of electrocompetence of each batch of cells. This information is valuable when troubleshooting, since few or no transformants may result from low concentrations of DNA, DNA of low quality, or because cells are not sufficiently competent for a particular application.

1. Dilute a 50 μg/mL stock solution of pUC19 (or other suitable small plasmid) to 5 mg/mL with H_2O, and transform 2 μL (10 ng) into cells (Section 3.2.).
2. After the 1-h incubation at 37°C, dilute the transformed cell suspension 10^4- and 10^5-fold in SOC, plate 100 μL of each dilution, and incubate overnight at 37°C.

3. Count the number of transformants on one or both plates (=N). The efficiency or number of transformants/µg = N × dilution × 10 × 100. The factor of 10 accounts for the fact that the total volume of transformed cells is 1.0 mL, but only 0.1-mL aliquots are plated. The factor of 100 accounts for ratio between 1 µg (used to express the efficiency) and 10 ng used to transform the cells. Therefore, cells with an efficiency of 5×10^9/µg will yield 500 colonies on the 10^4 dilution plate and 50 colonies on the 10^5 dilution plate.

4. Notes

1. Most strains of *E. coli* used as plasmid hosts are recA⁻ (which prevents most homologous recombination). However, when plasmids contain many or long repeated regions, strains with different recA⁻ alleles (perhaps in concert with other markers, such as *gyrA*) may yield different levels of recombination. For example, the mammalian expression plasmid, pMSG (Pharmacia, Piscataway, NJ) has repeated SV40 mRNA splicing and polyadenylation signal sequences, and is unstable during long-term growth in DH5α *(recA1 gyrA)*, but it is completely stable in HB101 *(recA13)* (J. N., unpublished observations). Try different strains if you suspect that problems are the result of restriction barriers or DNA rearrangements. Another useful marker pair is *lacZYA* and Δ(*lacZ*)M15, which complements the *lac* α-fragment encoded by many vectors and produces blue colonies on plates containing X-GAL and the *lac* inducer IPTG.
2. Carbenicillin, used in place of ampicillin at an equivalent concentration, eliminates satellite colonies.
3. Using 50 µL of X-GAL/plate will give colonies with an intense blue color. However, X-GAL is expensive, and the color obtained with 40 µL/plate is usually sufficient. Also, the color intensity will increase with time, especially if plates are transferred to 4°C after being incubated overnight at 37°C.
4. For optimum transformation, cells should be harvested at early to midlog phase. For *E. coli* strains HB101 and DH5α, an A_{600} of 0.4 is optimum. However, other strains may have different optima. Thus, it is prudent to perform a simple growth experiment with new strains to ensure that the cells are at an appropriate stage of growth when harvested. The results from a sample growth experiment are pictured in Fig. 1A. Briefly, 2.5 mL of an overnight culture of *E. coli* strain HB101 was transferred to 250 mL of LB in a 500-mL flask and incubated at 37°C with agitation. The growth of the culture was monitored for several hours by measuring the A_{600}. The usual early lag in growth was followed by exponential growth and the beginning of a plateau phase. Four separate 250-mL cultures were initiated from the same overnight culture and harvested when the A_{600} was 0.4–0.85. Transformation efficiencies were highest from cultures harvested

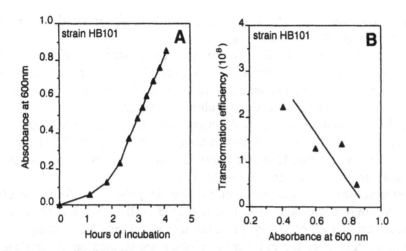

Fig. 1. Transformation efficiency as a function of cell density at the time of harvest. (**A**) Growth curve for HB101 cells in LB medium. (**B**) Transformation efficiency plotted against the optical density (A_{600}) prior to preparation of electrocompetent cells. Culture volumes were adjusted such that final cell concentrations were equal. Transformation efficiencies were measured after cells were frozen for 1 d at –80°C.

in early to midlog phase (Fig. 1B). Similar results were observed with DH5α cells grown in the same manner (data not shown). Typical cultures of DH5α or HB101 with an A_{600} of 0.4 yield transformation efficiencies in the range of $0.1–5 \times 10^9$ transformants/μg of pUC19 DNA. This cell density was chosen as a compromise between high transformation efficiency and total cell yield. The average cell yield using this protocol is $10^9–10^{10}$ viable cells/mL of frozen stock.

5. Some centrifuge bottles may be sterilized by autoclaving. If this is not possible, sterilize by rinsing once with 70% ethanol and then twice with sterile H_2O. Maintain cultures at 4°C during all subsequent steps until they are stored at –80°C. Transformation efficiencies are reduced (and less reproducible) if cells are allowed to reach room temperature during centrifugation and washing steps.

6. It is important to resuspend cell pellets completely by pipeting to remove all medium salts. Do not use a vortex mixer to resuspend the cells, since this may reduce cell viability. Cell pellets become increasingly loose after successive washes, so use care when removing supernatant from pelleted cells, preferably by aspiration instead of decanting.

7. Prior to freezing cells, a 40-μL aliquot of cells may be electroporated without DNA to test the conductivity of the cell suspension. If arcing occurs,

the cells should be washed once more with 10% glycerol and checked for excessive conductivity a second time.

8. The freezing medium, 10% glycerol, is effective with most *E. coli* strains. Some strains may transform more efficiently if 290 mos*M*/mL sucrose are added to the medium to alter the osmotic profile *(6)*. A difference in osmolality between the intra- and extracellular media may induce a flow of plasmid DNA molecules across the membrane and into the cell. Divalent cations (e.g., Mg^{2+} or Ca^{2+}) and various buffers (e.g., HEPES) have also been recommended. However, most *E. coli* strains do not require these additives for efficient transformation *(7)*.

9. Temperature is a critical parameter during the pulse. Transformation efficiency may decrease by 100-fold if pulsed at 20°C *(7)*, and there is an increased chance of arcing. In addition, a sharp increase in temperature immediately after the pulse may enhance transformation by creating a thermodiffusion gradient that carries the DNA into the cell *(6)*. The protocol outlined here takes advantage of this possibility, since the SOC medium is at room temperature when added to the cells immediately after the pulse.

10. The salt concentration of the final cell suspension to be pulsed should remain below 5 mEq (especially for high-voltage protocols required for transformation of bacteria). Thus, the DNA solution to be added to the cells should not increase significantly the conductivity of the suspension. TE buffer will not interfere with electrotransformation as long as it is diluted at least 10-fold when added to the cell suspension. If the DNA has been treated with restriction enzymes (i.e., to linearize unwanted molecules; *see* refs. *3,4,8)* or ligase, it should be desalted either by ethanol precipitation or, more rapidly, by spin-column chromatography *(9)*.

11. For a given cell concentration, transformation frequency (transformants per survivor) increases linearly with DNA concentration in the range of 10 pg/mL to 7.5 µg/mL *(1,7)*. In this range, DNA concentration determines the probability that a cell will be transformed *(1)*. However, at a high cell concentration, a relatively low DNA concentration will result in maximal transformation efficiency (colony-forming units/µg of DNA). Thus, the overall yield of transformants is dependent on the transformation frequency and the number of cells present. High efficiency with low frequency is an appropriate strategy for the construction of libraries, in which case cotransformants are to be avoided. DNA concentrations <10 ng/mL and cell concentrations >3 × 10^{10} cells/mL are recommended *(1)*. If high frequency is required, DNA concentrations from 1 to 10 µg/mL can transform a large proportion of the cells. For routine fragment subcloning, the concentration of intact, replicable recombinant molecules is usually a small fraction of the total DNA. As a rule, we use 0.025 µg of total DNA/

electroporation. This value can be increased or decreased empirically, or by estimating the efficiency of each step of a subcloning strategy and using the known transformation efficiency of the electrocompetent cells.

12. Plasmid size can be an important factor in electrotransformations. The number of transformants per mole of plasmid decreases as the size of the plasmid increases. Siguret et al. *(10)* obtained 2.9×10^9 transformants/µg with a 2.9-kbp plasmid, but only 0.5×10^8 transformants/µg with a 13.6-kbp plasmid. The quantity of DNA in the range of 10 pg to 2 ng/transformation did not influence these transformation efficiencies *(10)*.

13. Electrotransformation with linearized plasmids or linear genomes (such as phage λ) results in lower transformation efficiencies than observed with circular plasmids. For example, linear unit length λ (48 kbp) transforms with 0.1% of the efficiency observed for a small plasmid, such as pUC19 *(7)*. The uptake of linear vs circular (relaxed or supercoiled) DNA induced by a pulsed electric field is not different. However, the stability of the DNA in the cytoplasm depends on its topology. Linear DNA is degraded in *E. coli* rather quickly, which may account for the observed transformation efficiencies that are one to four orders of magnitude lower than those for circular molecules *(11)*. This difference is the basis for site-directed mutagenesis procedures involving comutagenesis of restriction sites *(3,4)* and efficient subcloning strategies *(12)*.

14. DNA introduced by pulsed electric fields is more vulnerable to restriction endonucleases than DNA introduced by other methods. Restriction barriers can cause dramatic decreases in transformation efficiency. When subcloning, it is therefore prudent to use DNA isolated from the same strain used for electrotransformation and amplification of plasmids of interest. Alternatively, it is possible to overcome such barriers by using high concentrations of DNA.

15. Transform cells using cuvets with either 0.1- or 0.2-cm electrode gaps. The narrower gap will produce higher field strengths for a given voltage, which may improve transformation efficiency, but it also increases the chance of arcing. Using 20 or 30 µL of cells instead of 40 µL conserves competent cells, but gives slightly lower transformation efficiencies and increases the chance of arcing.

16. If arcing occurs, the time constant will be <1.0, and the transformation should be repeated since few or no transformants will result. If a particular DNA sample always causes arcing, but no arcing occurs when DNA is omitted, either desalt the DNA sample (*see* Note 10), add less DNA, switch from a 0.1- to a 0.2-cm electrode gap, or use a lower voltage. However, any of these measures except desalting may reduce transformation efficiency. If the time constant is too short, the sample may be too con-

ductive. If the time constant is too long, check the capacitance and resistance settings.

17. Highest efficiencies are achieved using rich medium after electroporation (i.e., SOC; *1*). If efficiency is not important, LB can be used in place of SOC. Also, after SOC is added, the cells may be incubated for 1 h in the electroporation cuvet. However, gentle shaking in a larger test tube is recommended.

18. If pure supercoiled plasmid DNA is used, transformation frequencies will be very high (usually >10^9 transformants/µg). If only one or a few transformants are required, a loopful of the cell suspension can be streaked on one-quarter of a plate, and a single plate can be used to recover four independent transformants. If transformation efficiencies are expected to be low, it is best to distribute the entire cell suspension on one or more plates. It is possible to plate 200–250 mL on a single plate if the excess liquid is evaporated by incubating upright at 37°C for 5–10 min without a lid. Alternatively, some or all of the cells can be plated on a single plate if they are first collected by centrifugation at 3000g for 10 min and then resuspended in 100–200 mL of the supernatant. However, if the quantity of cells is too great, "pools" of dead cells may form on selective plates and prevent growth of transformants. This problem is common if a large volume of cells are electroporated or electrocompetent cells are highly concentrated.

Acknowledgments

Our research is supported by NIH training grant CA 09078 (E. M. M.), and NIH grants CA 54079, CA 62058, and CA 55302 (J. A. N.) from the National Cancer Institute.

References

1. Dower, W. J., Miller, J. F., and Ragsdale, C. W. (1988) High efficiency transformation of *E. coli* by high voltage electroporation. *Nucleic Acids Res.* **16,** 6127–6145.
2. Zell, R. and Fritz, H.-J. (1987) DNA mismatch-repair in *Escherichia coli* counteracting the hydrolytic deamination of 5-methyl-cytosine residues. *EMBO J.* **6,** 1809–1815.
3. Deng, W. P. and Nickoloff, J. A. (1992) Site-directed mutagenesis of virtually any plasmid by eliminating a unique site. *Anal. Biochem.* **200,** 81–88.
4. Ray, F. A. and Nickoloff, J. A. (1992) Site-specific mutagenesis of almost any plasmid using a PCR-based version of unique-site elimination. *Biotechniques* **13,** 342–346.
5. Sambrook, J., Fritsch, E. F., and Maniatis, T. (1989) *Molecular Cloning: A Laboratory Manual,* 2nd ed. Cold Spring Harbor Laboratory, Cold Spring Harbor, NY.
6. Antonov, P. A., Maximova, V. A., and Pancheva, R. P. (1993) Heat shock and osmotically dependent steps by DNA uptake in *Escherichia coli* after electroporation. *Biochim. Biophys. Acta.* **1216,** 286–288.

7. Shigekawa, K. and Dower, W. J. (1988) Electroporation of eukaryotes and prokaryotes: a general approach to the introduction of macromolecules into cells. *Biotechniques* **6,** 742–751.

8. Ray, F. A., Miller, E. M., and Nickoloff, J. A. (1994) Adapting unique site elimination site-directed mutagenesis for marker rescue and domain replacement. *Anal. Biochem.* (in press).

9. Nickoloff, J. A. (1994) Sepharose spin column chromatography: a fast, nontoxic replacement for phenol:chloroform extraction/ethanol precipitation. *Mol. Biotechnol.* **1,** 105–108.

10. Siguret, V., Ribba, A.-S., Cherel, G., Meyer, D., and Pietu, G. (1994) Effect of plasmid size on transformation efficiency by electroporation of *Escherichia coli* DH5α. *Biotechniques* **16,** 422–424.

11. Conley, E. C. and Saunders, J. R. (1984) Recombination-dependent recircularization of linearized pBR322 plasmid DNA following transformation of *Escherichia coli. Mol. Gen. Genet.* **194,** 211–218.

12. Taghian, D. G. and Nickoloff, J. A. Subcloning strategies and protocols, in *Basic DNA and RNA Protocols* (Harwood, A., ed.), Humana, Totowa, NJ (in press).

CHAPTER 9

Electrotransformation in *Salmonella*

Kenneth E. Sanderson, P. Ronald MacLachlan, and Andrew Hessel

1. Introduction

Enteric bacteria are not naturally competent for DNA-mediated transformation, but methods derived from the Ca^{2+}-shock method of Mandel and Higa *(1)* permit uptake of DNA at levels adequate for molecular cloning. Transformation frequencies using Ca^{2+}-shock methods with selected strains of *E. coli* K-12 range from 10^6 to 10^8 transformants/µg of covalently closed circular plasmid DNA, but efficiency is frequently lower because these methods are very sensitive to purity of reagents, cleanliness and quality of glassware and plasticware, and methods of cell handling *(2)*.

Transformation of plasmids by Ca^{2+}-shock methods was demonstrated in *Salmonella typhimurium* LT2 by Lederberg and Cohen *(3)*, although the frequencies observed were low. In addition, transfection by free DNA of the bacteriophage P22 was detected in *S. typhimurium* by the same methods *(4)*. "Rough" strains lacking the galactose-4-epimerase activity *(galE)*, which produced lipopolysaccharide (LPS) lacking side chains and part of the LPS core, were efficiently transfected, but transfection of "smooth" strains was significantly lower. LPS structure also affected the frequencies of plasmid transformation in *S. typhimurium* with Ca^{2+}-shock methods; *galE* and *rfaF* mutants gave transformation frequencies up to 10^6 transformants/µg DNA, but smooth strains and strains with other LPS-deficient mutants were typically 100-fold lower (Fig. 1) *(5)*. Trans-

From: *Methods in Molecular Biology, Vol. 47: Electroporation Protocols for Microorganisms*
Edited by: J. A. Nickoloff Humana Press Inc., Totowa, NJ

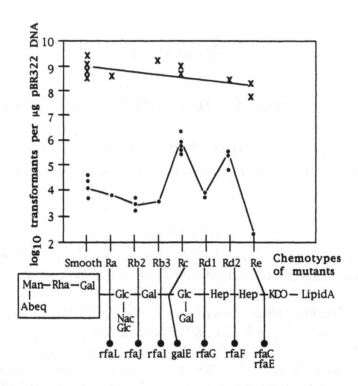

Fig. 1. Effect of LPS in the recipient cell on electrotransformation and Ca^{2+}-shock transformation in *S. typhimurium*. Numbers of Ap^R (ampicillin-resistant) transformants with strains of the LPS chemotypes indicated are shown in the figure. The frequencies for electrotransformation (indicated with "X") were reported by Binotto et al. *(12)*, whereas the frequencies for transformation by Ca^{2+}-shock (indicated by "·") were reported by MacLachlan and Sanderson *(5)*. The chemical composition of the LPS and the genotypes of the strains for genes affecting LPS synthesis (shown in italics) are at the bottom. Each mutant blocks the incorporation of sugars into the LPS from the site of mutation toward the left; thus, smooth strains have the intact LPS, whereas the *rfaG* mutant (chemotype Rd1) has LPS terminating with heptose, lacking glucose and the other sugars to the left. The strains used in electroporation are as follows. Where there is more than one strain for a chemotype, they are in order of reducing frequency of electrotransformation. Smooth, CL4419 (as SL1027, but also *ilv-452 hsdL6 hsdSA29*); SL3770 (prototrophic *rfa⁺*); SL1027 (*metA22 metE551 trpC2 H1-b H2-e,n,x fla-66 rpsL120 xyl-404*) (cured of Fels 2); LB5000 (as SL1027 but also *ilv-452 hsdL6 hsdSA29 hsdSB121 leu-3121*). Chemotype Ra, SA3749 (as SL3770 but *rfaL446*). Chemotype Rb3 (as SL3770 but *rfaI432*). Chemotype Rc, LB5010 (as LB5000 but *galE856*); SL1306 (as SL3770 but also *galE503*). Chemotype Rd2, SL3789 (as SL3770 but *rfaF511*). Chemotype Re, SL1102 (as SL1027 but also *rfaE543*); SA1377 (*rfaC630* [P22⁺]).

fer of plasmids between genera, such as *E. coli* and *Salmonella,* by Ca^{2+}-shock methods is usually possible only if host restriction mutants are available in the recipient strain.

Electrotransformation is a very efficient way to transfer plasmids between strains of *S. typhimurium* LT2. Unlike Ca^{2+}-shock methods, transformation frequencies are little affected, if at all, by the LPS composition of the cell envelope, so smooth strains (with LPS side chains), which are the strains most commonly used in experimental studies, are efficiently transformed. We also obtain efficient intergeneric electrotransfer between *E. coli* K-12 and *S. typhimurium* LT2, even when the strains have the wild-type alleles of *hsd* genes for DNA restriction. The methods of cell preparation and of electrotransformation described here are derived from those reported by Dower et al. *(6,7).*

2. Materials
2.1. Plasmids and Bacterial Strains

Electrotransformation frequencies can be determined using the 4.3-kb plasmid pBR322; other plasmids, especially chimeric plasmids, such as pBR322 or pBluescript carrying insertions of DNA of *S. typhimurium,* yield results that are qualitatively similar *(8,9).* The bacterial strains are derivatives of *S. typhimurium* LT2 (many isolated by Roantree et al. *[10],* described in the legend to Fig. 1) or of *E. coli* K-12 (*see* Footnote b of Table 1). Cells are grown in Luria broth at 37°C, unless stated otherwise, and the strains are stored frozen at –76°C in broth containing 15% glycerol. Prior to use, each strain is single-colony isolated, and in some specific situations, the phenotype for lipopolysaccharide (LPS) expression is confirmed by bacteriophage sensitivity tests *(11).* The strains and plasmids are available on request from the Salmonella Genetic Stock Centre (SGSC, c/o K. E. Sanderson, Department of Biological Sciences, University of Calgary, 2500 University Drive N.W., Calgary, Alberta, Canada, T2N 1N4, Tel.: (403) 220-6792; Fax: (403) 289-9311; e-mail: Kesander@acs.ucalgary.ca.).

2.2. Solutions and Reagents for Electrotransformation

1. TE: 10 m*M* Tris-HCl, pH 8.0, and 1 m*M* Na$_2$EDTA.
2. Luria broth: 10 g bacto-peptone, 5 g bacto-yeast extract, 5 g NaCl, 1 L H$_2$O, and 3.5 mL 1*M* NaOH. Sterilize by autoclaving, allow to cool, and add 2.5 mL of 40% filter-sterilized D-glucose.

Table 1
Electrotransformation of pBR322 DNA Between Strains
of *S. typhimurium* LT2 and *E. coli* K-12[a]

| Recipient strain | Genotype[b] | Ap^R transformants/µg pBR322 DNA | |
		DNA isolated from *S. typhimurium* LT2[c]	DNA isolated from *E. coli* K-12[c]
S. typhimurium			
SL1027	*hsdL$^+$ hsdSA$^+$ hsdSB$^+$*	9.1×10^8	5×10^6
CL4419	*hsdL6 hsdSA29 hsdSB$^+$*	3.7×10^9	6×10^7
LB5010	*hsdL6 hsdSA29 hsdSB121*	8.2×10^8	2×10^8
E. coli			
AB1157	*hsd$^+$*	8.7×10^5	2.1×10^8
C600	*hsd$^+$*	6.9×10^6	7.3×10^8
LE392	*hsdR514*	2.6×10^9	1.1×10^9

[a]Electrotransformation used cells from the large-scale method described in the text.

[b]Full genotype of the *S. typhimurium* strains is in the legend to Fig. 1; genotypes of the *E. coli* strains are as follows: AB1157, *thi-1 argE3 proA2 leu-8 thr-4 his lacY mtl xyl-5 ara-14 gal-2 tsx rpsL31* F$^-$; C600, *thr-1 leu-6 thi-1 supE lacY tonA21* F$^-$; LE392, *hsdR514 supE44 supF58 lacY1 galK2 galT22 metB1 trpR55* F$^-$.

[c]Source and preparation of the DNA are described in the text.

3. HEPES buffer (1 m*M* HEPES, pH 7.0): add 10 mL of 100 m*M* HEPES stock (23.8 g HEPES, 1 L H$_2$O) to 990 mL water, and adjust pH to 7.0.
4. 10% Glycerol (ultra-pure grade is recommended).
5. SOC broth: 20 g bacto-tryptone, 5 g bacto-yeast extract, 10 m*M* NaCl, 2.5 m*M* KCl, 10 m*M* MgCl$_2$, 10 m*M* MgSO$_4$, and 20 m*M* glucose.
6. Luria agar plus ampicillin: 15 g bacto-agar added to 1 L Luria broth (described above). Sterilize by autoclaving, cool to 50°C, and add 2.5 mL of 20 mg/mL ampicillin stock solution (1 g of ampicillin in 50 mL of 40% ethanol, store at −20°C) for a final concentration of 50 mg/L ampicillin.

3. Methods

3.1. Preparation of Cells for Electrotransformation

3.1.1. Preparation of Large Numbers of Cells

1. Prepare a fresh overnight broth culture by inoculating 10 mL of Luria broth with a single colony of the bacterial strain to be used. Grow overnight at 37°C. Inoculate 1 L of L-broth with 10 mL of the fresh overnight culture and grow at 37°C with aeration to an A_{640} of 0.75 (*see* Note 1).
2. Chill the cells on ice for at least 15 min and then collect the cells by centrifugation (4000*g*, 10 min) at 4°C.

3. Resuspend the cell pellet in 1 vol (1 L) of cold 1-mM HEPES buffer, pH 7.0, recentrifuge to pellet the cells, suspend the cell pellet in 1/2 vol (0.5 L) of cold 1-mM HEPES buffer, and centrifuge again (*see* Note 2).

4. Suspend the pellet in 1/50 vol (20 mL) of 10% glycerol, centrifuge, and then resuspend the final pellet in 1/100 vol (10 mL) of 10% glycerol. This 100-fold concentrated suspension can be used immediately, or it can be distributed in small aliquots, frozen in a dry ice–ethanol bath, and stored at –76°C (*see* Notes 3 and 4).

3.1.2. Rapid Methods to Produce a Smaller Number of Cells

The above procedure produces 10 mL of cell suspension; 40 µL of cells are needed for each electroporation. Therefore up to 250 electroporations can be done with one preparation of cells. It is often convenient to produce smaller volumes of cells, such as when a specific strain will be used only a few times.

For example, the procedure above may be scaled down using a 30-mL starting culture volume to produce 300 µL of competent cells, sufficient for about seven electroporations. The following procedure yields enough cells for a single electroporation.

1. Grow about 4 mL of cells to an A_{640} of 0.75.
2. Centrifuge 1.5 mL of cells in a 1.5-mL microfuge tube, discard the supernatant, refill the tube with cells and recentrifuge, and repeat so that a pellet from about 4 mL of cells is produced.
3. Wash the cells three times with 1.5 mL of cold 1 mM HEPES buffer each time, wash once with 80 µL of 10% glycerol, and resuspend the cells in 40 µL of 10% glycerol.

3.2. Preparation of DNA

Highly purified pBR322 DNA may be isolated from *S. typhimurium* LT2 by alkaline lysis methods followed by double banding in CsCl, suspended in TE and held in small aliquots at –76°C *(5,12)*. DNA concentration is determined by either of two methods that give very similar results: measurement of A_{260}, or linearization of the plasmid DNA by a restriction endonuclease, separation on agarose, and comparison with standard amounts of DNA after ethidium bromide staining. However, for routine electroporation, DNA of very high quality is not needed (*see* Note 5).

3.3. Procedures for Electrotransformation

1. Thaw 40 µL of frozen cells on ice, or transfer 40 µL of freshly prepared cells to a chilled 1.5-mL microfuge tube, and add DNA in a small volume (2 µL) of TE buffer.

2. Transfer the entire contents of the above tube to a chilled electroporation cuvet with a 0.2-cm electrode gap using a Pasteur pipet, and tap the cuvet on a table to distribute the suspension evenly between the electrodes at the bottom of the cuvet. Alternatively, transfer the DNA to the wall of the cuvet, add the cells to the drop of DNA, and then tap the cuvet as above. Keep the cuvets on ice until time of electroporation.

3. Carefully remove all moisture from the outside surface of the cuvet with a tissue. Pulse once with a field strength of 12.5 kV/cm, 25 µF capacitance, and 200 Ω resistance, yielding a pulse length of 5 ms (*see* Note 6).

4. Immediately following the pulse, add 1 mL of chilled SOC broth, transfer the mixture to a 17×100 mm polypropylene tube, and incubate at 37°C for 1 h (*see* Note 7).

5. Plate 0.1 mL of the cell suspension onto plates with appropriate antibiotics to select the electrotransformants.

4. Notes

1. Although we routinely use cultures grown to a cell density corresponding to an A_{640} of 0.75, electroporation frequencies are not strongly dependent on the cell density of the starting culture, and cultures with a broad range of cell densities yield good transformation efficiencies.

2. The purpose of the washes is to remove salts that increase the conductivity of the transformation mixture and may cause arcing in the cuvet. Some experimenters use cold double-distilled water for washing rather than HEPES buffer and report no significant differences. There are two common causes of arcing: the presence of salts in the DNA solution and insufficient washing of the cells. Excess salt in the DNA stock solutions can be removed by ethanol precipitation followed by one or more washes with cold 70% ethanol or by drop dialysis (a piece of washed dialysis membrane is floated on the surface of the buffer solution, a drop of the DNA stock solution is carefully applied to the membrane; after a few minutes of incubation at room temperature, the drop is then carefully removed). If salts are from the cell suspension itself, an additional wash with 10% glycerol normally prevents arcing. However, even if arcing does occur, the experiment should be completed since a useful number of transformants may still be produced.

3. Although we routinely use a 100X concentrated cell suspension for electroporation (1 L of cell suspension concentrated to 10 mL of cells in glycerol), final concentration factors from 50–300X have been used successfully.

4. The use of cells that have been frozen and stored rather than fresh cells usually reduces cell survival and subsequent electrotransformation formation by about 50%.

5. DNA used for electroporation need not be highly purified. Highly purified DNA maximizes the frequency of transformants/μg DNA, but DNA that is isolated by alkaline lysis or boiling and purified by phenol-extraction and alcohol washing as described by Sambrook et al. *(13)* generally yields an acceptable number of electrotransformants. However, DNA preparations containing residual salts will frequently cause arcing across the cuvet electrodes (*see* Note 2).

6. Exponentially decaying electric pulses may be generated by a Gene Pulser apparatus (Bio-Rad, Richmond, CA) connected to a Pulse Controller unit (Bio-Rad), containing a high-power, 20 Ω resistor in series with the sample, and a selection of resistors of 100–1000 connected to Ω in parallel with the sample. The cells may be electroporated in Potter-type cuvets (Bio-Rad) with an electrode gap of 0.2 cm. The standard conditions used, unless otherwise stated, are a field strength of 12.5 kV/cm (the highest field strength achievable with this equipment and cuvet) and a pulse length of 5 ms (using 200 Ω resistance and 25 μF capacitance).

7. The period of incubation prior to plating on selective media is dependent on the antibiotic resistance that is being selected. Very little or no incubation for gene expression is needed for tetracycline resistance; 1 h is adequate for ampicillin resistance; several hours may be needed for optimal kanamycin resistance. Some investigators suggest that the electroporated cells are fragile and should not be shaken during incubation. It is also acceptable to incubate the cell suspension in the cuvet rather than transferring to another tube.

8. Effect of DNA concentration on efficiency: A wide range of DNA concentrations is effective in electrotransformation. The recovery of transformants increases linearly with DNA in the electroporation cuvet at concentrations from 10 ng/mL to 10 μg/mL *(12)*. Using *E. coli* K-12 recipients, others report linearity over the range of 10 pg DNA/mL to 7.5 μg DNA/mL *(6)*.

9. Effect of LPS in the cell envelope on efficiency: Figure 1 illustrates the structure of the LPS of *S. typhimurium,* the genes that affect synthesis of the LPS, and the frequencies of transformation by Ca^{2+}-shock methods and electrotransformation methods for mutants in each gene *(5,12)*. Transformation frequencies with smooth strains of *S. typhimurium* range from 4 × 10^8 to 3.4 × 10^9/μg pBR322 DNA; strains with LPS of chemotypes Ra, Rb3, Rc, Rd2, and Re give similar frequencies. Ca^{2+} shock treatment is less efficient, and the frequency of transformation of some mutants is barely detectable. Smooth and most rough strains give only 10^4 transformants/μg DNA, strains with Rc *(galE)* and Rd2 *(rfaF)* LPS give up to 2 × 10^6, but Re chemotype strains give only 10^2 transformants/μg DNA (Fig. 1). Whereas transformation of *S. typhimurium* strains by Ca^{2+}-shock methods is strongly influenced by the LPS in the cell envelope, electrotransformation is little affected, if at all, and gives high frequencies with

all the strains tested. Thus, the physical barrier that the cell envelope presents to DNA is less relevant in electrotransformation than in Ca^{2+}-shock methods in *S. typhimurium*. However, the cell envelope influences electrotransformation in some gram-positive organisms, for in *Streptococcus* species, gentle digestion of cell walls gives optimal uptake of plasmid DNA in electrotransformation *(14)*.

10. Electrotransformation between species: *S. typhimurium* LT2 has three host-restriction systems, *hsdSA, hsdSB,* and *hsdL,* each capable of restricting entry of foreign DNA; *E. coli* K-12 has one known system, *hsdK (15).* Host-restriction systems reduce the frequency of F-mediated conjugational transfer of plasmids, such as F'*lac*+, from *E. coli* K-12 to *S. typhimurium* LT2 by 100- to 1000-fold *(16)*. The host-restriction systems of *S. typhimurium* are very efficient in preventing transformation of plasmids from *E. coli* K-12 to *S. typhimurium* LT2 by Ca^{2+}-shock methods, since transformants are rarely recovered unless host-restriction mutants of *S. typhimurium* are used (P. R. MacLachlan and K. E. Sanderson, unpublished data). Electrotransformation is much more efficient than Ca^{2+}-shock methods for intergenus transfer. pBR322 DNA isolated from *S. typhimurium* gives about 10^9 transformants/µg DNA when electrotransformed into either *S. typhimurium* or host-restriction-deficient strains of *E. coli* K-12; restriction-proficient strains of *E. coli* produce transformation frequencies 100- to 1000-fold lower. Similarly, pBR322 DNA isolated from *E. coli* K-12 gave 10^8–10^9 transformants when electrotransformed into *E. coli* K-12 and similar frequencies into *S. typhimurium* LT2 restriction-deficient lines, but the frequency of electrotransformation into SL1027, an Hsd+ line, is reduced about 100-fold. Thus, the reciprocal barrier for plasmid electrotransformation between *E. coli* K-12 and *S. typhimurium* LT2 reduces transformation by 100- to 1000-fold, and as expected, the barrier is absent in host-restriction deficient mutants. However, despite the presence of a wild-type restriction barrier, electrotransformation of plasmids between host-restriction-proficient (Hsd+) members of these two genera is quite practical, even when hybrid plasmids with inserts of heterologous chromosomal DNA are used; this was proven by the transfer of pBR322 and pBluescript vectors carrying inserts of *S. typhimurium* DNA in both directions between the two genera *(8,17)*.

Acknowledgments

We acknowledge the assistance of J. Binotto and D. M. Sirisena. This work was supported by an Operating Grant from the Natural Sciences and Engineering Research Council of Canada and by grant RO1AI34829 from the National Institute of Allergy and Infectious Diseases.

References

1. Mandel, M. and Higa, A. (1970) Calcium dependent bacteriophage DNA infection. *J. Mol. Biol.* **53**, 157–162.
2. Hanahan, D. (1987) in *Escherichia coli and Salmonella typhimurium: Cellular and Molecular Biology* (Low, K. B., Magasanik, B., Schaechter, M., and Umbarger, H. E., eds.), American Society for Microbiology, Washington, DC, pp. 1177–1183.
3. Lederberg, E. M. and Cohen, S. N. (1974) Transformation of *Salmonella typhimurium* by plasmid deoxyribonucleic acid. *J. Bacteriol.* **119**, 1072–1074.
4. Bursztyn, H., Sgaramella, V., Ciferri, O., and Lederberg, J. (1975) Transfectability of rough strains of *Salmonella typhimurium. J. Bacteriol.* **124**, 1630–1634.
5. MacLachlan, P. R. and Sanderson, K. E. (1985) Transformation of *Salmonella typhimurium* with plasmid DNA: differences between rough and smooth strains. *J. Bacteriol.* **161**, 442–445.
6. Dower, W. J., Miller, J. F., and Ragsdale, C. W. (1988) High efficiency transformation of *Escherichia coli* by high voltage electroporation. *Nucleic Acids Res.* **16**, 6127–6145.
7. Shigekawa, K. and Dower, W. (1988) Electroporation of eukaryotes and prokaryotes: a general approach to the introduction of macromolecules into cells. *Biotechniques* **6**, 742–750.
8. MacLachlan, P. R., Kadam, S. K., and Sanderson, K. E. (1991) Cloning, characterization, and DNA sequence of the *rfaLK* region for lipopolysaccharide synthesis in *Salmonella typhimurium* LT2. *J. Bacteriol.* **173**, 7151–7163.
9. Sirisena, D. M., Brozek, K. A., MacLachlan, P. R., Sanderson, K. E., and Raetz, C. R. H. (1992) The *rfaC* gene of *Salmonella typhimurium. J. Biol. Chem.* **267**, 18,874–18,884.
10. Roantree, R. J., Kuo, T.-T., and MacPhee, D. G. (1977) The effect of defined polysaccharide core defects upon antibiotic resistances of *Salmonella typhimurium. J. Gen. Microbiol.* **103**, 223–234.
11. Wilkinson, R. G., Gemski, P., Jr., and Stocker, B. A. D. (1972) Non-smooth mutants of *Salmonella typhimurium:* differentiation by phage sensitivity and genetic mapping. *J. Gen. Microbiol.* **70**, 527–554.
12. Binotto, J., MacLachlan, P. R., and Sanderson, K. E. (1991) Electrotransformation in *Salmonella typhimurium* LT2. *Can. J. Microbiol.* **37**, 474–477.
13. Sambrook, J., Fritsch, E. F., and Maniatis, T. (1989) *Molecular Cloning: A Laboratory Manual,* 2nd ed., Cold Spring Harbor Laboratory, Cold Spring Harbor, NY.
14. Powell, I. B., Acken, M. G., Hillier, A. J., and Davidson, B. E. (1988) A simple and rapid method for genetic transformation of lactic streptococci by electroporation. *Appl. Environ. Microbiol.* **54**, 655–660.
15. Bullas, L. R. and Ryu, J.-I. (1983) *Salmonella typhimurium* LT2 strains which are r–m+ for all three chromosomally located systems of DNA restriction and modification. *J. Bacteriol.* **156**, 471–474.
16. Sanderson, K. E., Janzer, J., and Head, J. (1981) Influence of lipopolysaccharide and protein in the cell envelope on recipient capacity in conjugation of *Salmonella typhimurium. J. Bacteriol.* **148**, 283–293.
17. Schnaitman, C. A., Parker, C. T., Klena, J. D., Pradel, E. L., Pearson, N. B., Sanderson, K. E., and MacLachlan, P. R. (1991) Physical maps of the *rfa* loci of *Escherichia coli* K-12 and *Salmonella typhimurium. J. Bacteriol.* **173**, 7410,7411.

Chapter 10

Electrotransformation of *Pseudomonas*

Jonathan J. Dennis and Pamela A. Sokol

1. Introduction

The family *Pseudomonadaceae* incorporates a broad range of species
that are common inhabitants of soil, fresh water, and marine environments, where they are active in organic matter mineralization. Some species can cause diseases of plants, with various host specificities, whereas
others are important opportunistic human and animal pathogens. Many
species are metabolically diverse, capable of utilizing a variety of organic
compounds as the sole source of carbon and energy. The type species of
this family is *Pseudomonas aeruginosa (1)*. *Pseudomonas* species have
previously been reported to be transformed genetically by conjugation,
transfection, or chemical transformation. In addition, several investigators have described the electroporation of *P. aeruginosa* and *P. putida (2–5)*.

Electroporation uses the electrical discharge of a capacitor to form a
high-intensity electric field that generates pores in the membranes of
cells, which allows the entry of exogenous DNA molecules. Although a
large proportion of the cells are killed during this electrical pulse, many
cells undergo reversible permeability, provided the electric field strength
does not exceed a critical level *(6)*. Electroporation has become a valuable technique for the introduction of plasmids into bacterial cells for
which other genetic systems have not been developed. In addition,
electroporation has permitted the routine manipulation of DNA in many
different *Pseudomonas* species. This chapter describes a protocol for successfully electroporating *Pseudomonas*.

From: *Methods in Molecular Biology, Vol. 47: Electroporation Protocols for Microorganisms*
Edited by: J. A. Nickoloff Humana Press Inc., Totowa, NJ

Table 1
Plasmid Vectors Used in Electroporation Experiments

Plasmid	Description	Antibiotic concentration in selection media
pUCP18	pUC18 derivative-Ap^r	Pa.—750 µg/mL Cb Pa. (m)—750 µg/mL Cb Pf.—750 µg/mL Cb Pp.—750 µg/mL Cb
pUCP18c	pUCP18 + Cm^r cassette-Ap^r, Cm^r	Pa.—250 µg/mL Cm Pc.—80 µg/mL Cm
pUCP26	pUCP18 derivative-Tc^r	Pa.—250 µg/mL Tc Pa. (m)—250 µg/mL Tc
pKT230	Broad host range-Sm^r, Km^r	Pa.—700 µg/mL Km

Abbreviations: Pa.—*P. aeruginosa*; Pa. (m)—*P. aeruginosa*-mucoid; Pc.—*P. cepacia*; Pf.—*P. fluorescens*; Pp.—*P. putida*; Ap—Ampicillin; Cb—Carbenicillin; Cm—Chloramphenicol; Km—Kanamycin; Sm—Streptomycin; Tc—Tetracycline.

2. Materials
2.1. Plasmid Vectors

Suitable plasmid vectors are described in Table 1. pUCP18 was extensively used for the comparison of transformation efficiencies between methods and between species. pUCP18 was constructed by H. P. Schweizer (7) by inserting the origin of replication from the broad host-range vector pRO1600 into a nonessential region of pUC18. To construct a chloramphenicol-resistant plasmid capable of replicating in *Pseudomonas* species, a 1.6-kb *Bam*HI fragment from pUC18CM (8) was ligated into the multiple cloning site of pUCP18. The tetracycline resistant plasmid pUCP26 was also derived from pUCP18 (9). In addition, the electroporation of plasmid pKT230 is described to demonstrate the transformation efficiency of a large, medium-copy number plasmid in *Pseudomonas* (10) (*see* Note 4). Plasmid DNA may be isolated by the alkaline-lysis method, CsCl gradient-purified, and dialyzed against TE buffer (10 m*M* Tris pH 8.0, 1 m*M* EDTA) (11).

2.2. Pseudomonas *Strains*

Table 2 lists some of the strains that have been successfully transformed. *P. fluorescens* NMR1336 and *P. putida* PpG1 are available from M. Vasil (Department of Microbiology and Immunology, University of Colorado Medical School, Denver, CO 80262). In addition, we tested a

Table 2
Pseudomonas Strains Used in Electroporation Experiments

Species	Strain	Source or reference
P. aeruginosa	PAO1	*12*
P. aeruginosa-mucoid	PAO-muc	*13*
P. cepacia	249-2	*14*
	Pc715j	*15*
	K63-3	*15*
P. fluorescens	NMR1336	Vasil
P. putida	PpG1	Vasil

mucoid variant of *P. aeruginosa* PAO1 *(13)* for its ability to be transformed by electroporation (*see* Note 6). These species of *Pseudomonas* require only low-level containment procedures. Autoclave all contaminated materials prior to disposal.

2.3. Solutions and Reagents for Electroporation

1. LB (Luria-Bertani), Miller, (Difco Laboratories, Detroit, MI): Dissolve 25 g in 1 L distilled water. Autoclave and store at room temperature. To make LB agar, add 15 g agar/L of LB. For the selection of transformants, add the appropriate antibiotic for the respective plasmid as listed in Table 1. Store at 4°C (*see* Note 5).
2. Magnesium electroporation buffer (MEB): 1 mM $MgCl_2$ and 1 mM HEPES. Add 240 mg/L HEPES and 203 mg/L $MgCl_2 \cdot 6H_2O$. Adjust the pH to 7.0, and autoclave. Chill to 4°C before use. Store at 4°C (*see* Notes 3, 8, and 9).
3. Sucrose magnesium electroporation buffer (SMEB): 300 mM sucrose, 1 mM $MgCl_2$, and 1 mM HEPES. Prepare MEB as above, and add 102.7 g/L sucrose before autoclaving (*see* Notes 1 and 2).
4. SOC medium: Add 10 g bacto-tryptone, 2.5 g bacto-yeast extract, 0.29 g NaCl, 93.2 mg KCl, 1.0 g $MgCl_2 \cdot 6H_2O$, and 0.60 g $MgSO_4$ to a final volume of 500 mL in distilled water. Autoclave, cool, and add 3.5 mL filter-sterilized 50% glucose. Store at room temperature.

3. Methods

1. Inoculate a flask containing 10 mL of LB with *Pseudomonas,* and grow overnight at 37°C with shaking. Use 1 mL of this culture to inoculate a 50-mL culture of fresh LB, and incubate at 37°C with shaking until late-logarithmic phase (OD_{600} = 0.5 – 1.0; approx 5 h).
2. Pipet 20 mL of this culture into a sterile 40-mL centrifuge tube, and chill on ice. Centrifuge the cells at 2300*g* for 10 min at 4°C.

3. Remove and discard the supernatant, and gently resuspend the cell pellet in 20 mL of ice-cold MEB. Centrifuge as above (*see* Notes 1, 2, 3, 8, and 9).

4. Perform a second wash step as described above. Remove the MEB wash supernatant, gently resuspend the cell pellet in 20 mL of ice-cold MEB, and recentrifuge.

5. Remove the MEB wash supernatant, and resuspend the cell pellet in 1 mL of ice-cold MEB. Transfer this cell suspension to a sterile 1.5-mL tube. Centrifuge the cells at 2300*g* at 4°C for 5 min.

6. After carefully removing the MEB wash supernatant, gently resuspend the cell pellet in 100 μL of MEB. Mix 150 ng of plasmid DNA with 100 μL of cells, and allow the cell–DNA mixture to incubate in an ice bath for 5 min before the electroporation is performed (*see* Note 7).

7. Place the DNA–cell mixture in a prechilled (4°C), sterile electroporation cuvet with a 0.2-cm electrode gap (Bio-Rad, Richmond, CA), and electroporate. We use a Gene Pulser with pulse-controller unit (Bio-Rad) with the electrical settings as follows: 2.5 kV (12.5 kV/cm), 25 μF, 200 Ω (*see* Notes 10 and 11).

8. Immediately following the discharge of the capacitor through the cells, add 900 μL of room-temperature SOC medium directly to the cuvet, and gently mix with the cells. Transfer the cells back to the original microfuge tube, and incubate at 37°C for 1 h.

9. Plate the cells onto prewarmed LB agar plates containing the appropriate antibiotic, and incubate for 24 h at 37°C.

4. Notes

1. The critical factor in the high-efficiency electroporation of *Pseudomonas* species in general, and *Pseudomonas aeruginosa* in particular, is the selection of the wash/electroporation buffer. Some investigators studying the electroporation of *P. aeruginosa* have indicated that 300 m*M* sucrose is necessary in the wash buffer to prevent cell lysis *(5)*. However, the inclusion of osmotic agents, such as glycerol and sucrose, in the electroporation buffer is probably only required for the stabilization of protoplasts, rather than whole cells *(6)*.

2. Several investigators have demonstrated increased rates of transformation in *Pseudomonas* species by including 1 m*M* MgCl$_2$ in the wash/electroporation buffer, in addition to 300 m*M* sucrose *(2,4)*. In this regard, *P. aeruginosa* is unlike *E. coli*, which can be washed and electroporated in very low-ionic-strength buffers, such as sterile water or 1 m*M* of HEPES *(16)*. This enhanced electroporation efficiency reported for *P. aeruginosa* contrasts with the electroporation of *Camplyobacter jejuni*, where 1 m*M* Mg^{2+} strongly inhibits electrotransformation *(17)* and the electroporation of *E. coli*, where it has no effect *(6)*. Both sucrose and sucrose-magne-

Table 3
Transformation Efficiencies of *P. aeruginosa* PAO1
Using Magnesium vs Sucrose-Magnesium Electroporation Buffer

Buffer used	No. of cells/mL[a]	Transformants/μg DNA
MEB	3.6×10^{11}	6.5×10^5
SMEB	3.3×10^{11}	7.0×10^4

[a]Number of prepulse cells available to be transformed following wash steps with electroporation buffers.

sium electroporation buffers result in transformation efficiencies in *P. aeruginosa* approaching 10^5 CFU/μg DNA *(2,4)*. However, it is unclear which buffer constituent is most important in *Pseudomonas* electroporation.

3. When the sucrose is omitted from the sucrose-magnesium electroporation buffer, the transformation efficiency of *P. aeruginosa* is significantly increased. There is an approx 10-fold increase in transformation efficiency with *P. aeruginosa* when using MEB vs SMEB (Table 3). This increase in transformation efficiency is not the result of an increase in cell viability when using MEB, since viable cell counts are approximately equal in both SMEB and MEB electroporation experiments. Using MEB, *P. aeruginosa* transformation efficiencies approach 10^6 CFU/μg DNA.

The increase in transformation efficiency when using MEB probably reflects the importance of Mg^{2+} ions in maintaining the structural integrity of the *P. aeruginosa* LPS. Studies have indicated that *P. aeruginosa* is extremely sensitive to divalent cation chelators, such as EDTA, which remove Mg^{2+} ions from the LPS and destabilize the *P. aeruginosa* outer membrane *(18)*. Washing and electroporating *P. aeruginosa* in low-ionic-strength buffers probably destabilize the *P. aeruginosa* outer membrane and interferes with membrane reformation following electroporation. The inclusion of a low concentration of $MgCl_2$ in the electroporation buffer reduces this destabilization. Substituting a low-ionic-strength buffer (1 m*M* HEPES, 300 m*M* sucrose) for MEB in the third wash step reduces the *P. aeruginosa* transformants obtained from 9.1×10^5 to 1.5×10^5/μg DNA.

4. The transformation efficiency of *P. aeruginosa* using different types of plasmids with different origins of replication and antibiotic resistance markers in strain PAO1 is shown in Table 4. After adjusting the plasmid DNA concentrations to give equimolar amounts of plasmid molecules, there is a general, inverse relationship between plasmid size and electroporation efficiency. However, this effect may also be caused by differences in plasmid characteristics, such as promoter strength, antibiotic marker efficiency, or plasmid topology.

Table 4
Transformation Efficiencies
of *P. aeruginosa* PAO1 Using Different Plasmids

Plasmid	Size, kb	Electroporation efficiency[a]
pUCP18	4.6	5.6×10^5
pUCP26	5.0	2.4×10^5
pUCP18c	6.2	1.0×10^5
pKT230	11.9	3.4×10^4

[a]Expressed in number of transformants/0.02 pmol DNA.

5. Because *P. aeruginosa* is able to generate spontaneous resistance mutants against a variety of antibiotics, we employ antibiotic concentrations considerably higher than that reported elsewhere *(3,4)*. Increased antibiotic concentrations should prevent the overestimation of transformation efficiency. The results from the experiments described above also suggest that it is difficult to compare electroporation efficiencies using different methods, strains, and plasmids unless these variables are tested simultaneously.

6. Under certain conditions, *P. aeruginosa* produces an alginate capsule, and its morphological phenotype is described as mucoid. The *P. aeruginosa* strain PAO1 utilized in the above experiments is a nonmucoid form, which is typical for laboratory strains. However, under certain growth conditions, PAO1 will convert to a mucoid form *(13)*. The electroporation of mucoid *P. aeruginosa* occurs at a very low frequency, even when MEB is used as the wash/electroporation buffer. Mucoid PAO1 transformants were obtained with both pUCP18 and pUCP26, although at low frequencies (6.6×10^2 CFU/µg DNA and 4.0×10^2 CFU/µg DNA, respectively). The low transformation efficiency is not the result of increased cell death, since the number of cells surviving electroporation was similar to nonmucoid controls. However, increasing the field strength of the electroporation event by increasing the size of the resistor in parallel to 400 or 600 Ω had little effect on the transformation efficiency of mucoid *P. aeruginosa*.

7. Freezing and storing *P. aeruginosa* cells at –70°C in SMEB buffer prior to electroporation result in only a small decrease in transformation efficiency. The capacity of cells to be frozen, thawed, and electroporated allows a large batch of cells to be washed, concentrated, and stored in aliquots for future electroporation experiments, increasing the convenience of electroporating *P. aeruginosa*.

8. The increase in transformation efficiency observed with *P. aeruginosa* when using MEB vs SMEB is not observed with other species of *Pseudomonas*. Both MEB or SMEB buffers are adequate for obtaining transfor-

Table 5
Electroporation Efficiencies
of *P. cepacia* Strains Transformed with pUCP18c

P. cepacia strain	Transformation efficiency[a]
249-2	4.3×10^4
Pc715j	7.4×10^3
K63-3	3.7×10^3

[a]Expressed in number of transformants/µg DNA.

mation efficiencies of approx 10^4 CFU/µg DNA with *P. fluorescens* or *P. putida*. For example, using pUCP18, 1.3×10^4 and 1.4×10^4 transformants were obtained in *P. fluorescens* using MEB and SMEB, respectively. Similarly, in *P. putida* 1.3×10^4 transformants were obtained using MEB, whereas 1.5×10^4 transformants were obtained with the same plasmid using SMEB. These experimental values are similar to those reported by others *(4)*.

9. Several strains of *P. cepacia* have also been transformed in experiments using MEB electroporation buffer. This protocol is relatively efficient at producing transformants in well-described laboratory strains of *P. cepacia* (249-2), as well as with clinical isolates (i.e., Pc715j and K63-3). Because most *P. cepacia* strains are resistant to β-lactams, the plasmid used in these experiments was a pUCP18 derivative containing a chloramphenicol cassette (pUCP18c). The results shown in Table 5 indicate that clinical isolates of *P. cepacia* may be slightly more difficult to transform by electroporation than strains passed frequently on rich growth media. Similar low transformation efficiencies have been reported for clinical isolates of *P. aeruginosa (2)*.

10. The manufacturers of electroporation devices do not recommend the use of electrically conducting material, such as $MgCl_2$, in electroporation buffers when using a pulse-controller unit. However, the inclusion of 1 m*M* $MgCl_2$ in the electroporation buffer was important to the recovery of *P. aeruginosa* transformants and did not alter the performance of the electroporation device.

11. Cuvets with different sizes of electrode gaps are available commercially. The 0.2-cm electrode gap cuvet was used exclusively in the *Pseudomonas* transformations described in this chapter. The 0.1-cm electrode gap cuvet can also be used to electroporate *Pseudomonas*. However, the amount of DNA added to the cells must be substantially reduced to prevent arcing. In addition, if the DNA sample contains too much salt, the cell–DNA sample will be too conductive and cause arcing. The 0.1-cm gap cuvets appear to be more sensitive to the purity of the DNA preparation. The manufacturers of the Gene Pulser

recommend that when using 0.1-cm cuvets, the voltage setting be decreased from 2500 to 1800 V to reduce field strength from 25 to 18 kV/cm.

Although we routinely use new cuvets for each transformation, it is possible to reuse cuvets. Electroporation cuvets can be cleaned by soaking the electrode gap overnight in 70% ethanol, washing with soapy water, rinsing the electrode gap thoroughly with distilled water, blowing the excess water out of the electrode gap with pressurized air, and autoclaving for 10 min. Cuvets can be reused approximately five times or until cracks appear in the plastic.

Acknowledgments

This work was supported by a grant from the Canadian Cystic Fibrosis Foundation.

References

1. Krieg, N. R. (ed.) (1984) *Bergey's Manual of Systematic Bacteriology,* 9th ed. Williams & Wilkens, Baltimore, MD.
2. Diver, J. M., Bryan, L. E., and Sokol, P. A. (1990) Transformation of *Pseudomonas aeruginosa* by electroporation. *Anal. Biochem.* **189,** 75–79.
3. Farinha, M. A. and Kropinski, A. M. (1990) High efficiency electroporation of *Pseudomonas aeruginosa* using frozen cell suspensions. *FEMS Microbiol. Lett.* **70,** 221–226.
4. Fiedler, S. and Wirth, R. (1988) Transformation of bacteria with plasmid DNA by electroporation. *Anal. Biochem.* **170,** 38–44.
5. Smith, A. W. and Iglewski, B. H. (1989) Transformation of *Pseudomonas aeruginosa* by electroporation. *Nucleic Acids Res.* **17,** 10,509.
6. Shigekawa, K. and Dower, W. J. (1988) Electroporation of eukaryotes and prokaryotes: a general approach to the introduction of macromolecules into cells. *BioTechniques* **6,** 742–751.
7. Schweizer, H. P. (1991) *Escherichia-Pseudomonas* shuttle vectors derived from pUC18/19. *Gene* **97,** 109–112.
8. Schweizer, H. P. (1988) The pUC18CM plasmids: a chloram-phenicol resistance gene cassette for site-directed insertion and deletion mutagenesis in *Escherichia coli. BioTechniques* **8,** 612–616.
9. West, S. E. H., Schweizer, H. P., Dall, C., Sample, A. K., and Runyen-Janecky, L. J. (1994) Construction of improved Escherichia-Pseudomonas shuttle vectors derived from pUC18/19 and nucleotide sequence of the region required for their replication in *Pseudomonas. Gene* **148,** 81–86.
10. Bagdasarian, R. L., Lurz, R., Ruckert, B., Franklin, F. C. H., Bagdasarian, M. M., Frey, J., and Timmis, K. N. (1981) Specific-purpose plasmid cloning vectors. II. Broad host range, high copy number, RSF1010-derived vectors, and a host-vector system for gene cloning in *Pseudomonas. Gene* **16,** 237–247.
11. Sambrook, J., Fritsch, E. F., and Maniatis, T. (1989) *Molecular Cloning: A Laboratory Manual,* 2nd ed. Cold Spring Harbor Laboratory, Cold Spring Harbor, NY.

12. Holloway, B. W., Krishnapillai, V., and Morgan, A. F. (1979) Chromosomal genetics of *Pseudomonas. Microbiol. Rev.* **43,** 73–102.
13. Woods, D. E., Sokol, P. A., Bryan, L. E., Storey, D. G., Mattingly, S. J., Vogel, H. J., and Ceri, H. (1991) *In vivo* regulation of virulence in *Pseudomonas aeruginosa* associated with genetic rearrangement. *J. Infect. Dis.* **163,** 143–149.
14. Beckman, W. and Lessie, T. G. (1979) Response of *Pseudomonas cepacia* to B-lactam antibiotics: utilization of penicillin G as the carbon source. *J. Bacteriol.* **140,** 1126–1128.
15. McKevitt, A. I. and Woods, D. E. (1984) Characterization of *Pseudomonas cepacia* isolates from patients with cystic fibrosis. *J. Clin. Microbiol.* **19,** 291–293.
16. Dower, W. J., Miller, J. F., and Ragsdale, C. (1988) High efficiency transformation of *E. coli* by high voltage electroporation. *Nucleic Acids Res.* **16,** 6127–6245.
17. Miller, J. G., Dower, W. J., and Tompkins, L. S. (1988) High-voltage electroporation of bacteria: genetic transformation of *Campylobacter jejuni* with plasmid DNA. *Proc. Natl. Acad. Sci. USA* **85,** 856–860.
18. Haque, H. and Russell, A. D. (1974) Effect of ethylene-diamine-tetraacetic acid and related chelating agents on whole cells of gram-negative bacteria. *Antimicrob. Agents Chemother.* **5,** 447–452.

CHAPTER 11

Electroporation of *Xanthomonas*

Teresa J. White and Carlos F. Gonzalez

1. Introduction

Species within the genus *Xanthomonas* demonstrate pathogenesis to a variety of plant types, including rice, crucifers, cotton, wheat, peppers, tomatoes, and geraniums. However, *Xanthomonas* species do not respond well to chemical treatments that induce competence necessary for transformation studies, and a highly reproducible natural transformation system has yet to be developed. Electrotransformation, a technique by which bacteria are transformed by electroporation, offers an alternative to natural transformation systems *(1)*. Our studies have primarily been concerned with *Xanthomonas oryzae* pv. *oryzae*, the causal agent of bacterial leaf blight in rice *(2)*; however, information on other species is included in our discussion.

Electroporation involves the application of a short, high-voltage pulse across a suspension of cells that results in transient membrane permeability and subsequent uptake of DNA *(1)*. The most critical parameters controlling electroporation are the initial electric field strength and the pulse duration. Compensation between electric field strength and pulse duration provides a general relationship between these parameters, and optimization of this relationship allows for the development of an efficient electrotransformation system *(3,4)*. In the development of a protocol for *Xanthomonas*, five US strains of *X. o. oryzae* and five different pathovars of *Xanthomonas campestris* were subjected to electrotransformation. However, electrical parameters were optimized for only *X. o. oryzae* strain X37-2, resulting in 3.6×10^9 transformants/µg plasmid DNA *(5)*. Transformation efficiencies and frequencies for the additional

From: *Methods in Molecular Biology, Vol. 47: Electroporation Protocols for Microorganisms*
Edited by: J. A. Nickoloff Humana Press Inc., Totowa, NJ

Table 1

Effects of X37-2-Optimized Electrical Parameters

on Electroporation of Other Strains of *X. o. oryzae*[a]

Strain	Survival, %	Efficiency, transformants/μg DNA[b]	Frequency, transformants/survivor
X37-2	42	3.6×10^9	1.8×10^{-2}
X1-5	32	1.8×10^7	2.9×10^{-4}
X7-2D	44	1.3×10^7	2.2×10^{-4}
X1-8	76	2.7×10^7	4.1×10^{-4}
X8-3	41	7.8×10^6	1.8×10^{-4}

[a]Electrical parameters: 12 kV/cm, 5-ms pulse duration. Efficiency and frequency were determined after an expression period of 1 h. Table presents averaged data from duplicate trials *(5)*.
[b]Plasmid pUFR027 *(11)* was used for all experiments.

Table 2

Effects of X37-2-Optimized Electrical Parameters

on Electroporation of Different Pathovars of *X. campestris*[a]

X. campestris pathovar	Host	Survival, %	Efficiency, transformants/μg DNA[b]	Frequency, transformants/survivor
campestris	Cruncifers	52	1.3×10^6	4.6×10^{-6}
malvacearum	Cotton	27	7.6×10^4	9.3×10^{-4}
pelargonii	Geraniums	2	1.2×10^6	8.0×10^{-4}
translucens	Wheat	25	2.1×10^7	4.3×10^{-3}
vesicatoria	Peppers, tomatoes	37	0	$<10^{-9}$

[a]Electrical parameters: 12 kV/cm, 5-ms pulse duration. Efficiency and frequency were determined after an expression period of 1 h. Table presents averaged data from duplicate trials *(5)*.
[b]Plasmid pUFR027 *(11)* was used for all experiments.

strains of *X. o. oryzae* (Table 1) and the pathovars of *X. campestris* (Table 2) were lower than those obtained for the optimized strain. Therefore, it may be essential that electrical parameters be individually determined for each bacterial strain to yield optimal efficiencies and frequencies of transformation. The parameters presented provide an effective starting point for optimization trials on these strains and other xanthomonads.

2. Materials

1. Growth medium: NBY broth: 8.0 g/L of nutrient broth (Difco, Detroit, MI), 2.0 g/L of yeast extract (Difco), 2.0 g/L of K_2HPO_4 (anhydrous), and 0.5 g/L of KH_2PO_4 *(6)*. Autoclave and cool. Add 20 mL of 50% sterile

glucose and 1.0 mL of $1M$ MgSO$_4$. Store at room temperature. For plate medium, add 20 g agar/L of broth, autoclave, and add glucose and MgSO$_4$ solutions. Mix well before pouring. Store hardened plates at 4°C.

2. Electroporation buffer (EPB): 10% sterile glycerol in sterile deionized distilled H$_2$O (ddH$_2$O).

3. Vector selection and preparation: Several desirable features for a suitable vector include a broad host-range origin of replication, an antibiotic resistance selection marker, a *lacZ*α cassette containing a multiple cloning site, and a partition locus (*see* Note 1).

   Purified plasmid DNA is essential for efficient transformation. CsCl-ethidium bromide density-gradient ultracentrifugation is a proven method for purifying DNA. However, there are purification columns currently available that significantly reduce the amount of time, equipment, and chemicals required, while providing inexpensive purified plasmid DNA. Following the final precipitation of the DNA pellet, rinse with ice-cold 70% ethanol to remove residual salt from the precipitated DNA, air-dry, and resuspend in ddH$_2$O. Store at –20°C or at 0°C on ice.

3. Methods
3.1. Cell Preparation

1. Strains of *X. o. oryzae* and pathovars of *X. campestris* can be routinely cultured on nutrient yeast extract (NBY) agar *(6)* plates and incubated at 28°C for 48 h (*see* Note 2). Working cultures are stored at 4°C and transferred at least every 2 wk. Stock cultures are maintained at –20°C in NBY containing 20% glycerol or as freeze-dried stocks.

2. Grow bacterial strains for 48 h at 28°C on NBY agar. Using a sterile loop, inoculate one loopful of bacterial cells from a fresh culture into a 250-mL sterile Erlenmeyer flask containing 55 mL of NBY broth without glucose or MgSO$_4$ (*see* Note 3). Incubate the culture at 28°C on a rotary shaker for 16–18 h at 225 rpm, and grow cells to midlog-growth phase ($A_{640} = 0.5$–0.65).

3. The following cell preparation protocol is a modification of the procedure of Dower et al. *(7)*. Concentrate a 50-mL sample by centrifugation ($3800g$) in a sterile centrifuge tube at 4°C for 10 min. Wash the cell pellet in 0.5 and 0.25 vol of cold sterile ddH$_2$O. Vortex well to resuspend cells before centrifuging. Resuspend cells in 10 mL of cold EPB, concentrate by centrifugation, and resuspend in 1.0 mL of cold EPB.

4. Transfer the cell suspension to a sterile microcentrifuge tube, centrifuge at $13,600g$ for 1.5 min at 4°C, and resuspend cell pellet to a volume of 0.5 mL in cold EPB to yield a final cell concentration of ca. 1–5×10^{11} colony-forming units (CFU)/mL. Hold cells on ice and use for electrotransformation within 2–4 h (*see* Note 4).

3.2. Electroporation Protocol

1. Transfer a 60-μL aliquot of concentrated cells to a sterile, cold 0.6-mL microcentrifuge tube containing 1–3 μL of plasmid DNA (*see* Note 5), and mix by pipeting.
2. Load a 40-μL sample of the cell/DNA mixture between prechilled electrodes (0.56-mm gap).
3. Pulse with 667 V to generate an electric field strength of 12 kV/cm (*see* Note 6), and a 5-ms pulse duration (*see* Note 7).
4. Immediately add 1.0 mL of complete NBY broth, and transfer to a 17 × 100 mm round-bottom, sterile polypropylene tube. Gently mix the sample, and transfer a 100-μL aliquot to a prechilled polypropylene tube to assess survival of the pulsed cells. Incubate the remaining sample at 28°C with constant shaking (225 rpm) for 1–3 h to allow for expression of antibiotic resistance.

3.3. Recovery of Transformants

1. Following the recovery/expression period, prepare a dilution series of the pulsed sample in sterile ddH$_2$O. Plate 100-μL aliquots onto NBY agar and NBY agar supplemented with the appropriate concentration of antibiotic (*see* Note 1).
2. Incubate the cultures at 28°C for 4 d. Visible colonies can be individually picked using a sterile inoculating loop or a toothpick, and restreaked to confirm resistance. Selective and nonselective plating allows for the determination of the transformation efficiency (transformants/μg DNA) and frequency (transformants/survivor).

4. Notes

1. Depending on the copy number of the chosen vector, preliminary experiments should include a series of platings to determine the minimum inhibitory concentration of antibiotic necessary to ensure confident selection of transformants. For example, we found that NBY agar amended with 25 μg/mL of kanamycin allowed for selection of transformants containing plasmid pUFR027 and that 50 μg/mL of spectinomycin were necessary to select transformants harboring plasmid pHM1, a derivative of pRI40 *(8)*. Reference material is available for most commonly used antibiotics *(9)*.

    The partition function provides for distribution of plasmid copies at cell division *(10)*. Unstable inheritance of plasmids that replicate at low or moderate copy number per cell can be corrected with an efficient partition system. In the development and optimization of this protocol, plasmid pUFR027 *(11)*, which is a broad host-range vector that contains the *lacZα* gene, and a partition locus were used during all electrotransformation experiments. Members of this plasmid and cosmid group have been successfully introduced into other strains of *X. o. oryzae (5,12,13)* and pathovars of *X. campestris (5,11,14)*. Using a protocol similar to the one described here and vector pUFR027, Choi and Leach *(13)* electroporated Philippine strains of *X. o. oryzae* at frequencies ranging from 10^{-8} to 10^{-3} transformants/survivor.

2. The inability to electrotransform cultures of *Xanthomonas* successfully could be the result of inherent restriction-modification (R-M) systems. Wang et al. *(15)* purified two sequence-specific type II endonucleases from *X. o. oryzae*, and the modifying methyltransferase gene from one of these systems has recently been identified and characterized *(12)*. The presence of these R-M systems in strains of *X. o. oryzae* can be verified by the resistance of genomic DNAs to digestion by endonucleases *Pst*I (*Xor*I isoschizomer) or *Pvu*I (*Xor*II isoschizomer) *(12,13)*.

3. Xanthomonads produce copious amounts of extracellular polysaccharides that interfere with uptake of plasmid DNA during electrotransformation experiments. To reduce this problem, always inoculate broth cultures from a freshly grown agar plate and do not add glucose to the culture since this enhances the production of extracellular polysaccharides.

4. Following cell preparation, some researchers quick-freeze concentrated cells in liquid nitrogen and store them at $-70°C$ *(12,14)*. Before use, cells should be thawed slowly on ice. However, when using previously frozen cells of *X. o. oryzae*, surviving cells were reduced by 50–60%, and transformation efficiencies fell noticeably. Electroporation of freshly prepared cells provided consistently higher transformant yields.

5. A linear relationship exists between the proportion of cells transformed and the DNA concentration. Using optimized electrical parameters with *X. o. oryzae* strain X37-2, the transformation frequency increased from 2.4×10^{-7} to 5.3×10^{-2} transformants/survivor as plasmid DNA concentration increased from 140 pg/mL to 20 µg/mL. Even at the highest plasmid DNA concentration, a saturation level was not observed *(5)*.

6. Optimization trials for *X. o. oryzae* strain X37-2 revealed a large window of transformability with transformation efficiencies ranging from 2.5×10^8 to 3.6×10^9 transformants/µg DNA for field strengths of 10.3–17.1 kV/cm with a 5-ms pulse duration *(5)*. Transformation efficiencies for other strains of *X. o. oryzae* and for different pathovars of *X. campestris* averaged ca. 100-fold less when the optimized parameters were used *(5,14)*.

7. It is important to prechill the sterile electroporation chamber well before using. We use the BTX T100 apparatus (Biotechnologies and Experimental Research, Inc., San Diego, CA) in conjunction with the Flatpack chamber. With this equipment, place the BTX Flatpack chamber on an

ethanol-sterilized aluminum sample weigh boat or a piece of aluminum foil, and chill thoroughly. Be sure to remove any excess condensation from the outer walls of the chamber before placing it in the holder. With other enclosed chamber designs, simply place the chamber directly on ice.

Also, contrary to most manufacturers' instructions, sample chambers can be reused if rinsed thoroughly with sterile ddH$_2$O and sterilized with 95% ethanol. However, with continued reuse, the probability of arcing increases. Arcing can also result from salt contamination of DNA, which increases the conductivity of the sample mixture.

Acknowledgments

We wish to thank S. H. Choi and J. E. Leach of Kansas State University for contributing in-press data.

References

1. Mercenier, A. and Chassy, B. M. (1988) Strategies for the development of bacterial transformation systems. *Biochimie* **70**, 503–517.
2. Jones, R. K., Barnes, L. W., Gonzalez, C. F., Leach, J. E., Alvarez, A. M., and Benedict, A. A. (1989) Identification of low virulence strains of *Xanthomonas campestris* pv. *oryzae* from rice in the USA. *Phytopathology* **79**, 984–990.
3. Calvin, N. M. and Hanawalt, P. C. (1988) High-efficiency transformation of bacterial cells by electroporation. *J. Bacteriol.* **170**, 2796–2801.
4. Chassy, B. M., Mercenier, A., and Flickinger, J. (1988) Transformation of bacteria by electroporation. *Trends in Biotechnol.* **6**, 303–309.
5. White, T. J. and Gonzalez, C. F. (1991) Application of electroporation for efficient transformation of *Xanthomonas campestris* pv. *oryzae*. *Phytopathology* **81**, 521–524.
6. Vidaver, A. K. (1967) Synthetic and complex media for the rapid detection of fluorescence of phytopathogenic pseudomonads: effect of the carbon source. *Appl. Microbiol.* **15**, 1523,1524.
7. Dower, W. J., Miller, J. F., and Ragsdale, C. W. (1988) High efficiency transformation of *E. coli* by high voltage electroporation. *Nucleic Acids Res.* **16**, 6127–6145.
8. Innes, R. W., Hiroes, M. A., and Kuemple, P. L. (1988) Induction of nitrogen-fixing nodules on clover requires only 32 kilobase pairs of DNA from the *Rhizobium trifolii* symbiosis plasmid. *J. Bacteriol.* **170**, 3793–3802.
9. Sambrook, J., Fritsch, E. F., and Maniatis, T. (1989) *Molecular Cloning: A Laboratory Manual.* Cold Spring Harbor Laboratory, Cold Spring Harbor, NY.
10. Nordstrom, K. (1985) Replication, incompatibility and partition, in *Plasmids in Bacteria* (Helinski, D. R., Cohen, S. M., Clewell, D. B., Jackson, D. A., and Hollaender, A., eds.), Plenum, New York, pp. 119–123.
11. DeFeyter, R., Kado, C. I., and Gabriel, D. W. (1990) Small, stable shuttle vectors for use in *Xanthomonas*. *Gene* **88**, 65–72.
12. Choi, S. H. and Leach, J. E. (1994) Identification of the *Xor*II methyltransferase gene and a *vsr* homolog from *Xanthomonas oryzae* pv. *oryzae*. *Mol. Gen. Genet.* **224**, 383–390.

13. Choi, S. H. and Leach, J. E. (1994) Genetic manipulation of *X. oryzae* pv. *oryzae*. *Int. Rice Res. Notes* **19,** 31,32.
14. Shaw, J. J. and Khan, I. (1993) Efficient transposon mutagenesis of *Xanthomonas campestris* pathovar *campestris* by high-voltage electroporation. *BioTechniques* **14,** 556–558.
15. Wang, R. Y. H., Shedlarski, J. G., Farber, M. B., Kuebbing, D., and Ehrlich, M. (1980) Two sequence specific endonucleases from *Xanthomonas oryzae*. *Biochim. Biophys. Acta.* **606,** 371–385.

CHAPTER 12

Transformation of *Brucella* Species with Suicide and Broad Host-Range Plasmids

John R. McQuiston, Gerhardt G. Schurig, Nammalwar Sriranganathan, and Stephen M. Boyle

1. Introduction

The six species that make up the genus *Brucella* infect a wide variety of animals and humans *(1)*. This bacterial species is gram-negative and classified as a facultative intracellular pathogen (reviewed in ref. 2). In the livestock industry of the United States and numerous countries throughout the world, the major economic impact of *Brucella* is its ability to induce abortions in cattle *(3)*. *B. abortus* is able to infect and replicate within macrophages *(4–6)*, trophoblasts *(7–9)*, and a variety of other cells *(10–12)*. It is not clearly understood how virulent strains of *B. abortus* survive in the bovine host for extended periods of time in the face of a detectable cell-mediated immunity *(3)*. In order to determine the genetic basis of the virulence of *Brucella*, i.e., to replicate and live inside professional and nonprofessional phagocytic cells, it is necessary to identify virulence genes as well as characterize these genes by complementation and mutational studies. Thus, the ability to transform *Brucella* with suicide plasmids (to introduce transposons or cause allelic exchanges) or replicating plasmids (to perform complementation studies) is paramount in implementing molecular approaches to unravel the genetic basis for *Brucella* virulence determinants.

From: *Methods in Molecular Biology, Vol. 47: Electroporation Protocols for Microorganisms*
Edited by: J. A. Nickoloff Humana Press Inc., Totowa, NJ

The DNA transformation procedures described below for introducing plasmids into *B. abortus* are based on methods originally described by Dower et al. *(13)* and modified for *Brucella* by Lai et al. *(14)* and Kovach et al. *(15)*.

2. Materials
2.1. Plasmids

The plasmid pSUP2021 is a 13.3-kb replicon that contains the Tn5 element and can autonomously replicate only in *Escherichia coli (16)*. The Tn5 element contains kanamycin and streptomycin resistance genes, which can be used as selective markers for transposition. pSUP2021 contains the chloramphenicol resistance gene, which can be used to monitor replication. The plasmid pBBR1MCS is a 4.7-kb broad host-range, mobilizable replicon *(15)* that contains a chloramphenicol resistance gene, a multiple cloning site, and T3 and T7 promoters to allow for inducible expression of cloned genes. For purposes of electroporation, both plasmids can be prepared by the alkaline lysis procedure of Ish-Horowicz and Burke *(17)*.

2.2. Bacteria

1. *B. abortus* 2308 can be obtained from Norman Cheville, Agriculture Research Service, National Animal Disease Center, P.O. Box 70, Ames, IA 50010 (*see* Note 1 regarding use of BL-3 pathogen).
2. *E. coli* DH5α can be purchased from Bethesda Research Laboratories, Life Technologies Inc. (Gaithersburg, MD).

2.3. Solutions and Reagents for DNA Transformation

1. Trypticase soy broth (TSB) (Becton-Dickinson, Cockeysville, MD).
2. Cold distilled, deionized H_2O (ddH_2O, 4°C).
3. SOC-B medium: 6% TSB (w/v), 10 mM NaCl, 2.5 mM KCl, 10 mM MgCl$_2$, 10 mM MgSO$_4$, and 20 mM glucose.
4. TSB agar plates containing kanamycin (100 µg/mL) or chloramphenicol (15 µg/mL).
5. Plasmid DNA: 0.5 µg/uL in water.

3. Methods
3.1. Preparation of Competent Cells (see Note 2)

1. Inoculate a single colony of *B. abortus* 2308 into 2 mL of TSB medium. Grow at 37°C with shaking at 180 rpm for 30–33 h. Based on these growth conditions, the Klett units after 48 h of incubation should be around 130–180 (*see* Note 3).

2. Inoculate 2 mL of the culture into 380 mL TSB medium into a 1-L flask with a side arm for measuring Klett units. Incubate at 37°C with shaking at 180 rpm until Klett units reach 70–75.
3. Chill the culture on ice 15–30 min, and transfer it to 250-mL prechilled, sterile centrifuge bottles. Cells should be kept ice cold or at 4°C for all subsequent steps.
4. Centrifuge cells 20 min at 5000g at 4°C.
5. Pour off the supernatant, resuspend the pellet in 10–15 mL of cold ddH$_2$O (4°C), add cold ddH$_2$O to 380 mL, and mix well.
6. Centrifuge cells for 15 min at 5000g at 4°C.
7. Repeat steps 5 and 6.
8. Pour off the supernatant, and resuspend the pellet in 20 mL cold ddH$_2$O, mix well, and transfer to 25-mL prechilled centrifuge tubes.
9. Centrifuge cells 10 min at 5000g at 4°C.
10. Pour off supernatant immediately, and resuspend the pellet in remaining liquid using a glass Pasteur pipet. Measure the volume of the cell suspension, and add cold ddH$_2$O, if necessary, to reach a final volume of 650 µL.
11. Transfer the cell suspension to a prechilled microcentrifuge tube, add plasmid at a ratio of 1–5 µL of 0.5 µg DNA/µL to 100 µL of cell suspension, mix well, and store the cell–DNA mixture on ice for up to 30 min.

3.2. DNA Electroporation Protocol

1. Add 100 µL cell–DNA mixture between electrodes separated by 0.5 cm.
2. Pulse at 625 V for 10 ms (equivalent to electric field strength 12.5 kV/cm), 0.4 µF (*see* Notes 4–6).
3. Transfer the electroporated cells to a 15-mL tube after adding 1-mL of SOC-B media (*see* Notes 7 and 8).
4. Incubate the electroporated cells for 24 h at 37°C with shaking at 180 rpm.
5. Transfer 250-µL aliquots of electroporated cell culture onto TSB/Kn plates (100 µg/mL) in the case of pSUP2021 or TSB/Cm plates (15 µg/mL) in the case of pBBR1MCS; incubate at 37°C for 4 d (*see* Note 9).
6. Observe and record the number of Kn^R or Cm^R clones (*see* Notes 10–13).

4. Notes

1. Because *B. abortus* is a pathogen (*18*) capable of infecting humans and animals, all the above procedures must be carried out in a Biosafety Level 3 (BL-3) facility. All culture manipulations as well as the electroporation protocols must be performed inside a Class II biocontainment cabinet to reduce the risk of infection from aerosols.
2. All materials and reagents coming into contact with bacteria must be sterile.
3. A convenient and very safe manner to grow *Brucella* without having to remove the culture is to use a flask with a side arm (Bellco Glass Co.,

Vineland, NJ) that can be inserted into a KLETT-SUMMERSON Photo-electric colorimeter (Fisher Scientific) to measure the cell density. Theoretically, the starting Klett value should be about 1.0; it takes about 16–20 h to reach a Klett value of 75. The final cell density should be around 3.0×10^{11} cells/mL.

4. If using nondisposable electrodes (BTX #474—0.5-mm gap), sterilize the electrodes and cuvets with 70% ethanol for 15 min or more, and dry under a class 2 containment hood.

5. With some species or strains of *Brucella* that seem to transform less efficiently (e.g., rough *B. abortus* strain RB51, and *B. melitensis*), repeating the wash step (Section 3.1., step 5) up to five times can greatly improve transformation efficiency (up to 10-fold).

6. Keeping anything coming in contact with the cells (i.e., tubes, water, centrifuge rotors) on ice is probably the most critical parameter. It has been observed that even transferring the tubes from the centrifuge to the hood at room temperature can dramatically reduce efficiencies.

7. SOC-B media can be replaced with TSB plus 10 mM MgCl$_2$ and 10 mM MgSO$_4$; this medium is easier to make.

8. After the electrodes are pulled out from the cuvet, some cells may remain in the cuvet; wash the cuvet with SOC-B medium to recover as many cells as possible, but be sure not to change the final volume of the culture and do not use the same cuvet again. Sometimes, electroporated cells can become very sticky and difficult to remove from electrodes completely. When this happens, SOC-B medium may be added between the electrodes to dilute the cell suspension and facilitate recovery of all cells. The electrodes must be rinsed thoroughly with cold sterile ddH$_2$O (e.g., 10 times with 1.0 mL of water using a glass Pasteur pipet) to remove all ions. Otherwise, arcing may occur and cause damage to the electrodes when the next electroporation is performed.

9. If the incubation time following plating on the TSB/kanamycin plates is longer than 4 d, spontaneous kanamycin-resistant (Kn^R) mutants will appear on negative control plates (i.e., with cells electroporated without DNA). Generally, the difference between spontaneous Kn^R mutants and true Kn^R clones bearing Tn5 can be differentiated by colony size. Usually, the colonies of spontaneous Kn^R mutants are much smaller than colonies bearing Tn5.

10. In the case of *B. abortus,* electroporation of pSUP2021 or a nonreplicating plasmid containing a *Brucella* gene disrupted by an antibiotic resistance marker, approx $1–5 \times 10^2$ Kn^R CFU/µg of plasmid are routinely obtained.

11. In the case of *B. melitensis,* using the above parameters, approx 10-fold less Kn^R CFU/µg of plasmid are routinely obtained (unpublished observations).

12. In the case of *B. abortus,* electroporation of the broad host-range plasmid pBBR1MCS, approx 2×10^4 Ap^R CFU/µg of plasmid are routinely obtained (unpublished observations).
13. In the case of nonreplicating plasmids described in Notes 10 and 11, cells that grow on the antibiotic selective medium reflect allelic exchange, i.e., recombination as opposed to transformation.

References

1. Alton, G. G., Jones, L. M., Angus, R. D., and Verger, J. M. (1988) *Techniques for the Brucellosis Laboratory.* Institute National de la Reserche Agronomique, Paris, France.
2. Smith, L. D. and Ficht, T. A. (1990) Pathogenesis of *Brucella. Crit. Rev. Microbiol.* **17,** 209–230.
3. Nicoletti, P. (1984) The epidemiology of brucellosis in animals. *Dev. Biol. Stand.* **56,** 623–628.
4. Cheers, C. (1984) Pathogenesis and cellular immunity in experimental murine brucellosis. *Dev. Biol. Stand.* **56,** 237–246.
5. Harmon, B. G., Adams, L. G., and Frey, M. (1988) Survival of rough and smooth strains of *Brucella abortus* in bovine mammary gland macrophages. *Am. J. Vet. Res.* **49,** 1092–1097.
6. Harmon, B. G., Adams, L. G., Templeton, J. W., and Smith, R. (1989) Macrophage function in mammary glands of *Brucella abortus*-infected cows and cows that resisted infection after inoculation of *Brucella abortus. Am. J. Vet. Res.* **50,** 459–465.
7. Anderson, T. D. and Cheville, N. F. (1986) Ultrastructural morphometric analysis of *Brucella abortus*-infected trophoblasts in experimental placentitis. Bacterial replication occurs in rough endoplasmic reticulum. *Am. J. Pathol.* **124,** 226–237.
8. Anderson, T. D., Cheville, N. F., and Meador, V. P. (1986) Pathogenesis of placentitis in the goat inoculated with *Brucella abortus.* II. Ultrastructural studies. *Vet. Pathol.* **23,** 227–239.
9. Meador, V. P. and Deyoe, B. L. (1989) Intracellular localization of *Brucella* abortus in bovine placenta. *Vet. Pathol.* **26,** 513–515.
10. Holland, J. J. and Pickett, M. J. (1956) Intracellular behavior of *Brucella* variants in chick embryo cells in tissue culture. *Proc. Soc. Exp. Biol. Med.* **93,** 476.
11. Hatten, B. A. and Sulkin, S. E. (1986) Intracellular production of *Brucella* L forms. II. Induction and survival of *Brucella abortus* L forms in tissue culture. *J. Bacteriol.* **91,** 14–20.
12. Detilleux, P. G., Deyoe, B. L., and Cheville, N. F. (1990) Penetration and intracellular growth of *Brucella abortus* in nonphagocytic cells *in-vitro. Infect. Immun.* **58,** 2320–2328.
13. Dower, W. J., Miller, J. F. and Ragsdale, C. W. (1988) High efficiency transformation of *E. coli* by high voltage electroporation. *Nucleic Acids Res.* **16,** 6127–6145.
14. Lai, F., Schurig, G. G., and Boyle, S. M. (1991) Electroporation of a suicide plasmid bearing Tn5 into *Brucella abortus. Microbial Pathogenesis* **9,** 363–369.

15. Kovach, M. E., Phillips, R. W., Elzer, P. H., Roop, R. M., II, and Peterson, K. M. (1994) pBBR1MCS: a broad-host-range cloning vector. *BioTechniques* **16,** 800,801.
16. Simon, R., Priefer U., and Punhler, A. (1983) A broad host range mobilization system for *in-vivo* genetic engineering: transposition mutagenesis in gram negative bacteria. *BioTechnology,* **November,** 784–791.
17. Ish-Horowicz, D. and Burke, J. F. (1981) Rapid and efficient cosmid cloning. *Nucleic Acids Res.* **9,** 2989–2998.
18. Richmond, J. Y. and McKinney, R. W. (1993) in *Biosafety in Microbiological and Biomedical Laboratories.* US Government Printing Office, Washington, DC, pp. 85,86.

CHAPTER 13

Electroporation
of *Francisella tularensis*

Gerald S. Baron, Svetlana V. Myltseva,
and Francis E. Nano

1. Introduction

Francisella tularensis is the gram-negative cocco-bacillus that is the etiologic agent of the highly infectious zoonosis, tularemia. In North America, tularemia presents as an acute febrile lymphadenitis that can progress into a life-threatening pneumonic illness *(1)*. The disease is usually acquired through the bite of an arthropod vector or from handling contagious wild rabbits. In Europe and Asia, tularemia is often acquired from consumption of contaminated water or inhalation of contaminated dust. The course of disease for European and Asian tularemia is much milder than the North American disease, with rare fatal infections.

The differences seen in the North American and Eurasian diseases are the result of the difference in the infecting strains. The predominant North American strain has been designated *F. tularensis* type A (or biotype *tularensis*), and is capable of causing a fatal infection in rabbits as well as in many rodents. The Eurasian strain, designated type B or biotype *palaeartica,* is not able to cause a fatal infection in rabbits, but causes a fatal infection in mice and guinea pigs. The type B strain is present in North America, but is less often associated with human disease than the type A biotype. Both type A and type B strains are highly infectious, being able to initiate infections with an inoculum of <10 organisms *(2)*.

At least two other forms of *Francisella* can be found in nature, but their exact distribution and contribution to human disease are poorly

From: *Methods in Molecular Biology, Vol. 47: Electroporation Protocols for Microorganisms*
Edited by: J. A. Nickoloff Humana Press Inc., Totowa, NJ

understood. *F. tularensis* biotype *novicida* has been isolated from three different locations in the US and has caused disease in two individuals *(3,4)*. Its DNA relatedness to strains of *F. tularensis* is as high as the interrelatedness of type A and type B strains, i.e., approx 90% *(4)*. The *novicida* biotype is highly virulent in mice, but apparently noninfectious for healthy humans. Another species, *F. philomiragia,* has been found to infect people with chronic granulomatous disease or other defects in their immune system, and is widespread in nature *(4,5)*.

Studies with *F. tularensis* biotypes A, B, and *novicida,* have demonstrated that these strains cause infections characterized by parasitism of host macrophages *(6)*. The immunology of clearance of *F. tularensis* infections is Class II restricted and dependent on the presence of γ-interferon, and is therefore consistent with the view that *F. tularensis* grows primarily intracellularly *(7–11)*. In vitro studies have demonstrated the ability of *F. tularensis* to grow inside macrophages *(12)*.

2. Materials

1. Strains: The *F. tularensis* live vaccine strain ("LVS," ATCC 29684) is the most widely studied strain and has been proven to be safe to use in the laboratory using Biosafety Level 2 (BL2) containment and procedures. *F. tularensis* biotype *novicida* U112 (ATCC 15482) is the fastest-growing *Francisella* strain, with a generation time of 1 h, and is also safe to use under BL2 conditions.
2. Plasmids and transposons: Plasmids pFEN504-3 and pLA68-11 (*see* Note 1) should be used as positive controls for the electroporation of the LVS and the *novicida* biotypes, respectively *(13)*.
3. Growth media: For electroporation experiments, *Francisella* strains are grown in trypticase soy broth (Difco, Detroit, MI) supplemented with 0.1% cysteine HCl (TSB-C; Sigma, St. Louis, MO) (*see* Notes 6–8). Transformants are selected on cystine heart agar (Difco) supplemented with 5% defibrinated horse blood (PML Microbiological) and kanamycin sulfate (Sigma) (5 µg/mL for LVS and 15 µg/mL for the *novicida* biotype).
4. Electroporation wash buffer: 500 m*M* sucrose (Anachemia, Champlain, NY).

3. Methods
3.1. Cell Preparation

1. Grow cells in TSB-C to 100 Klett units (A_{600} ~0.2) using the number 47 filter (~540 nm).
2. Centrifuge 40 mL of culture at 10,000*g* for 5 min, and resuspend the culture in 40 mL of 500 m*M* sucrose. This should be repeated, and then the cell

suspension should be centrifuged and resuspended in 4 mL of sucrose solution (*see* Note 3). Aliquot 1 mL into each of four screw-cap microfuge tubes, then replace caps. Dip the microfuge tubes in disinfectant, and centrifuge for 5 min. Resuspend the pellet in 40 μL of the sucrose solution by pipeting the pellet up and down repeatedly. Incubate the cell suspension on ice until needed for electroporation.

3.2. Electroporation

1. Add 1 μL (~1 μg) of highly purified pFEN504-3 or pLA68-11 DNA to 40 μL of cells and transfer to a 0.1-cm gap cuvet.
2. Electroporate at 1.5 kV at 400 Ω, with capacitance set at 25 μF (*see* Note 2).
3. Add 1 mL TSB-C, and grow cells for 6 h (for LVS) or 2 h (for biotype *novicida*) at 37°C with shaking at 225 rpm.
4. Plate on blood agar plates supplemented with kanamycin (5 μg/mL for LVS and 15 μg/mL for *novicida* biotype), and incubate at 37°C. Transformants of the *novicida* biotype should appear in 1–2 d, and those of the LVS biotype should appear in 3 d (*see* Notes 4 and 5).

4. Notes

1. Plasmids pFEN504-3 and pLA68-11 should be used as positive controls in the electroporation of the LVS and *novicida* biotypes, respectively. Plasmid pFEN504-3 (ATCC 40849) is a pUC18-derived recombinant that has an LVS insert of approx 9 kbp, with a kanamycin resistance (Km) cassette located in the insert DNA. Plasmid pLA68-11 *(13)* contains a biotype *novicida* DNA insert of approx 6 kbp together with a Km cassette. To date, the only antibiotic marker that has consistently given antibiotic resistant transformants is a Km cassette derived from Tn*903 (14)*.
2. All manipulations that create aerosols, such as vortexing, should be carried out in a type 2 biosafety cabinet. Centrifuge tubes should be closed tightly and disinfected on the outside before centrifugation. Centrifuge tubes should be opened only in a biosafety cabinet. Centrifugation of highly virulent strains should be done in aerosol containment cannisters. For electroporation, the electrode/cuvet holder portion of the electroporation apparatus should be moved into the biosafety cabinet, since sparks sometimes dislodge the cap and generate an aerosol. Although the LVS and *novicida* biotypes of *F. tularensis* usually do not cause disease in healthy humans, special caution should be taken when working with aerosols, since inhalation is the most dangerous route of infection. Those choosing to work with the highly virulent *F. tularensis* strains should be vaccinated. Furthermore, research with virulent strains should be performed in a Level 3 biocontainment facility, and use of HEPA-filtered respirators during electroporation experiments should be considered.

3. The carbohydrate-rich capsule covering the LVS and most other strains of *F. tularensis* appears to be a hindrance to efficient electroporation; capsule negative strains are transformed at higher efficiencies more consistently. Fortunately, the capsule can be removed with relatively mild treatments. For example, vortexing cells in sucrose containing 1% *N*-hexadecyl-*N*,*N*-dimethyl-3-ammonio-1-propanesulfonate (Calbiochem, La Jolla, CA) for 30 s increases the electroporation efficiencies *(13)*. We have observed that growing *Francisella* in trypticase soy broth yields cells that are transformed better than cells grown in Chamberlain's defined medium *(15)*. It is possible that the large amount of glucose present in Chamberlain's medium stimulates the production of the capsule.

    The use of a capsule-negative strain permits more efficient electroporation *(16)*. At present, only the capsule-deficient variant of the LVS strain has been used in electroporation studies. Capsule-negative variants of highly virulent strains were identified several years ago, and such strains may be available from individual researchers.

4. The events involved in transformation of *Francisella* are poorly understood. The *novicida* biotype transforms approx 1000-fold better by chemical means than by electroporation, whereas the LVS biotype transforms at least 1000-fold better by electroporation. The only type of DNA that has successfully been introduced into *Francisella* is that which contains *Francisella* sequence, for example, recombinant clones with *Francisella* DNA inserts that are interrupted by a Km cassette. Different regions of the *Francisella* chromosome integrate into the chromosome at widely different rates. Autonomously replicating plasmids have been introduced into the *novicida* biotype, but this occurred only when the plasmids contained *Francisella* inserts; these plasmids transformed cells at very low efficiencies. Linear chromosomal DNA can be introduced into the *novicida* biotype by chemical transformation techniques, but cannot be electroporated into the LVS biotype. There is no evidence for a restriction/modification system in either the LVS or the *novicida* biotype.

5. When using a new sample of DNA for electroporation, a positive control should always be included. The transformation frequencies of different integrating loci can vary by a factor of 10,000 or more, and the location of the selectable marker in the locus strongly affects the transformation frequency.

6. The growth of the LVS biotype may be inhibited by small amounts of contaminants in the glassware or in media components. Each lot of horse blood should be evaluated for its ability to support growth of the LVS biotype. Insufficiently rinsed glassware may contain detergents that inhibit the growth of the LVS.

7. Cultures of the LVS should contain approx 10^5 cells/mL in order to start growth quickly; apparently the accumulation of a siderophore is required before growth can commence *(17)*.

8. Strains can be stored at $-76°C$ after mixing bacterial cultures 1:1 with 2.6% sterile gelatin.

References

1. Evans, M. E., Gregory D. W., Schaffner, W., and McGee, Z. A. (1985) Tularemia: a 30-year experience with 88 cases. *Medicine* **64,** 251–269.

2. Fortier, A. H., Slayter, M. V., Ziemba, R., Meltzer, M. S., and Nacy, C. A. (1991) Live vaccine strain of *Francisella tularensis*: infection and immunity in mice. *Infect. Immun.* **59,** 2922–2928.

3. Larson, C. L., Wicht, W., and Jellison, W. L. (1955) A new organism resembling *P. tularensis* isolated from water. *Public Health Rep.* **70,** 253–258.

4. Hollis, D. G., Weaver, R. E., Steigerwalt, A. G., Wenger, J. D., Moss, C. W., and Brenner, D. J. (1989) *Francisella philomiragia* comb. nov. (formerly *Yersinia philomiragia*) and *Francisella tularensis* biogroup *novicida* (formerly *Francisella novicida*) associated with human disease. *J. Clin. Microbiol.* **27,** 1601–1608.

5. Wenger, J. D., Hollis, D. G., Weaver, R. E., Baker, C. N., Brown, G. R., Brenner, D. J., and Broome, C. V. (1989) Infection caused by *Francisella philomiragia* (formerly *Yersinia philomiragia*). *Ann. Intern. Med.* **110,** 888–892.

6. White, J. D., Rooney, J. R., Prickett, P. A., Derrenbacher, E. B., Beard, C. W., and Griffith, W. R. (1964) Pathogenesis of experimental respiratory tularemia in monkeys. *J. Infect. Dis.* **114,** 277–283.

7. Anthony, L. S. D. and Kongshavn, P. A. L. (1988) H-2 Restriction in acquired cell-mediated immunity to infection with *Francisella tularensis* LVS. *Infect. Immun.* **56,** 452–456.

8. Surcel, H.-M., Ilonen, J., Poikonen, K., and Herva, E. (1989) *Francisella tularensis*-specific T-cell clones are human leukocyte antigen class II restricted, secrete interleukin-2 and gamma interferon, and induce immunoglobulin production. *Infect. Immun.* **57,** 2906–2908.

9. Anthony, L. S. D., Ghadirian, E., Nestel, F. P., and Kongshavn, P. A. L. (1989) The requirement for gamma interferon in resistance of mice to experimental tularemia. *Microb. Pathog.* **7,** 421–428.

10. Leiby, D. A., Fortier, A. H., Crawford, R. M., Schreiber, R. D., and Nacy, C. A. (1992) In vivo modulation of the murine immune response to *Francisella tularensis* LVS by administration of anticytokine antibodies. *Infect. Immun.* **60,** 84–89.

11. Anthony, L. S. D., Morrissey, P. J., and Nano, F. E. (1992) Growth inhibition of *Francisella tularensis* live vaccine strain by IFN-γ-activated macrophages is mediated by reactive nitrogen intermediates derived from L-arginine metabolism. *J. Immunol.* **148,** 1829–1834.

12. Anthony, L. S. D., Burke, R. D., and Nano, F. E. (1991) Growth of *Francisella* spp. in rodent macrophages. *Infect. Immun.* **59,** 3291–3296.

13. Anthony, L. S. D., Gu, M., Cowley, S. C., Leung, W. W. S., and Nano, F. E. (1991) Transformation and allelic replacement in *Francisella* spp. *J. Gen. Microbiol.* **137,** 2697–2703.

14. Way, J. C., Davis, M. A., Morisato, D., Roberts, D. E., and Kleckner, N. (1984) New Tn*10* derivatives for transposon mutagenesis and for constructing of *lac*Z operon fusions by transposition. *Gene* **32,** 369–379.

15. Chamberlain, R. E. (1965) Evaluation of live tularemia vaccine prepared in a chemically defined medium. *Appl. Microbiol.* **13,** 232–235.

16. Sandstrom, G., Lofgren, S., and Tarnvik, A. (1988) A capsule-deficient mutant of Francisella tularensis LVS exhibits enhanced sensitivity to killing by serum but diminished sensitivity to killing by polymorphonuclear leukocytes. *Infect. Immun.* **56,** 1194–1202.

17. Halmann, M., Benedict, M., and Mager, J. (1967) Nutritional requirements of *Pasteurella tularensis* for growth from small inocula. *J. Gen. Microbiol.* **49,** 451–460.

CHAPTER 14

A Simple and Rapid Method for Transformation of *Vibrio* Species by Electroporation

Hajime Hamashima, Makoto Iwasaki, and Taketoshi Arai

1. Introduction

Vibrios are facultative anaerobes. They are gram-negative, oxidase-positive rods, and they are distributed in seawater and in the water at the mouths of rivers. According to Bergey's *Manual of Systematic Bacteriology*, there are 20 species in the genus *Vibrio*, including the human pathogens *V. cholerae*, *V. parahaemolyticus*, *V. mimicus*, *V. fluvialis*, *V. vulnifficus*, *V. fetus*, and others. Members of the species *V. cholerae* are divided into O antigen groups from 1 to 140, and all strains that cause diarrhea as a major symptom belong to the O-1 group. Members of the O-1 group secrete extracellular enterotoxin and can be further divided into two groups by biotype, namely, classical and El Tor. Strains of *V. cholerae* that do not agglutinate in O-1 serum, but secrete enterotoxin and cause diarrhea are designated *V. cholerae* non-O-1 strains. All of the previous six pandemics from 1817 to 1923 were caused by the classical biotype of *V. cholerae*, but the present seventh pandemic from 1961 is caused by the El Tor biotype. Some strains of *V. cholerae* non-O-1, which agglutinate in O-139 serum (synonym Bengal), are causative agents of a cholera-like disease that is currently sweeping India and Bangladesh *(1,2)*. The cause of the change in biotype of the pathogens is not well understood.

From: *Methods in Molecular Biology, Vol. 47: Electroporation Protocols for Microorganisms*
Edited by: J. A. Nickoloff Humana Press Inc., Totowa, NJ

Strains of *V. parahaemolyticus* that have been isolated from patients are positive for the Kanagawa phenomenon, namely, hemolysis by direct thermostable hemolysin *(3)*, but strains isolated from foods and other environments are negative *(4)*.

The causes of various phenomena, such as the halophilic character of *V. parahaemolyticus,* the resistance of *V. cholerae* to alkali, the change in biotype of *V. cholerae,* and the short doubling time of *Vibrio* unlike that of Enterobacteriaceae, are of interest to researchers in microbiology and medicine. The genetic analysis of mutations or expression of genes in the genus *Vibrio* is necessary if we are to explain these phenomena. However, most current genetic investigations are confined to Enterobacteriaceae, and there are a few reports about *Vibrio.* We believe that the lack of studies is owing to the following features of *Vibrio.*

1. There are cryptic mini-plasmids in *Vibrio,* and few large plasmids, such as those found in enterobacterial strains, are found in *Vibrio (5,6).* Moreover, strains of *Vibrio* resistant to chloramphenicol (CP) are rarely observed in nature to our knowledge *(7),* one exception being that in the report by Aoki et al. *(8).*
2. The plasmids of enterobacterial strains are readily lost from *Vibrio* host cells without drug-selection pressure *(9,10).*
3. Most strains of *Vibrio* secrete extracellular DNase that is a barrier to transformation of *Vibrio (11).*

We have reported the transformation of three species of *Vibrio* by electroporation. The efficiency of electroporation is highly dependent on the strength of the electric field, the concentration of plasmid DNA, and the combination of plasmid and recipient strain *(12).* In our experiments, the efficiency of transformation of *V. parahaemolyticus* of which extracellular DNase activity was minimized was about two times greater than that with the wild-type strain (unpublished results). Kawagishi et al. reported that DNase activity of cells of *V. alginolyticus* was minimized by $CaCl_2$-EDTA treatment before electroporation *(13).* However, the inhibition of DNase activity by treatment with EDTA often resulted in the lysis of *V. cholerae* and *V. parahaemolyticus* cells. We describe a simpler and more effective electroporation method of *Vibrio* species in this chapter.

Because the production of β-lactamase encoded by a chromosomal gene has been observed in *Vibrio,* low concentrations of ampicillin cannot be used for a selective drug in the same way they can be used in *Escherichia coli. Vibrio* and in particular *V. parahaemolyticus* are normally grown in media containing 2–3% NaCl, which reduces the activity of tetracyclines (TCs). Therefore, TCs are unsuitable as the selective agent. As described above, few strains of *Vibrio* are resistant to CP and no multiple-resistance plasmids have been observed in *V. parahaemolyticus* apart from pSA55 *(7)*. Therefore, CP is suitable as a selective agent. The vector plasmids of *E. coli* K-12 and its variants are useful for transformation of *Vibrio*. However, plasmids from *E. coli* are unstable in *Vibrio (9,10)*, and therefore, the development of stable plasmids is clearly desirable. We have developed a "high frequency of transformation (HFT)" variant of *V. parahaemolyticus* that is able to harbor plasmids from *E. coli*.

The efficiency of electroporation of *Vibrio* is affected by the concentrations of Na^+, Mg^{2+}, and Ca^{2+} ions in the medium used for the growth of host cells, and it is decreased by the addition of 10% glycerol to the electroporation buffer, which is often used for the electroporation of species of Enterobacteriaceae.

2. Materials
2.1. Solutions

1. MH broth: 30% infusion from beef, 1.75% casamino acids, and 0.15% starch, pH 7.4 (*see* Note 3).
2. SMH broth: 30% infusion from beef, 1.75% casamino acids, 0.15% starch, and 2.5% NaCl, pH 7.4 (*see* Note 3).
3. EP buffer: 272 m*M* sucrose, 1 m*M* $MgCl_2$, 7 m*M* KH_2PO_4-Na_2HPO_4 buffer, pH 7.4 (*see* Notes 2 and 5).

2.2. Bacterial Strains and Plasmids

The following bacteria can be used as recipient strains: *V. parahaemolyticus* ST565 and SIH411, *V. cholerae* non-O-1 ST25, and *V. alginolyticus* ST223. These strains, apart from *V. parahaemolyticus* SIH411, were isolated from patients and from the environment in 1980 and 1981 *(7)* and maintained in a frozen store at –80°C. *V. parahaemolyticus* SIH411 is an HFT variant obtained from ST565 by mutagenesis with 1-methyl-3-nitro-1-nitrosoguanidine (unpublished results). The strains of *V. parahaemolyticus* and *V. alginolyticus* are grown in SMH broth, and the other strain is grown in MH broth. SMH- or MH-agar medium containing 6.25 µg/mL CP is used for the selection of transformed cells. Plasmids (pHSG398, pACYC184, and pBR325) can be isolated from *E. coli* by the procedure described by Birnboim and Doly *(14)*. Purification of plasmid DNA by CsCl density-gradient centrifugation in the presence of ethidium bromide is performed as described by Sambrook et al. *(15)*.

3. Methods

3.1. Preparation of Competent Cells for Electroporation

1. Incubate cells of the host strain with shaking at 37°C overnight in SMH broth, except in the case of *V. cholerae*. Use MH broth for the growth of *V. cholerae* (*see* Note 3).
2. Inoculate 1.0–1.5 mL of an overnight culture into 160 mL of fresh SMH broth or MH broth. Grow cells to midexponential phase, to an optical density (OD) at 600 nm of 0.5–0.6, with vigorous shaking at 37°C (*see* Note 4).
3. Divide the culture into 40-mL aliquots in 50-mL polypropylene tubes (e.g., Falcon 2070), and chill on ice.
4. Collect the cells by centrifugation at 2000–2700*g* for 20 min at 4°C in a swinging bucket rotor.
5. Wash the collected cells in each tube with 40 mL of ice-cold EP buffer with subsequent centrifugation at 2000–2700*g* for 20 min at 4°C.
6. Resuspend the washed cells in 8 mL of cold EP buffer at 20 times the concentration in the initial culture (*see* Note 5).
7. Divide this concentrated suspension into small aliquots, freeze on dry ice, and store at –80°C (*see* Note 5).

3.2. Electroporation

1. Thaw the frozen cells rapidly at 37°C (*see* Note 6) and add 1.0 µg of plasmid DNA to 200 µL of the competent cells (*see* Note 1). After thorough mixing, place the mixture on ice for 15 min (*see* Note 7).
2. Transfer the mixture of cells and DNA to a chilled cuvet, with a distance of 0.2 cm between electrodes.
3. Pulse samples with a time constant of 25 ms (25 µF capacitance, 1000 Ω), at 1.25–1.5 kV (field strength, 6.25–7.5 kV/cm) for *V. parahaemolyticus* and *V. alginolyticus,* and 1.5–1.75 kV (field strength, 7.5–8.75 kV/cm) for *V. cholerae* non-O-1.
4. Dilute the samples in 0.8 mL of prewarmed SMH broth or MH broth as quickly as possible, and then incubate at 37°C for 45–60 min (*see* Note 8). Plate 200-µL aliquots of the incubated samples on SMH agar or MH agar that contains 3–6 µg/mL CP. Incubate plates at 37°C for 24–36 h (*see* Note 9).

4. Notes

1. The extracellular DNase of *Vibrio* is a major barrier to transformation by osmotic-shock methods and also to electroporation. A large quantity of DNA is needed for the electroporation of most wild-type strains of *Vibrio* that secrete the extracellular DNase.

2. The highest efficiency of electroporation is obtained with 272 m*M* sucrose in the EP buffer. The efficiency is reduced by two- to fourfold higher concentrations of sucrose, such as is used with species having high intracellular pressure. We reported that it is necessary to use fourfold higher sucrose (1.1*M*) in EP buffer for the electroporation of *Staphylococcus aureus (16)*.
3. SMH broth or MH broth is suitable for the growth of host strains and for postpulse incubation; the efficiency is decreased 3- to 50-fold when LB medium is used *(16)*.
4. Optimally competent cells are obtained at an OD at 600 nm of 0.5–0.6. The efficiency is approximately equal in the case of a culture with an OD of 0.8 when the culture is diluted to an OD of 0.5.
5. Competent cells are concentrated 20-fold, and cells are suspended in EP buffer for storage. With 40-fold concentration, there is a twofold increase in yield of transformants, but it is difficult to handle such concentrated cell suspensions. Rapid freezing is necessary to keep cells in a competent state. The viability of stored cells is decreased when 10% glycerol is included in EP buffer. Therefore, glycerol is not recommended for the storage of competent cells. Competent cells can be stored for at least 6 mo at –80°C.
6. Frozen cells should be thawed immediately at 37°C. If thawing is performed slowly at room temperature or in an ice bath, the efficiency of electroporation decreases (by ~10-fold).
7. The optimal concentration of DNA depends on the characteristics of the host strain. The mixture of plasmid DNA and cells is incubated for 15 min on ice. When the duration of incubation is shorter or longer than 15 min, the efficiency decreases. These observations suggest that a binding step is necessary and that a nuclease is present in the suspension of cells.
8. The pulsed samples are diluted with prewarmed SMH broth or MH broth, as soon as possible after electroporation, and incubated at 37°C. If the samples are cooled on ice, the efficiency decreases. Incubation of samples for 45 min is sufficient. If the duration of incubation is <45 min, the efficiency decreases.
9. Using optimized conditions as described above, the highest frequency of electroporation of HFT variant strain of *V. parahaemolyticus* is 10^{-3}/surviving cell. Several plasmids were introduced into the wild-type strains of *Vibrio* species at frequency ranging from 10^{-6} to 10^{-4} transformants/surviving cell under these conditions. The electroporation conditions described here were determined as described *(12)*.

References

1. Albert, M. J., Siddique, A. K., Islam, M. S., Faruque, A. S. G., Amsarzzaman, M., Faruque, S. M., and Sack, R. B. (1993) Large outbreak of clinical cholera due to *Vibrio cholerae* non-O1 in Bangladesh. *Lancet* **341**, 704.

2. Shimada, T., Nair, G. B., Deb, B. C., Albert, M. J., Sack, R. B., and Takeda, Y. (1993) Outbreak of *V. cholerae* non-O1 in India and Bangladesh. *Lancet* **341,** 1347.

3. Honda, T., Taga, S., Takeda, Y., and Miwatani, T. (1975) Toxin produced by *Vibrio parahaemolyticus* with lethal effects in mice. *Jpn. J. Bacteriol.* **30,** 233 (in Japanese).

4. Sakazaki, R., Tamura, K., Kato, T., Obara, Y., and Yamai, S. (1968) Studies on the enteropathogenic, facultatively halophilic bacterium, *Vibrio parahaemolyticus.* 3. Enteropathogenicity. *Jpn. J. Med. Sci. Biol.* **21,** 325–331.

5. Guerry, P. and Colwell, R. R. (1975) Isolation of cryptic plasmid deoxyribonucleic acid Kanagawa-positive strains of *Vibrio parahaemolyticus. Infect. Immun.* **16,** 328–334.

6. Arai, T., Ando, T., Kusakabe, A., and Ullah, M. A. (1983) Plasmids in *Vibrio parahaemolyticus* strains isolated in Japan and Bangladesh. *Microbiol. Immunol.* **27,** 1021–1029.

7. Arai, T., Hamashima, H., and Hasegawa, H. (1985) Isolation of a new drug-resistance plasmid from a strain of *Vibrio parahaemolyticus. Microbiol. Immunol.* **29,** 103–112.

8. Aoki, T., Kitao, T., Watanabe, S., and Takeshita, S. (1984) Drug resistance and R plasmid in *Vibrio anguillorum* isolated in cultured ayu *(Plecoglossus altivelis). Microbiol. Immunol.* **28,** 1–9.

9. Kuwahara, S., Akiba, T., Koyama, K., and Arai, T. (1963) Transmission of multiple drug-resistance from *Shigella flexneri* to *Vibrio comma* through conjugation. *Jpn. J. Microbiol.* **7,** 61–67.

10. Yokota, T., Kasuga, T., Kaneko, M., and Kuwahara, S. (1972) Genetic behavior of factor *Vibrio cholerae. J. Bacteriol.* **109,** 440–442.

11. Marcus, H., Ketley, J. M., Kaper, J. B., and Holmes, R. K. (1990) Effects of DNase production, plasmid size, and restriction barriers on transformation of *Vibrio cholerae* by electroporation and osmotic shock. *FEMS Microbiol. Lett.* **68,** 149–154.

12. Hamashima, H., Nakano, T., Tamura, S., and Arai, T. (1990) Genetic transformation of *Vibrio parahaemolyticus, Vibrio alginolyticus,* and *Vibrio cholerae* non O-1 with plasmid DNA by electroporation. *Microbiol. Immunol.* **34,** 703–708.

13. Kawagishi, K., Okunishi, I., Honma, M., and Imae, Y. (1994) Removing of the periplasmic DNase before electroporation enhances efficiency of transformation in marin bacterium *Vibrio alginolyticus. Microbiology* **140,** (in press).

14. Birnboim, H. C. and Doly, J. (1979) A rapid alkaline extraction procedure for screening recombinant plasmid DNA. *Nucleic Acid Res.* **7,** 1513–1523.

15. Sambrook, J., Fritsch, E. F., and Maniatis, T. (1989) Transformation of *Escherichia coli* by high-voltage electroporation (electrotransformation), in *Molecular Cloning: A Laboratory Manual,* 2nd ed., book 1, Cold Spring Harbor Laboratory, Cold Spring Harbor, NY, pp. 21–52.

16. Hamashima, H., Tamura, S., Komaki, Y., and Arai, T. (1992) Transformation of *Staphylococcus aureus* and other staphylococcal species with plasmid DNA by electroporation. *Bult. Showa College of Pharm. Sci.* **28,** 227–234.

CHAPTER 15

Genetic Transformation of *Bacteroides* spp. Using Electroporation

C. Jeffrey Smith

1. Introduction

The greatest progress in development of methods for genetic manipulation of obligate anaerobic bacteria has been with the *Bacteroides* and *Clostridium,* which have become the model systems for gram-negative and gram-positive anaerobes, respectively. Advancement in the design of genetic systems for use with *Bacteroides* has followed a predictable pathway. These are predominant members of the gastrointestinal tract microflora of mammals, but many species also can be significant opportunistic pathogens. As pathogens, the *Bacteroides* are difficult to treat because of inherent resistance to many commonly used antibiotics and because of acquired resistance to some of the currently effective drugs. Thus, much of the work on *Bacteroides* genetics focused on analysis of transmissible antibiotic resistance determinants, and these resistance elements have in turn provided many of the plasmids and selective markers used in the development of *Bacteroides* genetics *(1–3).* It is important to note that the *Bacteroides* antibiotic resistance elements are in fact required components of genetic tools used in these organisms, because few if any genes or plasmids from other bacterial species function in the *Bacteroides (4).*

A key feature of any genetic system is the method used for the introduction of foreign DNA into the host cells of interest. With the *Bacteroides,* the method of choice has traditionally been to conjugate plasmid constructs from *Escherichia coli* to a *Bacteroides* recipient using an

From: *Methods in Molecular Biology, Vol. 47: Electroporation Protocols for Microorganisms*
Edited by: J. A. Nickoloff Humana Press Inc., Totowa, NJ

IncP conjugative helper plasmid *(5)*. The conjugation approach has proven to be versatile, independent of plasmid size, and generally applicable to all of the *Bacteroides* species tested. However, this methodology can be labor-intensive and time-consuming. In addition, the use of conjugation, especially for the transfer of plasmids between *Bacteroides* species, can result in simultaneous transfer of undesirable genetic elements that can interfere with genetic analyses. This is because *Bacteroides* have a multitude of uncharacterized, mobilizable elements that we are just now beginning to understand. Transformation can overcome many of these limitations and is now being used more often in the genetic manipulation of *Bacteroides*.

Previous studies using chemical methods to prepare cells for transformation either did not work or were restricted to a few *B. fragilis* strains (reviewed in ref. *2*). Electroporation, on the other hand, has proven to be a widely applicable transformation method for the *Bacteroides,* and the three electroporation methods described thus far have yielded relatively high transformation frequencies ($\sim 10^6$/µg plasmid DNA *[6–8]*). Although developed independently, each of the methods are similar, using cells harvested from mid- to late-logarithmic phase of growth and a 10% glycerol solution for electroporation. In addition, the *Bacteroides* display a high survival rate for the electroporated cells when electric pulses of up to 12.5 kV/cm are used. The similarity of each of the three electroporation methods and the results obtained are a testament to the utility and ease of electroporation. In this chapter, a standard electroporation protocol used for transformation of *Bacteroides* is described.

2. Materials

2.1. Bacterial Strains and Plasmids

Although chemical transformation and electroporation are not yet widely used in *Bacteroides* genetics, a variety of strains have been transformed successfully with many different plasmids (Table 1). The strain used most often as a host for transformation is the *B. fragilis* strain 638, which also is known as 638R *(9)* and TM4000 *(10)*. This is a rifampicin resistant mutant derived by M. Sebald from a plasmid-free clinical isolate originally obtained from the Michael Reese Hospital and Medical Center, Chicago *(11)*. Other *Bacteroides* and *Prevotella* species, listed in Table 1, have been electroporated successfully, but it is interesting to note that successful transformation of *B. thetaiotaomicron* has not yet been reported.

Table 1
Representative *Bacteroides* Strains and Plasmids
That Have Been Used Successfully for Electrotransformation

Strain or plasmid	Relevant characteristics[a]	Reference
B. fragilis		
638	Rf[r], plasmid-free	*6,8,18*
V531	Tc[r], Ap[r], two cryptic plasmids 3.7 and 10.2 kb	*6*
VPI-5785	Tc[r], three cryptic plasmids between 3 and 54 kb	*6*
B. uniformis		
V528	Rf[r], plasmid-free	*6*
1100	Tp[r], plasmid-free	*7*
B. ovatus		
VPI-0038-1 (V211)	Plasmid-free	*6*
P. ruminicola		
F101	Rf[r], plasmid-free	*7*
Plasmids		
pFD288	Cc[r]; 8.8-kb *Bacteroides/E. coli* shuttle vector	*6*
pBI191	Cc[r]; 5.4-kb *Bacteroides* vector	*6,19*
pBFTM10	Cc[r]; 14.6-kb *Bacteroides* R-plasmid	*6*
pIP417	Ni[r]; 7.7-kb *Bacteroides* R-plasmid	*8,19*
pFK707ΔH1	Cc[r]; 4.6-kb *Bacteroides* vector	*17,19*
pRR14	Tc[r]; 19.5-kb *P. ruminicola* R-plasmid	*7*
pDP1	Cc[r]; 19-kb *Bacteroides/E. coli* shuttle vector	*7*
Tn*4555*	Fx[r]; 12.5-kb circular transposon intermediate	*18*

[a]The relevant antibiotic resistance phenotypes are: Rf, rifampicin resistance; Tc, tetracycline resistance; Ap, ampicillin resistance; Cc, clindamycin (MLS) resistance; Ni, 5-nitro-imidazole (and metronidazole) resistance; Fx, cefoxitin resistance. The plasmid properties listed are the relevant properties in *Bacteroides* strains.

2.2. Bacteriological Media, Solutions, and Buffers

1. Hemin/menadione solution: 0.5 g hemin in 10 mL of 1*N* NaOH, 0.05 g menadione in 10 mL of 95% ethanol, and distilled H_2O to 1 L. The hemin and menadione solutions are prepared separately, and then added to the H_2O *(12)*.
2. Resazurin: 0.1% in H_2O.
3. Sodium bicarbonate: 10% in H_2O (filter-sterilized or autoclaved).
4. Antibiotic supplements: *see* Table 2.

Table 2
Antibiotic Supplements for Routine Selection of Transformants in BHIS Medium

Drug	Stock solution[a]	Final conc. in media
Clindamycin	25 mg/mL in H_2O	5 μg/mL
Cefoxitin	25 mg/mL in H_2O	15–20 μg/mL
Tetracycline	12.5 mg/mL in 50% ethanol	5 μg/mL
Metronidazole	25 mg/mL in H_2O	4 μg/mL
Rifampicin	20 mg/mL in methanol	20 μg/mL

[a]Antibiotic stocks are stored frozen except for clindamycin, which is maintained at 4°C.

5. BHIS medium: 37 g brain heart infusion broth (BHI; Difco, Detroit, MI), 1 g cysteine (free base), 10 mL hemin/menadione solution, 1 mL resazurin solution, 20 mL sodium bicarbonate solution, 15 g agar (when required), and H_2O to 1 L. The medium is prepared by dissolving the dehydrated BHI in 970 mL H_2O. Then the cysteine, hemin/menadione solution, and resazurin are added together with the agar (for solid media). The medium is autoclaved (121°C) immediately for 20 min, cooled to 55°C, and then sterile bicarbonate solution and antibiotic supplements are added. The bicarbonate concentration is calibrated for incubation in an atmosphere with 10% CO_2. Anaerobic atmospheres with different CO_2 concentrations may require adjustment of the sodium bicarbonate levels in order to obtain the pH 7.0–7.2 range preferred by most *Bacteroides* sp.
6. Electroporation buffer (EP1): 10% glycerol in 1 mM $MgCl_2$. This buffer is sterilized by filtration (0.2-μm filter), and stored at either room temperature or 4°C.
7. TE buffer: 10 mM Tris-HCl, pH 8.0, and 1 mM EDTA.

2.3. Equipment

1. Anaerobic chamber with 37°C incubator and an atmosphere of 10% H_2, 10% CO_2, and 80% N_2. We use a flexible anaerobic glove box from Coy Laboratories Inc. (Ann Arbor, MI).
2. GasPak jars with the appropriate gas-generating envelopes can be used as an alternative to the anaerobic chamber for incubation of Petri plates.

3. Methods

3.1. Preparation of Competent Cells

1. Inoculate 5 or 10 mL of BHIS containing the appropriate antibiotics (if required) from a fresh stock culture and incubate the culture overnight (~15–18 h) at 37°C in the anaerobic chamber.

2. Inoculate 200 mL of prewarmed (37°C) BHIS with 5 mL of the overnight culture, and incubate anaerobically until growth reaches midlogarithmic phase (OD_{550} = 0.65). This step may take several hours (*see* Note 1).
3. Harvest the culture by centrifugation at 4°C, 2600*g* for 10 min (*see* Note 2). Suspend the cell pellet in 200 mL of cold EP1 buffer, centrifuge as above, and then wash the cells again in 100 mL of cold EP1 (*see* Note 3).
4. After the last wash, suspend cells in 2.0 mL of cold EP1 buffer. At this point, the electrocompetent cells can be frozen for future use (*see* Note 4) or stored on ice for up to several hours.

3.2. Electrotransformation

1. Place a sterile 0.2-cm gap cuvet on ice for several minutes, and then add 100 µL of cell suspension to the prechilled cuvet.
2. Add 2–25 µL of DNA solution in sterile water or TE buffer to the cell suspension, and mix well by tapping the cuvet continuously.
3. Incubate the cuvet containing cells and DNA on ice for 5–10 min.
4. Place the cuvet in the electroporation chamber, and pulse the cell suspension at 2500 V (12.5 kV/cm) for between 5 and 10 ms. This range is generally achieved by setting the resistance to 400 or 600 Ω (*see* Note 5).
5. Immediately following the electroporation, place cuvets in the anaerobic chamber, remove the cell suspension, and add the cells to 1 mL of prewarmed (37°C) BHIS (*see* Note 6).
6. Incubate these cultures for 3 h to allow establishment and expression of plasmid genes.
7. Plate appropriate aliquots on selective media, and incubate these in the anaerobic chamber at 37°C. The plates are incubated in plastic bags to prevent drying, and colonies begin to appear in about 24 h. Incubation is generally carried out for a full 48 h.

   These electroporation conditions generally result in 75–90% cell survival and ~10^6 transformants/µg of supercoiled plasmid DNA. However, these frequencies depend on the strain used and the methylation state or source of the DNA (*see* Notes 7 and 8).

4. Notes

1. The precise electroporation conditions for *Bacteroides* are rather flexible and can be altered without serious consequences for most applications. We have found that cells in mid- to late-logarithmic growth phase (OD_{550} = 0.5–0.9) transform at similar frequencies. When cultures reach the higher end of this range, compensate by concentrating the cells only 50-fold rather than the 100-fold prescribed above.

2. The centrifugation values for the washing and harvesting steps are not critical, and in fact are dependent on the characteristics of the strains being used. The values presented above were designed for *B. fragilis* strain 638, which pellets very easily. In contrast, *B. uniformis* and *B. ovatus* strains make a significant amount of a loose, capsule-like material that makes them difficult to pellet. The centrifugation conditions must be altered appropriately for these strains.

3. The use of 1 mM MgCl$_2$ in the electroporation buffer can be eliminated as shown by others *(7,8)*. In our hands, the MgCl$_2$ seems to maintain integrity of the *Bacteroides* cells, and thus, we recommend its use; otherwise, the cells become difficult to wash and resuspend.

4. Electroporation procedures lend themselves quite nicely to the preparation of frozen competent cells, and the *Bacteroides* are no exception. Frozen competent cells are prepared exactly as described above, except that at the final step, aliquots are placed in tubes with a sealable top, flash-frozen in a dry ice/ethanol bath, and then stored at –70°C. In order to use these cells, they are removed from the freezer and allowed to thaw on ice. Once thawed, they are treated as in the standard procedure. We have noticed up to a 10-fold decrease in transformation frequency when using frozen cells, but this does not generally interfere with routine applications.

5. The electrical field strength and pulse length may be altered without serious effect on transformation. For example, in order to avoid the possibility of "arcing" with older cell suspensions, we often use lower field strengths (10 kV/cm) or shorter pulse times when performing simple plasmid transfers into strain 638. These conditions result in about 20-fold fewer transformants.

6. When working with anaerobic bacteria, a question that often arises is the need for strict anaerobic conditions during transformation. Although this has not been directly studied, *B. fragilis* is very aerotolerant, and the requirement for strict anaerobic conditions applies primarily to the expression period following electroporation. However, if it is practical, all procedures except for the actual electroporation should be performed inside the anaerobic chamber, which may lead to increased transformation efficiency (M. Sebald, personal communication). For anaerobes lacking the oxygen tolerance of *B. fragilis,* it is necessary to perform all steps of the procedure in the anaerobic chamber using prereduced buffers (*see* ref. *7*).

7. The most critical aspect of *B. fragilis* electroporation is the methylation state or source of the DNA. These organisms have an abundance of potent restriction/modification systems, and about six different restriction enzymes have been identified *(13,14)*. These restriction enzymes can significantly reduce the efficiency of transformation by both chemical *(15,16)*

Table 3

The Effect of Plasmid Source

on Transformation of *B. fragilis* with pFD288[a]

Source of PFD288[b]	Transformants/μg DNA[c]
B. fragilis	6.2×10^4 (100%)
B. uniformis	5.27×10^3 (8.5%)
B. thetaiotaomicron	8.06×10^3 (13%)
B. ovatus	4.96×10^3 (8%)
B. coli[d]	6.2×10^1 (0.1%)

[a]Transformations were performed at 10 kV/cm, 5 ms using frozen competent cells.

[b]pFD288 was purified by CsCl-ethidium bromide density-gradient centrifugation from each of the stains listed.

[c]The percent of transformants compared to transformation with pFD288 isolated from *B. fragilis* is in parentheses.

[d]Values obtained with DNA isolated from *E. coli* were more variable than those observed with DNA from other *Bacteroides* strains, and ranged from 0 to 1.2 transformants/μg DNA.

and electroporation methods (ref. *6* and Table 3). The decrease in efficiency can be so great that in cases where the plasmid DNA is prepared from an *E. coli* host, no transformants will be obtained unless all other parameters are optimal. Generally, we observe a 1000- to 10,000-fold decrease in the number of transformants when using DNA isolated from *E. coli*, but only about a 10-fold decrease when using DNA isolated from other *Bacteroides* strains (Table 3). To date, no restriction-less *B. fragilis* mutants or variants have been described.

8. Electroporation has found its most frequent use in the transfer of plasmids between *Bacteroides* strains. However, it has been used successfully to clone a *Bacteroides* chromosomal antibiotic resistance gene directly in *Bacteroides (17)*. The key to this procedure was first to use conjugation to mobilize the resistance gene into strain 638. This 638-modified DNA became the source of the antibiotic resistance gene for cloning. Then using a vector also purified from 638, it was possible to achieve high enough transformation frequencies to permit the isolation of the chromosomal antibiotic resistance gene. The success of these experiments depended on having DNA modified by the 638 host strain, and this underscores the importance of the *Bacteroides* restriction systems.

Acknowledgments

I am grateful to M. Sebald for sharing unpublished observations on *Bacteroides* electroporation. This work was supported in part by a grant

from the Biotechnology Research and Development Corporation and by
PHS AI-28884.

References

1. Salyers, A. A., Shoemaker, N. B., and Guthrie, E. P. (1987) Recent advances in *Bacteroides* genetics. *Crit. Rev. Microbiol.* **14,** 49–71.
2. Odelson, D., Rasmussen, J., Smith, C. J., and Macrina, F. L. (1987) Extrachromosomal systems and gene transmission in anaerobic bacteria. *Plasmid* **17,** 87–109.
3. Smith, C. J. (1989) Clindamycin resistance and the development of genetic systems in the *Bacteroides. Dev. Indust. Microbiol.* **30,** 23–33.
4. Smith, C. J., Rogers, M. B., and McKee, M. L. (1992) Heterologous gene expression in *Bacteroides fragilis. Plasmid* **27,** 141–154.
5. Guiney, D. G., Hasegawa, P., and Davis, C. E. (1984) Plasmid transfer from *Escherichia coli* to *Bacteroides fragilis*: differential expression of antibiotic resistance phenotypes. *Proc. Natl. Acad. Sci. USA* **81,** 7203–7206.
6. Smith, C. J., Parker, A., and Rogers, M. B. (1990) Plasmid transformation of *Bacteroides* spp. by electroporation. *Plasmid* **24,** 100–109.
7. Thomson, A. M. and Flint, H. J. (1989) Electroporation induced transformation of *Bacteroides* ruminicola and *Bacteroides* uniformis by plasmid DNA. *FEMS Microbiol. Lett.* **61,** 101–104.
8. Sebald, M., Reysset, G., and Breuil, J. (1990) What's new in 5-nitro-imidazole resistance in the *Bacteroides fragilis* group? in *Clinical and Molecular Aspects of Anaerobes* (Borriello, S. P., ed.), Wrightson Biomedical, Petersfield, pp. 217–225.
9. Privitera, G., Dublanchet, A., and Sebald, M. (1979) Transfer of multiple antibiotic resistance between subspecies of *Bacteroides fragilis. J. Infect. Dis.* **139,** 97–101.
10. Tally, F. P., Snydman, D. R., Shimell, M. J., and Malamy, M. H. (1982) Characterization of pBFTM10, a clindamycin-erythromycin resistance transfer factor from *Bacteroides fragilis. J. Bacteriol.* **151,** 686–691.
11. Stiffler, P. W., Keller, R., and Traub, N. (1974) Isolation and characterization of several cryptic plasmids from clinical isolates of *Bacteroides fragilis. J. Infect. Dis.* **130,** 544–547.
12. Holdeman, L. V., Cato, E. P., and Moore, W. E. C. (1977) *Anaerobe Laboratory Manual,* 4th ed. Virginia Polytechnic Institute and State University, Blacksburg.
13. Azeddoug, H., Reysset, G., and Sebald, M. (1992) Characterization of restriction endonuclease *Bfr*BI from *Bacteroides fragilis* strains BE3 and AIP 10006. *FEMS Microbiol. Lett.* **95,** 133–136.
14. Roberts, R. J. and Macelis, D. (1991) Restriction enzymes and their isochizomers. *Nucleic Acids Res.* **19(Suppl.),** 2077–2109.
15. Smith, C. J. (1985) Development and use of cloning systems for *Bacteroides fragilis*: cloning of a plasmid-encoded clindamycin resistance determinant. *J. Bacteriol.* **164,** 294–301.
16. Smith, C. J. (1985) Polyethylene glycol-facilitated transformation of *Bacteroides fragilis* with plasmid DNA. *J. Bacteriol.* **164,** 466–469.

17. Haggoud, A., Reysset, G., and Sebald, M. (1992) Cloning of a *Bacteroides fragilis* chromosomal determinant coding for 5-nitroimidazole resistance. *FEMS Microbiol. Lett.* **95,** 1–6.
18. Smith, C. J. and Parker, A. C. (1993) Identification of a circular intermediate in the transfer and transposition of Tn*4555*, a mobilizable transposon from *Bacteroides* spp. *J. Bacteriol.* **175,** 2682–2691.
19. Reysset, G., Haggoud, A., Su, W., and Sebald, M. (1992) Genetic and molecular analysis of pIP417 and pIP419: *Bacteroides* plasmids encoding 5-nitroimidazole resistance. *Plasmid* **27,** 181–190.

CHAPTER 16

Electrotransformation
of *Agrobacterium*

Jhy-Jhu Lin

1. Introduction

Agrobacterium tumefaciens and *Agrobacterium rhizogenes* are plant soil-borne pathogens. Plants infected by Agrobacterium develop symptoms of tumor formation or adventitious hairy roots *(1)*. These symptoms result from the integration of a fragment of DNA (T-DNA) from the bacterial tumor-inducing (Ti) plasmid into plant genomic DNA *(2)*. This unique feature of the transfer of a discrete sequence of DNA from a bacterial plasmid DNA into plant genomic DNA makes *Agrobacterium* a natural vector for plant researchers to establish transgenic plants *(3)*.

Although different procedures have been described to establish transgenic plants, *Agrobacterium*-mediated transformation has been used extensively. The establishment of transgenic plants using *Agrobacterium*-mediated transformation utilizes a binary vector system. The binary vector system consists of two plasmids, a shuttle plasmid, and a disarmed Ti plasmid (a Ti plasmid in which the T-DNA region has been deleted), which coexist autonomously in *Agrobacterium* after the introduction of the shuttle plasmid into *Agrobacterium (4)*. Traditionally, triparental mating is the most prevalent method for introducing a recombinant plasmid into *Agrobacterium* cells *(5)*. However, it requires three different bacteria cultured together, multiple incubation temperatures, several types of growth media, and a long incubation period (Table 1). Moreover, owing to the possibility of contamination of the transformants with

From: *Methods in Molecular Biology, Vol. 47: Electroporation Protocols for Microorganisms*
Edited by: J. A. Nickoloff Humana Press Inc., Totowa, NJ

Table 1
Comparison of Transformation
of *A. tumefaciens* by Triparental Mating and Electroporation

	Triparental mating	Electroporation
Time	4–6 d	2 d
Media	YM, YT, TY, LB	YM
Bacteria	HB101 (with RK2013), *E. coli* with recombinant T-DNA plasmid, *A. tumefaciens* LBA 4404	*A. tumefaciens* LBA 4404
Incubation temperatures	37, 30, and 28°C	30°C
Contamination of transformants	Considerable risk with *E. coli*; requires several rounds of screening on selective media	Pure colonies obtained with first screening of transformants
Confirmation of transformants	Plasmid isolation recommended	Plasmid isolation not necessary

E. coli cells, it is necessary to purify the plasmid DNA from the transformants *(6)*. These steps complicate the introduction of recombinant DNA into *Agrobacterium* cells. Directly introducing recombinant DNA into *Agrobacterium* cells using a freeze/thaw procedure has been described, but the transformation efficiency is only 10^3 transformants/µg *(7–9)*. It is critical to have highly transformable *Agrobacterium* cells in order to construct a plant genomic or a cDNA library directly in *Agrobacterium* and for the isolation of genes by complementation of plant mutants. Moreover, the isolation of plant enhancers and promoters requires the random cloning of DNA fragments into promoter probe vectors followed by selection in transformed plants *(10)*. The low efficiencies of triparental mating and freeze/thaw procedures are not well suited for these applications. In contrast, *Agrobacterium* cells have been transformed with an efficiency of 10^5–10^8 transformants/µg by electroporation *(6,11–14)*. In addition, electrocompetent *Agrobacterium* cells are commercially available. In this chapter, the preparation of electrocompetent cells of *A. tumefaciens* and the conditions of electroporation are described.

2. Materials

2.1. Bacterial Strains

The disarmed octopine *A. tumefaciens* LBA4404 strain is commercially available (Clonetech, Palo Alto, CA; Gibco-BRL, Gaithersburg, MD). The American Type Culture Collection (ATCC, Rockville, MD) also has a comprehensive collection of *Agrobacterium* strains. However, many disarmed *Agrobacterium* strains were developed by biotechnology companies (such as Calgene [Davis, CA] and Monsanto [St. Louis, MO]).

2.2. DNA

The binary vector pBI121 (Clonetech) and its derivative pBICAT (Gibco-BRL) have been successfully electroporated into *A. tumefaciens* LBA4404 cells. Other binary vectors have been developed by commercial companies and academic researchers, and may be obtained from these sources.

2.3. Solutions

1. YM medium: For 1 L of broth, add 0.4 g yeast extract, 10.0 g mannitol, 0.1 g NaCl, 0.2 g $MgSO_4 \cdot 7H_2O$, and 0.5 g $K_2HPO_4 \cdot 3H_2O$. Adjust to pH 7.0 and autoclave. To prepare YM plates, add 15 g of agar/1 L of YM broth, autoclave, and dispense into Petri plates. YM medium is also commercially available from Gibco-BRL.
2. 10% Glycerol: Autoclave and store at 4°C.
3. YM storage broth: Add 40 mL of glycerol to 60 mL of YM broth, mix well, and filter-sterilize through a 0.2-μm filter.
4. 1X TE buffer: 10 m*M* Tris-HCl, pH 8.0, and 1 m*M* EDTA. Autoclave and store at 4°C.

3. Methods

3.1. Preparation of Master Cultures

1. Streak *Agrobacterium* cells from a plate or a stab onto a YM plate, and incubate at 30°C for 36–48 h.
2. Use a well-separated colony to inoculate 50 mL of YM broth in a 500-mL flask, and grow cells with vigorous aeration at 30°C until $OD_{550} = 0.2$.
3. Transfer 5 mL of cultured cells into a chilled 15-mL Falcon tube containing 5 mL of YM storage medium, mix well, and chill on ice for 10 min.
4. Transfer 0.5 mL of mixed cells into an ice-chilled 1-mL Nalgene cryotube, and quickly freeze in a dry-ice/alcohol bath for 10 min.
5. Store the master cultures in a –70°C freezer.

3.2. Preparation of Electrocompetent Cells

1. Streak *Agrobacterium* cells from a master culture onto a YM plate, and incubate at 30°C for 36–48 h.
2. Use a well-separated colony to inoculate 50 mL of YM broth in a 500-mL flask, and grow cells with vigorous aeration at 30°C until $OD_{550} = 0.2$.
3. Inoculate 100 µL of the cultured cells into 1.5 L of YM broth in a 2.8-L flask, and incubate the cells with vigorous aeration at 30°C, until $OD_{550} = 1.0$ (*see* Note 1).
4. To harvest, transfer the cells into 500-mL centrifuge bottles, and centrifuge in a cold rotor at 2600*g* for 10 min.
5. Wash the cell pellets by resuspending cells in 500 mL of ice-cold 10% glycerol. Centrifuge the cells at 2600*g* for 10 min, and carefully pour off the supernatant.
6. Repeat step 5.
7. Resuspend the cell pellet in ice-cold 10% glycerol to a final volume of 1.5 mL. Usually, cells can be resuspended in the 10% glycerol that remains in the centrifuge bottle. The final cell concentration should be 5×10^{10} cells/mL (*see* Note 2).
8. Cells can be used immediately, or can be frozen in a dry ice-ethanol bath and stored in a –70°C freezer for 6 mo.

3.3. Electroporation

1. Thaw the frozen *Agrobacterium* electrocompetent cells on ice immediately before use. For freshly prepared cells, keep the cells on ice before use.
2. Add 1.0 mL of YM broth to 15-mL Falcon tubes at room temperature.
3. Add 20–40 µL of the electrocompetent cells and 1–5 µL of DNA in TE (10 m*M* Tris-HCl and 1 m*M* EDTA, pH 8.0) buffer to an ice-cold 1.5-mL tube. Gently mix the samples by tapping the tube several times.
4. Transfer the samples to a microelectroporation chamber that is cooled to 4°C (*see* Notes 3 and 6).
5. Pulse with 16.7 kv/cm or higher with a 6-ms time constant (*see* Notes 5 and 6).
6. Transfer the cells to the YM broth in a 15-mL Falcon tube. Resuspend remaining cells in the microchamber with 200 µL of YM broth using a sterilized Pasteur pipet to remove as many cells as possible (*see* Note 7 and Table 2).
7. Incubate the cells at 30°C for 3 h with shaking at 250 rpm (*see* Note 8 and Table 3).
8. Plate 100 µL onto a selective YM plate. For unknown strains and freshly prepared electrocompetent cells, plate the cells on both selective and nonselective media using serial dilutions to determine percentage of the viable cells after electroporation and the transformation efficiency.

Table 2
Effect of the Expression Medium on the Transformation
Efficiency of *Agrobacterium tumefaciens* LBA4404
Using Electroporation

	Transformation efficiency[b]	
	YM plate	LB plate
Expression medium[a]		
YM broth	10.0	6.3
SOC broth	7.1	3.9
LB broth	0.96	0.25
EMC broth	5.3	3.2
M9 broth	0.36	0.38

[a]Medium components: LB broth: 1% bactotryptone, 0.5% yeast extract, 10 mM NaCl, pH 7.0. SOC broth: 2% bactotryptone, 0.5% yeast extract, 10 mM NaCl, 2.5 mM KCl, 20 mM ($MgCl_2 \cdot 6H_2O$ + MgSO · $7H_2O$), 20 mM glucose, pH 7.0. EMC broth: 2% bactotryptone, 1% yeast extract, 10 mM NaCl, 2.5 mM KCl, 1.5% succinic acid, pH 7.0. M9 broth: 1 mM $MgSO_4 \cdot 7H_2O$, 1.9 mM NH_4Cl, 4.2 mM NaH_2PO_4, 2.5 mM Na_2HPO_4, 0.4% glucose. YM broth: 0.04% yeast extract, 1% mannitol, 1.7 mM NaCl, 0.8 mM $MgSO_4 \cdot 7H_2O$, 2.2 mM $K_2HPO_4 \cdot 3H_2O$, pH 7.0.

[b]Transformation efficiency (CFU/µg × 10^6) was determined by pulsing 1 ng of pBI121 plasmid DNA into 20 µL of electrocompetent cells in a microelectroporation chamber at a field strength of 16.7 kV/cm. Ten microliters of the electroporated mixture were diluted to 1.0 mL using different expression media, cultured at 30°C for 3 h, and plated on either YM or LB plates containing 1.5% agar with 50 µg/mL kanamycin and 100 µg/mL streptomycin. The plates were incubated at 30°C for 2 d.

The optimization of transformation efficiency is dependent on the balance between the killing effect and the efficiency of introducing the DNA into cells. It is important to measure the total number of viable cells, especially when the conditions for electroporation have not been established for a particular strain.

4. Notes

4.1. Factors Affecting the Preparation of Electrocompetent Cells

1. In general, *Agrobacterium* cells grow well in a rich medium, such as LB broth. However, *Agrobacterium* cells are very mucoid and aggregate as the cell density increases to 10^8 cell/mL. For preparation of electrocompetent cells, it is difficult to obtain a high transformation efficiency if the cells are aggregated at the time of harvesting. Growth in YM broth

Table 3
Effect of Expression Period on the Transformation Efficiency

| Amount of pBI121 | Transformation efficiency, CFU/μg[a] | | |
	1-h expression	2-h expression	3-h expression
100 pg	1.6×10^7	2.4×10^7	3.3×10^7
1 ng	1.8×10^7	2.9×10^7	3.4×10^7
100 ng	1.7×10^7	2.9×10^7	3.8×10^7
Amount of pBI121	Total number of viable cells/reaction, CFU/mL[b]		
	1-h expression	2-h expression	3-h expression
100 ng	1.2×10^9	1.3×10^9	1.5×10^9

[a]Transformation efficiency was determined by electroporating various amounts of pBI121 into 20 μL of electrocompetent cells of *A. tumefaciens* LBA4404 at a field strength of 16.7 kV/cm, transferring 10 μL of mixture into 1 mL YM broth in a 15-mL Falcon tube, and incubating at 30°C, 250 rpm for different expression periods. Cells were plated onto YM medium with 50 μg/mL kanamycin and 100 μg/mL streptomycin, and the plates were incubated at 30°C for 2 d before counting the colonies. The results are the average number of three experiments and two duplicates for each experiment.

[b]Total number of viable cells were obtained as above except that cells were plated onto LB plates with 100 μg/mL streptomycin only.

eliminates aggregation when the cells reach a high density, and highly electrocompetent cells can be obtained consistently.

2. The concentration of cells is one of the most critical factors for obtaining a high transformation efficiency of *Agrobacterium*. To obtain a transformation efficiency $>5 \times 10^6$ transformants/μg DNA, it is important to have a concentration $\geq 5.0 \times 10^{10}$ cells/mL. Cells in log phase yield a higher electrocompetence than cells in stationary phase. However, it is impractical to prepare large amounts of electrocompetent cells at a concentration of 5.0×10^{10} cells/mL from log phase of cultures.

4.2. Factors Affecting the Electroporation

3. Use a pulse generator with the exponential decay wave form and a 0.1–0.15 cm microelectroporation chamber to generate a field strength >15.0 kV/cm for efficient transformation.

4. The conductivity of the electroporation medium is a critical factor. Although many media have been used for *E. coli* and *Agrobacterium*, we have found that 10% glycerol is an effective electroporation medium and a convenient cryoprotectant for the long-term storage of frozen electrocompetent cells.

5. Because *Agrobacterium* cells are small, a high field strength is required to achieve a high transformation efficiency. The transformation efficiency

increases as the field strength increases *(6)*. We are able to obtain consistently a transformation efficiency $>5 \times 10^6$ transformants/μg at a field strength of 16.7 kV/cm.

6. As with *E. coli*, low temperature has been found to improve the transformation efficiency of *Agrobacterium (15)*. Thawing the frozen cells at room temperature instead of on ice decreases the transformation efficiency.

7. The optimal broth for expression of *Agrobacterium* has been investigated. YM broth, a widely used medium in soil-borne bacteria, such as *Rhizobium*, has been shown to be the best expression medium for *Agrobacterium*. The difference between YM broth and LB broth can be as great as 40-fold *(16)*. Moreover, the use of YM broth for *Agrobacterium* eliminates the cell aggregation that is commonly observed when LB broth is used.

8. *Agrobacterium* cells require an incubation period for recovery after electroporation and for the expression of antibiotic resistance genes. In *A. tumefaciens* LBA4404, it has been shown that a 3-h expression period after electroporation results in a twofold increase in transformation efficiency compared to a 1-h expression period as in *E. coli (17,18)*.

References

1. Nester, W. E., Gordon, M. P., Amasino, R. M., and Yanofsky, M. F. (1984) Crown Gall: a molecular and physiological analysis. *Ann. Rev. Plant Physiol.* **35,** 387–413.
2. Chilton, M.-D., Saiki, R. K., Yadav, N., Gordon, M. P., and Quetier, F. (1980) T-DNA from *Agrobacterium* Ti plasmid is in the nuclear DNA fraction of Crown gall tumor cells. *Proc. Natl. Acad. Sci. USA* **77,** 4060–4064.
3. Miki, B. L., Fobert, P. F., Charest, P. J., and Iyer, V. N. (1993) Procedures for introducing foreign DNA into plants, in *Methods in Plant Molecular Biology and Biotechnology* (Glick, B. R. and Thompson, J. E., eds.), CRC, Boca Raton, FL.
4. Gruber, M. Y. and Crosby, W. L. (1993) Vectors for plant transformation, in *Methods in Plant Molecular Biology and Biotechnology* (Glick, B. R. and Thompson, J. E., eds.), CRC, Boca Raton, FL.
5. Ditta, G., Stanfield, S., Corbin, D., and Helinski, D. R. (1980) Broad host range DNA cloning system for gram-negative bacteria: construction of a gene bank of *Rhizobium meliotti. Proc. Natl. Acad. Sci. USA* **77,** 7347–7351.
6. Sigh, A., Kao, T.-H., and Lin, J.-J. (1993) Transformation of *Agrobacterium tumefaciens* with T-DNA vectors using high-voltage electroporation. *FOCUS* **15,** 84–87.
7. An, G., Ebert, P. R., Miltra, A., and Ha, S. B. (1988) Binary vectors, in *Plant Molecular Biology Manual* (Gelvin, S. B., Schilperoort, R. A., and Verma, D. P. S., eds.), Kluwer Academic, Dordrecht, The Netherlands, pp. A3: 1–19.
8. Holsters, M., De Waele D., Depicker, A., Messens, E., Van Montagu, C., and Schell, J. (1978) Transfection and transformation of *Agrobacterium tumefaciens. Mol. Gen. Genet.* **163,** 181–187.
9. Hofgen, R. and Willmitzer, L. (1988) Storage of competent cells for *Agrobacterium* transformation. *Nucleic Acids Res.* **16,** 9877.

10. Mersereau, M., Pazour, G. J., and Das, A. (1990) Efficient transformation of *Agrobacterium tumefaciens* by electroporation. *Gene* **90,** 149–151.
11. Wen-jun, S. and Forde, B. G. (1989) Efficient transformation of *Agrobacterium* spp. by high voltage electroporation. *Nucleic Acids Res.* **17,** 8385.
12. Nagel R., Elliott, A., Masel, A., Birch, R. G., and Manners, J. M. (1990) Electroporation of binary Ti plasmid vector into *Agrobacterium tumefaciens* and *Agrobacterium rhizogenes. FEMS Microbiol. Lett.* **67,** 325–328.
13. Mattanovich, D., Ruker, F., Machado, A. C., Laimer, M., Regner, F., Steinkellner, H., Himmler, G., and Katinger, H. (1989) Efficient transformation of *Agrobacterium* spp. by electroporation. *Nucleic Acids Res.* **17,** 6747.
14. Cangelosi, G. A., Best, E. A., Martinetti, G., and Nester, E. W. (1991) Genetic analysis of *Agrobacterium. Methods in Enzymol.* **204,** 384–397.
15. Dower, W. J. (1990) Electroporation of bacteria: a general approach to genetic transformation, in *Genetic Engineering—Principles and Methods,* vol. 12, pp. 275–296. Plenum, New York.
16. Lin, J.-J. (1994) A new expression medium for *Agrobacterium tumefaciens* using electroporation. *FOCUS* **16,** 18,19.
17. Hanahan, D. (1983) Studies on transformation of *Escherichia coli* with plasmids. *J. Mol. Biol.* **166,** 557–580.
18. Lin, J.-J. (1994) Optimization of the transformation efficiency of *Agrobacterium tumefaciens* cells using electroporation. *Plant Sci.* **101,** 11–15.

CHAPTER 17

Electroporation of *Helicobacter pylori*

Ellyn D. Segal

1. Introduction

The genus *Helicobacter* currently comprises over 15 different species with members described as microaerophilic gram-negative spiral rods with a G + C content of 35–38%. The most medically important *Helicobacter* is *Helicobacter pylori*, a human pathogen that has been causally linked to the development of gastritis, gastric ulcers, and gastric cancer *(1,2)*. The need to fulfill, at the molecular level, Koch's Postulates for any microbial pathogen investigated demands that the organism be applicable to all the molecular genetic techniques available. Ultimately one must be able to prepare DNA from the organism, clone and manipulate the DNA in the laboratory, and most importantly, insert DNA into the isolate. Only in this manner can isogenic mutants be created, allowing for the proper comparison of wild-type and mutant in an appropriate model to determine potential virulence factors.

Various techniques exist for producing DNA uptake into most prokaryotes: $CaCl_2$ or heat-shock-mediated transformation, transfection with bacteriophage, and electroporation. The latter has often been shown to be successful where other techniques have failed. Electroporation involves applying high-intensity electric fields of short duration to permeabilize biomembranes reversibly *(3)*. The amplitude (electric field strength) and duration (time constant) of the discharge waveform are important, and optimal values may depend on the species and strain being investigated.

Presented below is a method that has been proven successful for the transformation of *H. pylori* via electroporation *(4)*. This method appears

From: *Methods in Molecular Biology, Vol. 47: Electroporation Protocols for Microorganisms*
Edited by: J. A. Nickoloff Humana Press Inc., Totowa, NJ

to overcome previously reported restrictions for *H. pylori* transformation *(5,6)*, in which only freshly isolated clinical strains were found to be competent via natural transformation or that required the presence of an homologous cryptic plasmid in the recipient *H. pylori* strain. Reported transformation efficiency by electroporation for *H. pylori* ranges from 50/µg DNA *(7)* to 2×10^5/µg DNA *(4)*.

2. Materials
2.1. Growth and Passage of H. pylori

Helicobacter pylori is passaged in the laboratory on either 5% sheep blood plates (TSA II, BBL, Cockeysville, MD) or on brucella agar (Difco, Detroit, MI) plates that have had fetal bovine serum (FBS, Gibco, Grand Island, NY) added to a final concentration of 5% before pouring. Cultures should be grown in a BBL GasPak jar containing an anaerobic gas pack (without catalyst), or in a 5% CO_2 incubator.

2.2. Solutions and Reagents

1. Electroporation buffer (EPB): 272 mM sucrose, 15% glycerol, 2.43 mM K_2HPO_4, and 0.57 mM KH_2PO_4, pH 7.4. Filter-sterilize and store at 4°C.
2. Freezing media: Brain heart infusion media (Difco) plus 25% glycerol. Autoclave and store at room temperature.
3. DNA: DNA can be prepared in a variety of ways, from CsCl-purified to a one-step minipreparation. The most important factor is to remove all salts (to ensure low conductivity) and suspend the DNA in ddH_2O. This step is also necessary when electroporating DNA that has been ligated. Salt removal can be done by ethanol precipitation or by centrifuging the DNA through an ion-exchange minicolumn. Store at –20°C.

2.3. Drug Selection Levels

Table 1 lists the drug resistance markers that have been successfully transformed into and expressed in *H. pylori*. In some instances, the level of resistance varies; this might be strain-dependent, and researchers should keep in mind the necessity to determine the appropriate levels for their particular isolates.

2.4. Vector Design

A variety of vectors exist for delivering exogenous DNA into *H. pylori*. Although many are based on the *Campylobacter/E. coli* shuttle vector design *(12,13)*, it has been shown that such plasmids do not exist independently in *H. pylori*. Integration of an incoming plasmid into the chro-

Table 1
Drug Resistance Markers That
Have Been Expressed in *Helicobacter pylori*

	Concentration, µg/mL
Streptomycin	500–1000 *(4)*
Rifampicin	Taylor and Wang, unpub. results *(8)*
Metronidazole	60 *(9)*
Chloramphenicol	4–8 *(9,10)*
Kanamycin	5–20 *(5,11)*

mosome will occur if homologous *H. pylori* DNA is present on the plasmid, allowing the production of insertion mutations via shuttle mutagenesis strategies *(7,10,14)*. If the transforming plasmid contains DNA homologous to a cryptic plasmid present in the host strain, recombination will occur between the two plasmids producing a hybrid plasmid *(5)*.

3. Methods

3.1. Electroporation Protocol

1. The night before an experiment, grow *H. pylori* by inoculating a liquid culture of brucella broth + 5% FBS (*see* Note 1). Inoculation is done by removing at least a quarter of a 2- to 4-d-old plate of *H. pylori* with a sterile swab and suspending it into a 30–50 mL of culture in a 125–250 mL flask. The flask is placed into a GasPak jar containing a GasPak Anaerobic System Envelope (without a catalyst) and grown with agitation (80 rpm) at 37°C.
2. Check the overnight culture for growth of *H. pylori* and for contamination (*see* Note 2). For each electroporation, pellet 1×10^9 to 1×10^{10} bacteria by centrifugation (1500g, 3 min), wash three times in 500–1000 µL cold EPB, and suspend in a final volume of 40 µL EPB. The bacteria are put into prechilled 0.2-cm electroporation cuvets, and at least 2 µg of DNA (suspended in ddH$_2$O) are added (*see* Note 4). Electroporate using 2500-V, 25-µF capacitor and 200 Ω (in parallel with the sample chamber) (*see* Note 3).
3. After electroporation, the volume of each sample is brought up to 1 mL by the addition of phosphate-buffered saline (PBS), pH 7.4, and aliquots are plated onto 5% sheep blood agar plates (or brucella agar plates). Following an overnight expression period (*see* Note 5), the agar is cut out of the Petri dish and placed growth-side up on top of a brucella agar plate containing an appropriate antibiotic at a 2X concentration (*see* Note 6). Incubate the plate right-side up for 3–6 d to allow for outgrowth of resistant colonies (*see* Note 7).

3.2. Isolation of Resistant Colonies and Storage of Helicobacter

H. pylori grown on plates must be passaged every 3–5 d or viability will decline. Because of the length of time it often takes for resistant colonies to appear, it is advisable to pick the colonies as soon as they become large enough to transfer. This can be done with a sterile toothpick; a single colony large enough to see can be patched onto a fresh selection plate. Within 3–4 d, enough *H. pylori* will grow to produce a sufficient inoculum for either a liquid culture or to inoculate several plates.

Long-term storage of *H. pylori* at −70°C is accomplished by suspension in a solution of sterile brain heart infusion medium + 25% glycerol. We have found that strains that have been stored in this manner remain viable for at least 3 yr.

4. Notes

1. *H. pylori* is usually grown under microaerophilic conditions, using a GasPak jar containing an Anaerobic System without a catalyst. Many *H. pylori* strains can also be grown in a water-jacketed 5% CO_2 incubator, which provides additional space for the many plates that may be required for multiple experiments. Experiments have shown that there is no difference in electroporation efficiency between *H. pylori* grown under the two conditions (unpublished data). *H. pylori* grown on plates in a 5% CO_2 incubator will grow well when inoculated into a liquid culture and placed in a GasPak jar and agitated. *H. pylori* taken from plates in a 5% CO_2 incubator and put onto fresh plates in a GasPak system often shows retarded growth. The reason for this is unknown.

2. Check the overnight liquid culture of *H. pylori* for growth and contamination by performing a gram stain. Healthy *H. pylori* appear as spirals, gulls, or bent rods. If a high percentage of bacteria are coccoid in shape, the culture should not be used.

3. Electroporation conditions vary among reports in the literature, usually relating to the field strength. Resistance ranging from 200–400 Ω has been successfully used *(4,5)*. The individual researcher should consider this range when initially determining transformation efficiency.

4. As for many electroporation systems, the amount of DNA to be added follows the rule of "the more, the better." At least 2 µg of DNA should be used, with reported amounts as high as 30 µg of DNA for *H. pylori (5)*. It is important that there are no residual salts present in the DNA solution, especially when adding a larger amount of DNA to the electroporation cuvet (this will result in arcing owing to high conductivity during

electroporation). A volume of 1–10 µL is recommended. DNA should be stored in ddH$_2$O instead of TE (10 m*M* Tris, 1 m*M* EDTA).

5. The length of time for the nonselective growth period ranges from 12–48 h in the literature *(4,5,10)*. This parameter is most likely strain-dependent, relating to the relative growth rates of different *H. pylori* isolates. Our laboratory has not noticed any change in efficiency or increase in false positives on comparing a 24- vs 48-h growth period.

6. A common problem of *H. pylori* transformation is contamination of the plates, usually by mold. This is probably because of the numerous handling procedures that are needed during the outgrowth phase and the length of time needed for *H. pylori* to grow. Techniques that will help reduce contamination include:

   a. Sterilize the tools used to cut out the nonselective agar and to place it on the top of a selective plate. This can be done by dipping the tools in ethanol and flaming them after each plate is done.

   b. Wrap the circumference of each plate with surgical tape (3M Micropore); this will help prevent the spread of any airborne contaminants.

   c. If contamination does occur, it is sometimes possible to salvage most of the plate by cutting the contaminated portion away with a sterile scalpel and removing it from the plate.

7. The length of time for selective growth can also vary, with 3 d as a minimum and 6 d as a maximum. One problem with the longer time period is plate contamination, which was discussed above. A second problem is the growth of false positives. This later occurrence should be addressed by including a control transformation with no DNA added. This will provide a quantitative assay for false positive background levels. A selective plate on which colonies mainly appear on the periphery of the top agar level tends to indicate that the colonies are false positives.

References

1. Parsonnet, J., Friedman, G. D., Vandersteen, D. P., Chang, Y., Vogelman, J. H., Orentreich, N., and Sibley, R. K. (1991) *Helicobacter pylori* infection and the risk of gastric carcinoma. *New Engl. J. Med.* **325**, 1127–1131.

2. Nomura, A., Stemmerman, G. N., Chyou, P. H., Kato, I., Perez-Perez, G. I., and Blaser, M. (1991) *Helicobacter pylori* infection and gastric carcinoma among Japanese. *New Engl. J. Med.* **325**, 1132–1136.

3. Miller, J., Dower, W. J., and Tompkins, L. S. (1988) High-voltage electroporation of bacteria: genetic transformation of *C. jejuni* with plasmid DNA. *Proc. Natl. Acad. Sci. USA* **85**, 856–860.

4. Segal, E. D. and Tompkins, L. S. (1993) Transformation of *Helicobacter pylori* by electroporation. *Biotechniques* **14**, 225,226.

5. Tsuda, M., Karita, M., and Nakazawa, T. (1993) Genetic transformation in *H. pylori. Microbiol. Immunol.* **37**, 85–89.

6. Nedenskov-Sorensen, P., Bukholm, G., and Bovre, K. (1990) Natural competence for genetic transformation in *Campylobacter pylori*. *J. Infect. Dis.* **161,** 365.

7. Ferrero, R. L., Cussac, V., Courcoux, P., and Labigne, A. (1992) Construction of isogenic urease-negative mutants of *Helicobacter pylori* by allelic exchange. *J Bacteriol.* **174,** 4212–4217.

8. Taylor, D. E. (1992) Genetics of *Campylobacter* and *Helicobacter*. *Ann. Rev. Microbiol.* **46,** 35–64.

9. Taylor, D. E., Roos, K. P., and Wang, Y. (1993) Transformation of *H. pylori* by chromosomal metronidazole resistance and by a plasmid with a chloramphenicol resistance marker, in *Campylobacter Meets Helicobacter: The Joint Meeting*. Brussels. *Acta Gastroenterol. Belg.* **56(Suppl.),** 106.

10. Haas, R., Meyer, T. F., and van Putten, J. P. M. (1993) Aflagellated mutants of *H. pylori* generated by genetic transformation of naturally competent straits using transposon shuttle mutagenesis. *Molecular Microbiol.* **8,** 753–760.

11. Suerbaum, S., Josenhans, C., and Labigne, A. (1993) Cloning and genetic characterization of the *H. pylori* and *H. mustelae* flaB flagellin genes and construction of *H. pylori* flaA- and flaB-negative mutants by electroporation-mediated allelic exchange. *J. Bacteriol.* **175,** 3278–3288.

12. Labigne-Roussel, A., Harel, J., and Tompkins, L. S. (1987) Gene transfer from *E. coli* to *Campylobacter* species: development of shuttle vectors for genetic analysis of *C. jejuni*. *J. Bacteriol.* **169,** 5320–5323.

13. Labigne, A., Courcoux, P., and Tompkins, L. (1992) Cloning of *C. jejuni* genes required for leucine biosynthesis, and construction of *leu* negative mutant of *C. jejuni* by shuttle transposon mutagenesis. *Res. Microbiol.* **143,** 15–26.

14. Haas, R., Odenbreit, S., Heuermann, D., and Schmitt, W. (1993) Transposon shuttle mutagenesis as a general tool to identify and genetically define virulence determinants of *H. pylori*, in *Campylobacter Meets Helicobacter: The Joint Meeting*. Brussels. *Acta Gastroenterol. Belg.* **56(Suppl.),** 103.

CHAPTER 18

Electrotransformation of *Streptococci*

Robert E. McLaughlin and Joseph J. Ferretti

1. Introduction

The streptococci are a diverse genus of bacteria consisting of commensal and pathogenic organisms of human and veterinary origin, as well as industrially important species. Although many of the species are naturally transformable, several species do not enter a competent state. The ability to introduce exogenous DNA into these organisms is important in establishing genetic linkages to virulence and pathogenicity factors in some species, and the ability to stabilize or improve other phenotypic traits in other species.

There are several barriers that must be overcome for successful electrotransformation of streptococci. One of the most significant of these barriers is the cell wall. Earlier methods of electrotransformation included treatment of the cells with lysozyme *(1)* to help weaken the cell wall. Other methods included the addition of glycine or threonine to the growth medium as a cell-wall-weakening agent *(2–4)*. Although effective for many species and strains, the susceptibility of each strain to glycine needs to be determined independently, since high glycine concentrations can cause cell lysis during the electric pulse *(2)*. Simon and Ferretti noted that the amount of hyaluronic acid present on the cell surface may reduce the transformation frequency of some highly encapsulated strains *(5)*. The amount and specificity of endonucleases present in the various streptococci may also have a significant role in electrotransformation. Somkuti and Steinberg reported that the inability to transform one strain of *S. thermophilus* was presumably owing to high endonuclease activity *(4)*.

From: *Methods in Molecular Biology, Vol. 47: Electroporation Protocols for Microorganisms*
Edited by: J. A. Nickoloff Humana Press Inc., Totowa, NJ

A genetic approach to the study of *S. pyogenes* has been the focus of a number of laboratories. Thus, mechanisms to introduce DNA into this normally nontransformable organism has been a topic of high priority. Suvorov et al. *(6)* were the first to describe a procedure for the electrotransformation of *S. pyogenes*. Simon and Ferretti *(5)* studied the conditions for electrotransformation of the group A streptococci and found that strain NZ131 was able to be transformed at an efficiency of $>10^7/\mu g$. However, this high efficiency of electrotransformation was not characteristic of all strains, since 8 of 18 strains studied resulted in efficiencies of only 10^2–$10^4/\mu g$, whereas the remaining strains gave no transformants. Dunny et al. were able to transform *S. pyogenes* DW1009 with an efficiency of $\sim10^3/\mu g$ plasmid using a procedure optimized for *Enterococcus faecalis (2)*. A similar procedure has been used to transform *S. pyogenes* JRS14 and JRS75 as well as other strains, with transformation efficiencies of 10^3–$10^4/\mu g$ *(3,7,8)*. A wide range of efficiencies and strain-to-strain variability appear to be a common problem associated with the electrotransformation of this organism. A *recA* mutant of strain NZ131 has been recently constructed (L. Tao and J. Ferretti, unpublished data) and should be useful for studies in which chromosomal integration is not the desired outcome, e.g., in complementation, gene dosage, and expression studies.

The lactic acid streptococci are another group of streptococci that are not naturally transformable. Somkuti and Steinberg *(4)* were able to electrotransform a large number of strains with a procedure they optimized for the transformation of *S. thermophilus* ST128. Although strain-to-strain variability was also found, most strains yielded transformants at $\sim10^3/\mu g$ plasmid. Mollet et al. used a similar procedure to transform strain ST11 at an efficiency of 10^4–$10^5/\mu g$ with plasmids and $\leq10^2/\mu g$ with linear DNA *(9)*.

A recent report by Seroude et al. *(10)* describes the electrotransformation of a nontransformable strain of *S. pneumoniae* 801 and the parental strain Tr19 with plasmid DNA, with reported transformation frequencies of $\sim10^{-5}$.

The oral streptococci encompass *S. mutans, S. sanguis, S. gordonii*, and several other species. All have had DNA introduced into them by electrotransformation. Somkuti and Steinberg *(11)* transformed *S. gordonii* Challis at an efficiency of $\sim10^3/\mu g$. A *recA* mutant of *S. gordonii* has been reported by Vickerman et al. *(12)*. Fenno et al. *(13)*

transformed *S. parasanguis* FW213 and *S. gordonii* Challis, obtaining efficiencies of 10^2–10^3/µg with plasmid DNA, similar to that observed by Somkuti and Steinberg. Lee et al. *(1)* were able to transform *S. mutans* NG8 with linear DNA at an efficiency of <10^2. The successful electro-transformation of *S. sobrinus* 6715 was recently reported (Lassiter, personal communication) with an efficiency of 3.0×10^3/µg.

Ricci et al. have successfully electrotransformed *S. agalactiae,* obtaining frequencies with strain O90R (NCTC 9993) of >10^4/µg *(14).* Likewise, Framson (personal communication) was able to transform *S. agalactiae* at frequencies up to 10^5/µg plasmid.

This chapter provides a protocol for the electrotransformation of *S. pyogenes*, with variations and alternate protocols that have been used for the electrotransformation of other streptococci. In general, relatively high voltages in an osmotically stabilized electrotransformation buffer are essential for successful transformation.

2. Materials
2.1. Media and Reagents

1. Todd-Hewitt (TH) broth: 30 g/L (Difco, Detroit, MI). Add 15 g/L of agar for solid medium, and autoclave for 15 min at 121°C. Antibiotic(s) should be added to the medium once it has cooled to below 60°C, and the medium stored under conditions appropriate for the antibiotic (*see* Note 1).
2. Hogg-Jago (HJ) broth: 3% tryptone, 1% yeast extract, 0.5% KH_2PO_4, 0.5% beef extract supplemented with 1% lactose (HJL), and/or 1% glucose; used for the cultivation of *S. thermophilus (4).*
3. Several different electrotransformation buffers (EB) have been used by various investigators, and all report essentially the same transformation efficiencies. Whether each buffer will work equally well with all strains has not been determined. Therefore we present several buffers used in electrotransformation of streptococci (*see* Table 1). These buffers are used for the washing and storage of cells, unless otherwise noted. In general, most of the solutions contain hypertonic sucrose concentrations in low-ionic-strength buffer or water. The presence of 10–15% glycerol in the buffer allows for the storage of electrocompetent cells at –70°C. However, transformation efficiencies are generally lower with frozen cells.

2.2. DNA

1. Isolation of plasmid for the electrotransformation of streptococci has been done by a variety of methods, both from *E. coli (16)* and streptococci *(2,4).* Cesium chloride density gradient purification is generally performed to

Table 1
Solutions Used for the Electrotransformation of Streptococci

Bacteria	Electrotransformation buffer (EB)	Reference
S. pyogenes DW1009	0.625M sucrose 1 mM MgCl₂, pH 4.0	2
NZ131 (various strains)	0.5M sucrose in dH₂O	5
JRS 14, JRS75	15% Glycerol	7,8
(various strains)	272 mM glucose, 1 mM MgCl₂, pH 6.2	3
S. mutans NG8	7 mM potassium phosphate buffer, pH 7.4, 0.5M sucrose, 1 mM MgCl₂	1
S. parasanguis FW213	10 mM Tris-HCl, pH 6.0, 0.625M sucrose/ 10 mM Tris-HCl, pH 5.0, 0.625M sucrose^a	13
FW213	0.625M sucrose, 1 mM MgCl₂, pH 4.0	2
S. gordonii Challis DL1	10 mM Tris-HCl, pH 6.0, 0.625M sucrose/ 10 mM Tris-HCl, pH 5.0, 0.625M sucrose^a	13
	5 mM potassium phosphate buffer, pH 7.0/5 mM potassium phosphate buffer, pH 4.5, 1 mM MgCl₂, and 0.3M raffinose^a	11
S. pneumoniae 801, Tr19	0.5M sucrose in dH₂O	10
S. downeii MFe28	15% Glycerol	This laboratory
S. thermophilus (various strains)	5 mM potassium phosphate buffer, pH 7.0/5 mM potassium phosphate buffer, pH 4.5, 1 mM MgCl₂, and 0.3M raffinose^a	4
ST11	5 mM potassium phosphate buffer, pH 6.1, 0.5 mM MgCl₂, and 0.3M raffinose	9
S. sobrinus 6715	272 mM glucose, 1 mM MgCl₂, pH 4.0	M. O. Lassiter, personal communication
S. agalactiae (NCT 9993)	10% Glycerol	14
S. agalactiae H36B	0.625M sucrose, 1 mM MgCl₂, pH 4.0	2

^{a}Wash the cells one time in the first buffer (wash buffer), followed by one wash in the second buffer listed (EB).

maximize electrotransformation efficiencies, although we have obtained similar efficiencies using DNA prepared using commerically available kits (QIAprep-spin, Qiagen, Chatsworth, CA). Suspend DNA at a concentration of 1–2 µg/µL in a low-ionic-strength buffer (TE: 10 mM Tris-HCl [pH 7.5–8.5], 1 mM EDTA), or in distilled water [see Note 2]). Linear and nonreplicating homologous DNA may also be used for electrotransformation. This allows the generation of specific mutants by insertion inactivation following recombination (see Note 3).

3. Methods

3.1. Production of Electrocompetent Cells

1. Inoculate 5 mL of the appropriate growth medium with an aliquot of streptococci from a –70°C freezer stock culture. Incubate the culture at 37°C overnight (approx 16 h) without aeration (*see* Notes 4 and 5).
2. Dilute overnight culture 1:20 in 100 mL of fresh medium, and allow to grow without aeration at 37°C for an additional 2–3 h. Monitor the growth spectrophotometrically, and harvest cells when they reach the early log phase of growth (A_{560} = ~0.25, A_{600} = ~0.2) (*see* Note 6).
3. Collect cells by centrifugation at 8000g for 10 min at 4°C (*see* Note 7). All subsequent steps should be performed at 0–4°C using ice-cold reagents and prechilled centrifuge tubes to achieve maximal transformation efficiencies.
4. Carefully decant the medium, and suspend the pellet in 100 mL of ice-cold EB (or wash buffer) taking care to maintain sterility.
5. Pellet the cells by centrifugation at 8000g for 10 min at 4°C, and carefully decant the supernatant.
6. Repeat the washing procedure with a second 100-mL vol of EB, and pellet cells as step 5.
7. Resuspend the pellet in 0.5–1.0 mL of ice-cold EB, and transfer to prechilled 1.5-mL microfuge tubes. If cells are to be stored for later use, place them in a dry ice-ethanol bath to quick-freeze, and transfer to –70°C. Otherwise, cells should be kept on ice and used within 60 min of preparation.

3.2. Electrotransformation

1. Chill a sterile electroporation cuvet with a 0.2-cm electrode gap on ice or in the freezer for 5–10 min.
2. Add 1–2 µL DNA (~1.0 µg/µL) to an aliquot of electrocompetent cells (40–200 µL) (*see* Note 8). Mix by gently tapping the base of the microfuge tube or gently stir with a micropipet tip. A control without DNA should be included with each experiment. Incubate the mixture on ice for 2–3 min.
3. Transfer the cell mixture to the prechilled cuvet. Make sure that the suspension is evenly dispersed across the bottom of the cuvet.
4. Wipe any moisture from the surface of the cuvet, and place the cuvet into the holder. Apply a single pulse of 2.5 kV, capacitance at 25 µF and resistance at 200 Ω (*see* Note 9). The time constant will vary depending on the EB used and electrotransformation settings (*see* Note 10).
5. Add the appropriate medium to the cuvet and place on ice for at least 5 min. Incubate for 1.5–2 h at 37°C without aeration to allow cell recovery (*see* Notes 11 and 12).

6. Spread aliquots of the transformed cells on the appropriate medium supplemented with antibiotic(s) (*see* Note 13). If necessary, the cells can be diluted prior to plating. Incubate the plates at 37°C for 24–48 h.

4. Notes

1. TH medium may be supplemented to improve cell growth; for example, the addition of 5% (v/v) heat-inactivated horse serum to sterile medium (TH-HS) has been used for *S. pyogenes (5)* and *S. mutans (1)*. The addition of 0.2% yeast extract (THY) prior to sterilization has been used for the cultivation of *S. pyogenes* by Perez-Casal et al. *(7,8)*. The antibiotic concentration in selective media should be determined for each strain. This can be done by making a series of 1:2 dilutions of the antibiotic in broth medium. Inoculate each tube with the same amount of cells from an overnight culture, and incubate for 24–48 h. This first tube that has no growth has the approximate minimal inhibitory concentration (MIC) of the antibiotic. This concentration should then be verified in solid medium, since the MIC in liquid vs solid may vary.

2. Generally, 3–4 µg of plasmid/mL of *E. coli* for streptococcal integration vectors constructed with the pUC origin will be obtained. It may therefore be necessary to ethanol-precipitate DNA to obtain a concentration of 1 µg/µL. Alternatively, prepared cells can be added directly to an aliquot of precipitated plasmid DNA corresponding to 1–2 µg. DNA obtained from ligation reactions can also be used for electrotransformation. However, the mixture should be ethanol-precipitated and washed with 70% ethanol to concentrate the DNA and remove the salts. As above, the prepared cells can be added directly to the precipitated DNA pellet.

   Depending on the streptococcal species used, the size of the plasmid being used for transformation may affect efficiency. Simon and Ferretti reported similar transformation efficiencies of *S. pyogenes* strains with plasmids of 6 and 10 kbp *(5)*. However, Somkuti and Steinberg reported significant decreases in transformation efficiencies in *S. thermophilus* ST128 with plasmids of 7.6, 26.5, and 30.7 kbp and similar results in *S. gordonii* Challis *(4,11)*. As with other bacteria, smaller plasmids generally transform with higher frequencies.

3. Transformation with homologous DNA fragments has been accomplished with both linear DNA *(1,6,* M. G. Caparon, personal communication) and nonreplicating circular DNA. Simon and Ferretti reported that linear DNA transformed about 10^4 times less efficiently than circular plasmid DNA in the same strain *(5)*. Carefully purify DNA fragments from agarose following gel electrophoresis to remove extraneous DNA fragments. This can be done by several means *(15)* or by using a commercially available kit. Care

must be taken to ensure the absence of circular DNA in the transforming DNA preparation, since we have observed single crossover Campbell-type insertions occurring at a much higher frequency than double crossover events. Once transformants are obtained, the nature of the recombination event can be verified by Southern hybridization.

4. The addition of low concentrations of glycine or DL-threonine to the medium has been reported to improve the transformation of some strepto-cocci (*2–4*, P. E. Framson, personal communication). These agents act to weaken the cell wall, thereby making them more permeable. If glycine is used, it should be added to the autoclaved medium from a filter-sterilized stock solution. The concentration required for optimal transformation should be determined for each strain such that it causes an approx 80% inhibition in cell growth vs a control grown in the absence of glycine (*2*). Concentrations in the range of 20 m*M* glycine have been used successfully with *S. pyogenes* strains (*3,7,8*, M. G. Caparon, personal communication), whereas 80 m*M* glycine have been used with *S. agalactiae* (P. E. Framson, personal communication) and as high as 250 m*M* with some oral strepto-cocci (*2*, M. O. Lassiter, personal communication). The addition of 40 m*M* DL-threonine resulted in an increased transformation efficiency with some *S. thermophilus* strains (*4*).

5. *S. thermophilus* should be grown at 42°C.

6. *S. thermophilus* and *S. gordonii* Challis are reportedly not growth-phase-dependent, since cells harvested at stationary phase yield similar efficien-cies of transformation to those harvested from early stationary phase (*4,11*).

7. Perez-Casal et al. have reported an additional step that may increase the efficiency of electrotransformation with linear DNA. After harvesting the 100-mL cell culture, the cell pellet is resuspended in 5 mL of the spent culture medium. The cells are then subjected to heat shock in a 43°C water bath for 9 min, harvested as before, and washed twice with EB (*8*, M. G. Caparon, personal communication).

8. More than 1.0 µg of replicating plasmid DNA per transformation does not appear to increase the total number of transformants obtained. However, it is often necessary to use 5 or more micrograms of linear/nonreplicating DNA to obtain transformants.

9. Alternate parameters have been used to electrotransform streptococci successfully. The use of lower voltages (1.75–2.0 kV) with the same capacitance and resistance (*2,3,7*) or at 400 Ω (*8*, M. G. Caparon, personal communication) with a 0.2-cm electrode gap yielded similar transfor-mation frequencies for *S. pyogenes* as did a pulse of 2.05 kV, 25 µF, and 400 Ω for *S. thermophilus* (*4*). Systems lacking a resistance controller have also been successfully used to electrotransform *S. pyogenes, S. gordonii,*

S. mutans, and *S. agalactiae* at settings of 2.25–2.5 kV, 25 μF *(1,2,11),* and *S. thermophilus* at a setting of 1.6 kV, 25 μF using a 0.4-cm electrode gap *(4).* Ricci et al. reported the electrotransformion of *S. agalactiae* with settings of 1.8 kV, 25 μF, and 100 Ω using a 0.1-cm cuvet *(14). S. sobrinus* 6715 has been successfully transformed using the BRL Cellporator and booster at settings of 1.61 kV and 4 kΩ *(14).*

10. The use of the primary EB and settings listed in the protocol should result in a time constant of approx 4.7 ms *(5).* The use of a 15% glycerol solution and settings of 1.75 kV, 25 μF, and 400 Ω will generally produce time constants in the 7–8 ms range with *S. pyogenes* (M. G. Caparon, personal communcation) and approx 9 ms with *S. downeii* MFe28 (unpublished results).

11. The electrotransformed cells may also be transferred to a chilled 1.5-mL tube and incubated on ice for 30–60 min. The cells are then transferred to 10 mL of THY and incubated at 37°C for 1 h. Following incubation, harvest the cells (6500g, 10 min, 14°C), resuspend in 1 mL medium, and spread on selective medium (M. G. Caparon, personal communication).

12. After pulsing *S. thermophilus* (1.6 kV, 25 μF), add 0.1 mL of 10X HJL broth to the cells. The cells can then be maintained at 4–25°C for up to 24 h before plating on selective medium without loss of efficiency. A minimum of 4 h of incubation is required for phenotypic expression of erythromycin resistance to obtain maximal efficiency *(4).* Mollet et al. added 1 mL of 1.2 × M17 broth (Difco) to 200 μL of cells immediately after electrotransformation, and incubated the culture for 4 h at 42°C. Following incubation, the cells were mixed with soft agar (M17 + 1% sucrose *[10]* or HJ + 1% lactose *[4],* 0.6% agar) and plated on the same agar containing the appropriate antibiotic. Plates were incubated at 37–42°C for 48–72 h under anaerobic conditions.

13. We have observed growth of resistant streptococci after prolonged incubation on several antibiotics, particularly kanamycin and spectinomycin. It is therefore important to isolate transformants from the primary selection plate as soon as possible and streak them on fresh selective medium. Several antibiotics are bacteriostatic (i.e., the cells are not killed in the presence of the antibiotic, just inhibited from growing). Therefore, maintain transformants on selective medium for several passages to eliminate carryover of viable nontransformed cells.

References

1. Lee, S. F., Progulske-Fox, A., Erdos, G. W., Piacentini, D. A., Ayakawa, G. Y., Crowley, P. J., and Bleiweis, A. S. (1989) Construction and characterization of isogenic mutants of *Streptococcus mutans* deficient in major surface protein antigen P1 (I/II). *Infect. Immunol.* **57,** 3306–3313.

2. Dunny, G. M., Lee, L. N., and LeBlanc, D. J. (1991) Improved electroporation and cloning vector system for Gram-positive bacteria. *Appl. Environ. Microbiol.* **57,** 1194–1201.
3. Caparon, M. G. and Scott, J. R. (1991) Genetic manipulation of pathogenic streptococci. *Methods Enzymol.* **204,** 556–586.
4. Somkuti, G. A. and Steinberg, D. H. (1988) Genetic transformation of *Streptococcus thermophilus* by electroporation. *Biochemie* **70,** 579–585.
5. Simon, D. and Ferretti, J. J. (1991) Electrotransformation of *Streptococcus pyogenes* with plasmid and linear DNA. *FEMS Microbiol. Lett.* **82,** 219–224.
6. Suvorov, A., Kok, J., and Venema, G. (1988) Transformation of group A streptococci by electroporation. *FEMS Microbiol. Lett.* **56,** 95–100.
7. Perez-Casal, J., Caparon, M. G., and Scott, J. R. (1991) Mry, a *trans*-acting positive regulator of the M protein gene of *Streptococcus pyogenes* with similarity to the receptor proteins of two-component regulatory systems. *J. Bacteriol.* **173,** 2617–2624.
8. Perez-Casal, J., Caparon, M. G., and Scott, J. R. (1992) Introduction of the *emm6* gene into an *emm*-deleted strain of *Streptococcus pyogenes* restores its ability to resist phagocytosis. *Res. Microbiol.* **143,** 549–558.
9. Mollet, B., Knol, J., Poolman, B., Marciset, O., and Delley, M. (1993) Directed genomic integration, gene replacement, and integrative gene expression in *Streptococcus thermophilus. J. Bacteriol.* **175,** 4315–4324.
10. Seroude, L., Hespert, S., Selakovitch-chenu, L., Gasc, A.-M., Lefrancois, J., and Sicard, M. (1993) Genetic studies of cefotaxime resistance in *Streptococcus pneumoniae*: relationship to transformation deficiency. *Res. Microbiol.* **144,** 389–394.
11. Somkuti, G. A. and Steinberg, D. H. (1989) Electrotransformation of *Streptococcus sanguis* Challis. *Curr. Microbiol.* **19,** 91–95.
12. Vickerman, M. M., Heath, D. G., and Clewell, D. B. (1993) Construction of recombinant-deficient strains of *Streptococcus gordonii* by disruption of the *rec*A gene. *J. Bacteriol.* **175,** 6354–6357.
13. Fenno, J. C., Shaikh, A., and Fives-Taylor, P. (1993) Characterization of allelic replacement in *Streptococcus parasanguis*: transformation and homologous recombination in a "nontransformable" streptococcus. *Gene* **130,** 81–90.
14. Ricci, M. L., Nanganelli, R., Berneri, C., Orefici, G., and Pozzi, G. (1994) Parameters for electrotransformation of *Streptococcus agalactiae, in Pathogenic Streptococci: Present and Future* (Totolian, A., ed.), Lancer Publications, St. Petersburg, Russia, pp. 268,269.
15. Sambrook, J., Fritsch, E. F., and Maniatis, T. (1989) in *Molecular Cloning: A Laboratory Manual.* 2nd ed. Cold Spring Harbor Laboratory, Cold Spring Harbor, NY.

CHAPTER 19

Transformation of *Lactococcus* by Electroporation

Helge Holo and Ingolf F. Nes

1. Introduction

Lactococcus lactis strains are the principal organisms used in cheese-making, and their performance is of vital importance to the quality of the cheese. A goal for the dairy industry has always been to improve strains of *L. lactis* and to stabilize beneficial traits. In recent years, genetic engineering has become a useful tool for manipulating *L. lactis*. A critical step in this work is the introduction of DNA into the cells. The first successful transformations of *L. lactis* employed protoplast transformation in the presence of polyethylene glycol *(1)*. This method was not very reproducible and worked only for a few strains. More recently, electroporation proved to be an alternative to protoplast transformation in *L. lactis*. Harlander *(2)* was the first to employ electroporation in this species. Later work indicated that the cell wall is a physical barrier to the entering DNA not only during protoplast transformation, but also during electroporation. Powell et al. *(3)* used lysozyme treatment to obtain an increase in transformation efficiency, and van der Lelie et al. *(4)* obtained cells competent for electrotransformation by growing them in the presence of high concentrations of threonine to weaken their cell walls. However, the methods were not very efficient, and a number of strains could not be transformed.

In the method described here, high-efficiency transformation of *L. lactis* is achieved by growing the cells in osmotically stabilized media with glycine to weaken the cell wall *(5)*. With this method, we have

From: *Methods in Molecular Biology, Vol. 47: Electroporation Protocols for Microorganisms*
Edited by: J. A. Nickoloff Humana Press Inc., Totowa, NJ

obtained up to 10^8 transformants/μg plasmid DNA. The method has proven useful for all the strains of *L. lactis* tested, although at different efficiencies. The method has also proven useful for other gram-positive bacteria, e.g., *Enterococcus faecalis (6)*.

2. Materials

Sterilize all media by autoclaving.

1. M17: 5 g/L tryptone, 5 g/L soya peptone, 5 g/L meat digest, 2.5 g/L yeast extract, 0.5 g/L ascorbic acid, 0.25 g/L magnesium sulfate, and 19 g/L disodium glycerophosphate, pH 6.9.
2. GM17: M17 supplemented with 0.5% glucose *(7)*.
3. SGM17: GM17 containing 0.5*M* sucrose. Autoclave concentrated stock solutions of glucose (40%), sucrose (1*M*), and glycine (20%) separately, and after cooling, mix the proper components with the sterilized M17 medium at the correct concentrations (*see* Note 1).
4. SGM17MC: SGM17 containing 20 m*M* MgCl$_2$ and 2 m*M* CaCl$_2$.
5. SR plates *(8)*: 10 g/L tryptone, 5 g/L yeast extract, 200 g/L sucrose, 10 g/L glucose, 25 g/L gelatin, 15 g/L agar, 2.5 m*M* CaCl$_2$, and 2.5 m*M* MgCl$_2$, pH 6.8.
6. BSR (buffered SR) plates: SR plates supplemented with 19 g/L disodium glycerophosphate.
7. 0.5*M* sucrose containing 10% glycerol.
8. DNA dissolved in TE (10 m*M* Tris-HCl, 1 m*M* EDTA, pH 7.5).

3. Methods

3.1. Preparation of Competent Cells

1. Grow *L. lactis* in GM17 at 30°C to an OD at 600 nm of 0.5–0.8. Dilute cells 100-fold in SGM17 containing glycine and grow overnight (12–16 h) to an OD of 0.2–0.7 at 30°C (*see* Note 1).
2. Harvest cells by centrifugation at 5000*g* at 4°C and wash twice in ice-cold 0.5*M* sucrose containing 10% glycerol (see Note 2). Suspend cells in 1/100 culture volume of ice-cold 0.5*M* sucrose containing 10% glycerol. They can be used directly for transformation or stored in aliquots at –80°C until needed (*see* Note 3).

3.2. Electroporation

1. Thaw frozen competent cells on ice (*see* Note 3). In a polypropylene microcentrifuge tube, mix 40 μL of cells (*see* Note 4) with 1–2 μL of DNA (*see* Note 5). This mixture should then be carefully transferred to an ice-cooled electroporation cuvet (*see* Note 6). The formation of air bubbles should be avoided, and the mixture should be distributed evenly on the bottom of the cuvet, to reduce the risk of arcing during the electrical discharge.

2. Pulse once using the following settings: voltage, 2.0 kV; capacitance, 25 μF; resistance, 200 Ω. These settings should give a time constant of 4–5 ms (*see* Note 7).
3. Immediately following the discharge, add 0.96 mL of ice-cold SGM17 to the cuvet, transfer the mixture to a microcentrifuge tube, and incubate at 30°C for 2 h (*see* Note 8).
4. Plate the cell suspensions by spreading on SR or BSR plates (*see* Note 8) containing the appropriate selective agent (e.g., antibiotic) in order to select transformants. The most common marker genes used in lactococcal cloning vectors cause resistance to erythromycin or chloramphenicol (*see* Note 9). Erythromycin is used at 1 μg/mL and chloramphenicol at 5 μg/mL.
5. Incubate the plates at 30°C up to 3 d. The first colonies of transformants are usually visible after 1 d of incubation (*see* Note 10).

4. Notes

1. The glycine concentration needed for optimal transformation is strain-dependent, and we have found it to vary between 0.2 and 4% in the strains tested. If the glycine tolerance of the strain of interest is not known, it is advisable to grow the strain at several different glycine concentrations. The culture containing the highest concentration of glycine that permits growth overnight (OD 0.2–0.7) is chosen for the transformation.
2. The beneficial effect on the transformability of *L. lactis* of glycine and 0.5*M* sucrose was first demonstrated in strain BC101 *(5)*. The method was later used to transform a number of other strains of *L. lactis*. With one exception, strain IMN C-18, all of the strains were transformable by this procedure. Strain IMN C-18 could, however, be transformed when the cells were grown with glycine and 0.25*M* sodium succinate as osmoprotectant instead of sucrose. With all the strains tested, sodium succinate could replace sucrose in the growth medium, but usually the transformation efficiency was lower or similar to that of cells grown with sucrose.
3. An advantage of the method described here compared to other protocols for transformation of *L. lactis* is that frozen cells can be used. We have found no difference in competence between frozen and freshly prepared cells, and the cells can be stored for several months at −80°C without loss of transformability.
4. The efficiency of transformation is proportional to the concentration of cells and to the concentration of DNA in the electroporation cuvet *(5)*.
5. The method described here has been used for introducing plasmid DNA as well as ligation mixtures and bacteriophage DNA *(9)* into *L. lactis*. In plasmid transformation, the efficiency seems to be independent of plasmid size up to at least 20 kbp.

6. We use a Gene Pulser connected to Pulse Controller (Bio-Rad, Richmond, CA), usually with 2-mm electroporation cuvets (Bio-Rad). These cuvets are disposable, but we have found that they can be washed and reused many times without any loss of performance. After use, the cuvets are soaked in a neutral detergent solution, then rinsed with distilled water, and sterilized by soaking in 70% ethanol. Finally, the cuvets are air-dried.

7. The electroporation step in this procedure has been adapted from the protocol of Dower et al. *(10)* for the transformation of *Escherichia coli.* As with *E. coli,* the transformation efficiency is highly dependent on a sufficiently high field strength. We obtained the highest number of transformants at 12.5 kV/cm *(5),* which is the maximal field strength of the Gene Pulser when using 2-mm cuvets. We were unable to improve the transformation efficiency by using 1-mm cuvets to obtain even higher field strengths. At higher field strengths, there is an increased risk of arcing.

8. For many strains, a critical point in this transformation protocol appears to be the recovery of transformants. The SR medium, designed for the recovery of protoplasts *(8),* has proven superior to GM17 for growing the cells after transformation by the method described here. Although for many strains one does obtain transformants on GM17 plates, the yield is usually 30–90% lower than on SR plates. The number of transformants on SGM17 is the same as on GM17. For some strains, however, no transformants were obtained on GM17 plates, whereas on SR plates, more than 1×10^5 were found. For *L. lactis* NCDO 1986, we even observed that the transformants picked after growth on SR plates could be grown in SGM17, but not in GM17 broth. However, after a second transfer, the ability to grow without the osmostabilizer was restored.

9. We have observed an unacceptably high background of false positives when too many cells are spread onto SR plates containing 1 µg/mL erythromycin. We believe that this is because of the poor buffering capacity of the SR medium. When rare transformants are expected and thus as many cells as possible are spread on a single plate, we use BSR, which has the same buffering capacity as has GM17, but gives the same number of transformants as SR. The entire cell suspension from one electroporation experiment can be spread onto a BSR plate containing 1 µg/mL erythromycin without producing false positives.

10. By using this protocol and strain LM 0230, we have obtained up to 1×10^8 transformants/µg plasmid DNA, and the fraction of cells containing the plasmid can be as high as 5%. The transformation efficiency is, however, highly strain-dependent *(5).* The various lactococcal strains also differ with respect to glycine tolerance; the upper limit for growth is in the range 0.3–4% glycine in SGM17.

References

1. Kondo, J. K. and McKay, L. L. (1985) Gene transfer systems and molecular cloning in *N streptococci* a review. *J. Dairy Sci.* **68,** 2143–2159.
2. Harlander, S. K. (1987) Transformation of *Streptococcus lactis* by electroporation, in *Streptococcal Genetics* (Ferretti, J. J. and Curtiss, R., III, eds.), American Society for Microbiology, Washington, DC, pp. 229–233.
3. Powell, I. B., Achen, M. G., Hillier, A. J., and Davidson, B. E. (1988) A simple and rapid method for genetic transformation of lactic streptococci by electroporation. *Appl. Environ. Microbiol.* **54,** 655–660.
4. Van der Lelie, D., Van der Vossen, J. M. B. M., and Venema, G. (1988) Effect of plasmid incompatibility on DNA transfer to *Streptococcus cremoris. Appl. Environ. Microbiol.* **54,** 865–871.
5. Holo, H. and Nes, I. F. (1989) High frequency transformation, by electroporation, of *Lactococcus lactis* subsp. *cremoris* grown with glycine in osmotically stabilized media. *Appl. Environ. Microbiol.* **55,** 3119–3123.
6. Cruz-Rodz, A. L. and Gilmore, M. S. (1990) High efficiency introduction of plasmid DNA into glycine-treated *Enterococcus faecalis* by electroporation. *Mol. Gen. Genet.* **224,** 152–154.
7. Terzaghi, B. E. and Sandine, W. E. (1975) Improved medium for lactic streptococci and their bacteriophages. *Appl. Microbiol.* **29,** 807–813.
8. Okamoto, T., Fujita, Y., and Irie, R. (1983) Protoplast formation and regeneration of *Streptococcus lactis* cells. *Agric. Biol. Chem.* **47,** 259–263.
9. Lillehaug, D. and Birkeland, N. K. (1993) Characterization of genetic elements required for site-specific integration of the temperate lactococcal bacteriophage φ-LC3 and construction of integration-negative φ-LC3 mutants. *J. Bacteriol.* **175,** 1745–1755.
10. Dower, W. J., Miller, J. F., and Ragsdale, C. W. (1988) High efficiency transformation of *E. coli* by high voltage electroporation. *Nucleic Acid Res.* **16,** 6127–6145.

CHAPTER 20

Transformation
of *Lactobacillus* by Electroporation

Thea W. Aukrust, May B. Brurberg, and Ingolf F. Nes

1. Introduction

Species of the genus *Lactobacillus* are used worldwide in the production of fermented food or fodder from raw agricultural materials *(1)*. To obtain fermentation products that are reproducible and of high quality, fermentations are initiated by the addition of well-defined starter cultures. The techniques of molecular biology offer a controllable approach to improve commercially important lactic acid bacteria. An essential element of this approach for strain improvement is a method for introducing DNA into the cells.

Workers in a number of laboratories attempted to make *Lactobacillus* competent by partially removing the peptidoglycan cell wall, but this strategy was generally unsuccessful *(2* and references therein). A breakthrough was achieved when Chassy and Flickinger *(3)* discovered that *Lb. casei* could be efficiently and reproducibly transformed by electroporation. To date, it has become evident that many other *Lactobacillus* species also can be transformed with this technique, although the transformation efficiency is extremely variable, ranging from 10^2–10^7 transformants/µg of DNA.

Bearing in mind that the genus *Lactobacillus* presently comprises more than 60 highly diverse species *(4)*, the difference in transformability is perhaps not surprising. *Lactobacillus* species can be divided into three

From: *Methods in Molecular Biology, Vol. 47: Electroporation Protocols for Microorganisms*
Edited by: J. A. Nickoloff Humana Press Inc., Totowa, NJ

groups based on biochemical characteristics: the obligately homo-fermentative (group I), the facultatively heterofermentative (group II), and the obligately heterofermentative (group III) *(5)*. In the present chapter, we will describe protocols that we have successfully used for transformation of representative strains among the facultatively hetero-fermentative *Lactobacillus* species and the obligately heterofermentative *Lactobacillus* species.

Because of the heterogeneity within the genus *Lactobacillus*, electro-poration protocols need to be optimized for each species and, within the species, for each strain. It is therefore a challenging task to transform new, unknown strains with optimal efficiency. Here we present two basic protocols for electroporation that can be used to transform various lacto-bacilli and that form a good starting point for optimizing transformation of new strains. The first procedure is based on the use of sucrose and magnesium in the electroporation solution, and has been used for elec-troporation of *Lb. acidophilus (2,6)*, *Lb. bavaricus (7)*, *Lb. brevis (8)*, *Lb. casei (2)*, *Lb. curvatus (9)*, *Lb. delbrückii (8)*, *Lb. fermentum (2)*, *Lb. helveticus (10)*, *Lb. pentosus (11)*, *Lb. plantarum (2,12–14)*, *Lb. reuteri (2)*, and *Lb. sake (9)*. The second procedure is based on the use of poly-ethylene glycol (PEG) in the electroporation solution, and has been used for electroporation of *Lb. casei (15)*, *Lb. plantarum (14–16)*, *Lb. reuteri (17)*, and *Lb. sake (14)*. The PEG procedure also has been used for direct cloning of plasmid encoded traits in *Lb. sake (18)*.

2. Materials
2.1. Media

1. MRS broth *(19)*: 10.0 g peptone, 8.0 g meat extract, 4.0 g yeast extract, 20.0 g glucose, 1 mL monooleate (Tween 80), 2.0 g K_2HPO_4, 5.0 g sodium acetate · $3H_2O$, 2.0 g $(NH_3)_3$-citrate, 0.2 g $MgSO_4 · 7H_2O$, 0.05 g $MnSO_4 · 4H_2O$, distilled H_2O (dH_2O) to 1 L. Autoclave for 15 min. It is important to autoclave carefully in order to avoid browning, which may reduce the growth rate and also may have a negative effect on the transformation com-petence of the cells. MRS broth may be purchased from Oxoid Ltd. (Basingstoke, UK).
2. MRS plates: MRS broth solidified with 1.5% agar. Autoclave for 15 min.
3. MRS plates with antibiotics: Add antibiotics when medium cools to 60°C (*see* Note 1).
4. 20% (w/v) Glycine stock solution: 20 g glycine, dH_2O to 100 mL. Filter-sterilize through a 0.22-μm membrane. For alternative solutions, *see* Note 2.

5. MRSSM (MRS, 0.5M sucrose, 0.1M MgCl$_2$): concentrated MRS, 17.1 g sucrose, 2.0 g MgCl$_2$ · 6H$_2$O, dH$_2$O to 100 mL. Filter-sterilize as above.

2.2. Solutions for Electroporation

1. 1 mM MgCl$_2$: 0.20 g MgCl$_2$ · 6H$_2$O, dH$_2$O to 1 L. Autoclave.
2. 30% (w/v) PEG 1500 (polyethylene glycol, mol-wt range 1300–1600): 300 g PEG 1500, dH$_2$O to 1 L. Filter-sterilize as above.
3. SM: 326 g sucrose (952 mM), 0.71 g MgCl$_2$ · 6H$_2$O (3.5 mM), dH$_2$O to 1 L. Filter-sterilize as above.
4. DNA: Dissolve in TE (10 mM Tris-HCl, 1 mM EDTA, pH 7.5) to 0.10–1.0 µg/µL. Ligation mixtures should be precipitated by ethanol, washed with 80% ethanol, and dissolved in TE before electroporation (*see* Note 3 concerning choice of vector).

3. Methods

In the following, two alternative procedures for electroporation of lactobacilli are presented. In order to obtain optimum transformation efficiencies, both methods should be tested. Some strains are very sensitive to handling during the preparation of electrocompetent cells, so as a general precaution, all solutions should be chilled (4°C) and extreme care should be taken in all steps of the procedure when working with new strains (*see* Note 4).

3.1. Preparation of Competent Cells

1. Use a 25-mL preculture of lactobacilli in exponential growth phase to inoculate 100 mL of MRS or MRS supplemented with glycine. Inoculate to A_{600} = 0.25. Incubate the culture at 30°C until A_{600} = 0.6 (usually 2–4 h, depending on strain).
2. Harvest the cells by centrifugation, and decant the supernatant. Use the minimum speed and time sufficient to pellet the cells. These parameters will vary for different strains, but 1500g for 5 min has been used as a minimum.
3. Resuspend cells carefully (by pipeting the cells) in 100 mL of SM (procedure 1) or 1 mM MgCl$_2$ (procedure 2); first wash (*see* Note 5).
4. Pellet the cells by centrifugation, as in step 2. It may be necessary to increase centrifugation speed and time; this is strain-dependent.
5. Resuspend cells carefully as above in 100 mL of SM (procedure 1) or 30% PEG (procedure 2); second wash.
6. Pellet the cells by centrifugation, as in step 4.
7. Resuspend cells gently in 1 mL of SM (procedure 1) or 30% PEG (procedure 2). Pipet 100-µL aliquots (procedure 1) or 40-µL aliquots (procedure 2)

of the cells into separate polypropylene microcentrifuge tubes. Store on ice until needed for electroporation, which should be performed as soon as possible (*see* Note 6). The cell suspension should contain at least 10^9–10^{10} cells/mL.

3.2. Electroporation

1. Immediately before electroporation, add DNA in a volume ≤1/20 of the cell suspension volume. Mix with the pipet tip, and transfer carefully to an ice cold electroporation cuvet with a 2-mm electrode gap (*see* Note 7). Avoid trapping air bubbles, and make sure that the cell suspension is distributed evenly on the bottom of the cuvet.
2. Deliver the electric pulse using the following settings:
    Procedure 1 (electroporation in SM): voltage, 1.5 kV; capacitance, 25 μF; and resistance, 800 Ω. This should give a time constant of approx 10 ms.
    Procedure 2 (electroporation in PEG): voltage, 1.5 kV; capacitance, 25 μF; and resistance, 400 Ω. This should give a time constant of 9.0–9.5 ms (*see* Note 8).
3. Immediately following the discharge, add MRSSM up to 1 mL, and transfer the cell suspension to a microcentrifuge tube. Incubate at 30°C for 2 h.
4. Spread undiluted and serial dilutions of the cell suspension on MRS plates containing appropriate additives to select transformants. For evaluation of cell survival and transformation frequency (transformants/surviving cells), also spread serial dilutions of the cell suspension before and after electroporation on MRS plates. Incubate plates at 30°C for up to 3 d. The first colonies of transformants are usually visible after 1–2 d depending on the growth rate of the strain.

4. Notes

1. The most common marker genes in cloning vectors applicable to lactobacilli are genes causing resistance to erythromycin and chloramphenicol. Low concentrations of antibiotics should be used when selecting for transformants (i.e., 1 μg/mL of erythromycin and 5–10 μg/mL of chloramphenicol). In electroporation studies of *Lb. plantarum* strains with the plasmid pVS2 *(20)*, which has both an erythromycin and a chloramphenicol marker, we have observed for many strains that selection on chloramphenicol results in a transformation efficiency five- to sixfold higher than selection on erythromycin.
2. Addition of glycine to the growth medium inhibits formation of crosslinkages in the cell wall *(21)*, and this appears to enhance transformability of some lactobacilli strains. The glycine concentration needed for optimal transformation is strain-dependent, but to evaluate possible effects, we

often use 1% glycine in the medium *(13,14)*. To optimize a procedure for a specific strain, various concentrations should be tested. The positive effect of glycine may be increased further by growing the cells in osmotically stabilized media (i.e., $0.75M$ sorbitol) *(16)*. Threonine appears to have a similar effect as glycine and has been successfully applied by others *(11)*.

3. One should be aware that the choice of an appropriate cloning vector in some types of experiments can compensate for poor transformation efficiencies. A broad host-range origin of replication, or the combination of a *Lactobacillus* and an *E. coli* origin of replication in a shuttle vector allows amplification of the plasmid constructs in well-established and highly transformable host organisms, such as *B. subtilis, L. lactis,* or *E. coli,* before transformation into lactobacilli.

4. When optimizing an electroporation protocol, it is worthwhile testing whether the strain must be maintained at low temperatures to produce highly electrocompetent cells. For some strains, room-temperature solutions may be preferable *(15)*.

5. Many *Lactobacillus* strains produce slime (extracellular polysaccharides), which may have an adverse effect on transformation efficiency, most likely because of less efficient washing and concentrating of the cells. The presence of low concentrations of $MgCl_2$ in the first washing solution may help to circumvent this problem *(14)*. Usually 1 mM $MgCl_2$ is sufficient, but for heavy slime producers, concentrations from 10–100 mM may be useful. Addition of higher concentrations of $MgCl_2$ may markedly reduce the pulse length at the highest voltages (10–12.5 kV/cm), so these voltages should be avoided. In all cases, use the electroporation solution without any adjustments in the second wash step.

6. Some strains appear to lose their competence if they are kept too long on ice before electroporation, so when electroporating new strains, one should proceed immediately. For convenience, some strains may be stored in aliquots at –80°C without loss of electrocompetence for several weeks or months. This should be tested for each new strain.

7. Two-millimeter (or 1 mm) electroporation cuvets from Bio-Rad (Richmond, CA) are disposable, but may be washed and reused several times without loss of performance. After use, the cuvets should be soaked in a neutral detergent solution, rinsed well with distilled water, sterilized by soaking in 70% ethanol, and air-dried. To inactivate any remaining DNA, the cuvets can be treated by UVL-illumination (360 nm, 10 min) just before reuse.

8. A very important factor in electroporation is the electrical conditions used, such as voltage, field strength, and the length and shape of the pulse *(3,6,10,16,22)*. Despite differences in transformation efficiencies, strains we have tested show essentially the same response patterns to different

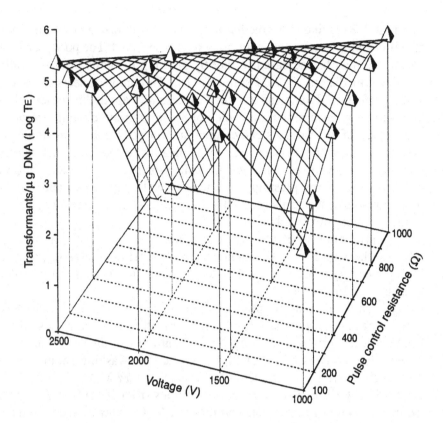

Fig. 1. Response surface plot showing transformation efficiency (trans-
formants/μg pVS2; log value) of *Lactobacillus plantarum* NC8 as a function of
voltage and pulse-control resistance using 2-mm electroporation cuvets and 40
μL of the nonionic 30% (w/v) PEG 1500 electroporation solution. The transformation efficiencies are indicated by arrows. For calculation of the response
surface plot, a second-degree polynomial with interactions in two variables was
fitted to data using the RSREG module of the SAS system (SAS Institute Inc.,
Cary, NC).

electrical pulses *(14)*. Figure 1 shows an example of a three-dimensional
response surface plot illustrating the combined effect of different voltage
and resistance combinations on transformation efficiency. Pulses with
time constants ranging from 2–20 ms resulted in successful transformation. The plot shows optimum efficiencies from high voltage/low resistance
(short pulses) to low voltage/high resistance (long pulses), demonstrating
the importance of specific combinations of voltage and resistance settings
for efficient transformation. At the highest voltages, only short pulses may

be used without extensive cell death and the resulting low number of transformants. At lower voltages, longer pulses are necessary to obtain the desired effect on the cells. The plot shown in Fig. 1 was obtained using an electroporation solution with low conductivity; hence, the resistance set by the apparatus was the major contributor to the length of the pulse. When more conductive electroporation solutions are used, increase the resistance of the apparatus.

The conditions used in the general procedures presented in this chapter were chosen from the least variable part of the response surface when different plots were compared. Although field strength (7.5 kV/cm) is identical in both procedures, a higher pulse-control resistance (800 Ω) is used in procedure 1 in order to compensate for the higher conductivity of the electroporation solution.

References

1. McKay, L. L. and Baldwin, K. A. (1990) Applications for biotechnology: present and future improvements in lactic acid bacteria. *FEMS Microbiol. Rev.* **87,** 3–14.
2. Luchansky, J. B., Muriana, P. M., and Klaenhammer, T. R. (1988) Application of electroporation for transfer of plasmid DNA to *Lactobacillus, Lactococcus, Leuconostoc, Listeria, Pediococcus, Bacillus, Staphylococcus, Enterococcus* and *Propionibacterium. Mol. Microbiol.* **2,** 637–646.
3. Chassy, B. M. and Flickinger, J. L. (1987) Transformation of *Lactobacillus casei* by electroporation. *FEMS Microbiol. Lett.* **44,** 173–177.
4. Hammes, W. P., Weiss, N., and Holzapfel, W. (1992) The genera *Lactobacillus* and *Carnobacterium,* in *The Procaryotes,* 2nd ed. (Bahlows, A., Trüper, H. G., Dworkin, M., Harder, W., and Schleifer, K. H., eds.), Springer-Verlag, New York, pp. 1535–1594.
5. Kandler, O. and Weiss, N. (1986) Regular, non-sporing Gram-positive rods, in *Bergey's Manual of Systematic Bacteriology,* vol. 2. (Sneath, P. H. A., Mair, N. S., Sharp, M. E., and Holt, J. G., eds.), Williams and Wilkins, Baltimore, pp. 1208–1234.
6. Luchansky, J. B., Kleeman, E. G., Raya, R. R., and Klaenhammer, T. R. (1989) Genetic transfer systems for delivery of plasmid deoxyribonucleic acid to *Lactobacillus acidophilus* ADH: conjugation, electroporation and transduction. *J. Dairy Sci.* **72,** 1408–1417.
7. Aymerich, M. T., Hugas, M., Garriga, M., Vogel, R. F., and Monfort, J. M. (1993) Electrotransformation of meat lactobacilli. Effect of several parameters on their efficiency of transformation. *J. Appl. Bacteriol.* **75,** 320–325.
8. Zink, A., Klein, J. R., and Plapp, R. (1991) Transformation of *Lactobacillus delbrückii* ssp. *lactis* by electroporation and cloning of origins of replication by use of a positive selection vector. *FEMS Microbiol. Lett.* **78,** 207–212.
9. Gaier, W., Vogel, R. F., and Hammes, W. P. (1990) Genetic transformation of intact cells of *Lactobacillus curvatus* Lc2-c and *Lact. sake* Ls2 by electroporation. *Lett. Appl. Microbiol.* **11,** 81–83.

10. Hashiba, H., Takiguchi, R., Ishii, S., and Aoyama, K. (1990) Transformation of *Lactobacillus helveticus* subsp. *jugurti* with plasmid pLHR by electroporation. *Agric. Biol. Chem.* **54**, 1537–1541.

11. Posno, M., Leer, R. J., Luijk, N. van, Giezen, M. J. F. van, Heuvelmans, P. T. H. M., Lokman, B. C., and Pouwels, P. H. (1991) Incompatibility of *Lactobacillus* with replicons derived from small cryptic *Lactobacillus* plasmids and segregational instability of the introduced vectors. *Appl. Environ. Microbiol.* **57**, 1822–1828.

12. Badii, R., Jones, S., and Warner, P. J. (1989) Sphaeroplast and electroporation mediated transformation of *Lactobacillus plantarum. Lett. Appl. Microbiol.* **9**, 41–44.

13. Aukrust, T. and Nes, I. F. (1988) Transformation of *Lactobacillus plantarum* with the plasmid pTV1 by electroporation. *FEMS Microbiol. Lett.* **52**, 127–131.

14. Aukrust, T. and Blom, H. (1992) Transformation of *Lactobacillus* strains used in meat and vegetable fermentations. *Food Res. Int.* **25**, 253–261.

15. Josson, K., Scheirlinck, T., Michiels, F., Platteeuw, C., Stanssens, P., Joos, H., Dhaese, P., Zabeau, M., and Mahillon, J. (1989) Characterization of a gram-positive broadhost-range plasmid isolated from *Lactobacillus hilgardii. Plasmid* **21**, 9–20.

16. Bringel, F. and Hubert, J.-C. (1990) Optimized transformation by electroporation of *Lactobacillus plantarum* strains with plasmid vectors. *Appl. Microbiol. Biotechnol.* **33**, 664–670.

17. Ahrné, S., Molin, G., and Axelsson, L. (1992) Transformation of *Lactobacillus reuteri* with electroporation: studies on the erythromycin resistance plasmid pLUL631. *Curr. Microbiol.* **24**, 199–205.

18. Axelsson, L., Holck, A., Birkeland, S.-E., Aukrust, T., and Blom, H. (1993) Cloning and nucleotide sequence of a gene from *Lactobacillus sake* Lb706 necessary for sakacin A production and immunity. *Appl. Environ. Microbiol.* **59**, 2868–2875.

19. Man, J. C. de, Rogosa, M., and Sharpe, M. E. (1960) A medium for the cultivation of lactobacilli. *J. Appl. Bacteriol.* **23**, 130–135.

20. Wright, A. von, Tynkkynen S., and Suominen, M. (1987) Cloning of a *Streptococcus lactis* subsp. *lactis* chromosomal fragment associated with the ability to grow in milk. *Appl. Environ. Microbiol.* **53**, 1584–1588.

21. Hammes, W., Schleifer, K. H., and Kandler, O. (1973) Mode of action of glycine on the biosynthesis of peptidoglycan. *J. Bacteriol.* **116**, 1029–1053.

22. Chassy, B. M., Mercenier, A., and Flickinger, J. (1988) Transformation of bacteria by electroporation. *Trends in Biotechnol.* **6**, 303–309.

CHAPTER 21

Electrotransformation of Staphylococci

Jean C. Lee

1. Introduction

Methods for genetic manipulation of Staphylococci have advanced considerably over the past decade. Transduction is the most common and simplest method for transferring plasmid DNA or chromosomal markers between strains of *Staphylococcus aureus*. However, some clinical isolates of *S. aureus* may be difficult or impossible to transduce because of restriction barriers. Moreover, staphylococcal phages are highly species-specific and cannot be propagated in species other than *S. aureus*.

With the advent of molecular microbiology, numerous staphylococcal genes were cloned and expressed in either *Escherichia coli* or *Bacillus subtilis*. Efforts to transfer these recombinant DNA molecules into the staphylococcus were hindered by restriction systems that prevented staphylococci from accepting DNA prepared from other bacterial genera *(1)*. Mutant *S. aureus* strains defective in one or more restriction systems were isolated *(1)*, so that recombinant DNA constructs could be returned to the staphylococcal genetic background. Before the development of electroporation-based transformation, uptake of DNA by staphylococci was accomplished by protoplast transformation. Protoplasts are prepared by treatment of the staphylococci with lysostaphin in a complex, osmotically stable medium. Transformation occurs when the protoplasts are "shocked" with polyethylene glycol in the presence of DNA. Transformants are selected by incubation of the mixture for several days on a complex cell-wall regeneration medium containing appropriate antibiotics.

Electroporation-based transformation has greatly facilitated the introduction of DNA molecules into the staphylococci *(2–5)*. The process of

From: *Methods in Molecular Biology, Vol. 47: Electroporation Protocols for Microorganisms*
Edited by: J. A. Nickoloff Humana Press Inc., Totowa, NJ

electroporation involves the brief exposure of exponential-phase bacterial cells to a high-voltage electrical field that facilitates DNA uptake by creating transient pores in the cell membrane. Electroporation of staphylococci is technically easier to perform than protoplast transformation. In addition, competent cells can be prepared and stored for months at –70°C, and transformants usually are recovered within 24 h.

2. Materials
2.1 Bacterial Strains

S. aureus RN4220 (R. P. Novick, New York University Medical Center) and SA113 (35556, American Type Culture Collection, Rockville, MD) are mutants of strain NCTC 8325 that lack restriction barriers; these strains are commonly used as recipients for electroporation of recombinant DNA molecules (*see* Note 1). Other staphylococcal strains vary considerably in their ability to be transformed with DNA prepared from *S. aureus* RN4220 or SA113. For example, DNA propagated in strain RN4220 (phage lytic group III) can be transferred readily to other members of this phage group. In contrast, strains belonging to phage type 94/96—members of lytic group V—represent a quite distinct class of *S. aureus* strains with rigid restriction barriers (*6*). It is difficult, but not impossible, to electroporate DNA isolated from strain RN4220 into members of this group.

2.2. Plasmids

Because small staphylococcal plasmids, like pC194 and pE194, are readily transformed into cells by electroporation, they serve as useful controls when the electroporation procedure is first being established in a laboratory. Numerous staphylococcal plasmids are available, some of which are shown in Tables 1 and 2. Shuttle vectors capable of replication in both staphylococci and *E. coli* have been constructed and include pLI50 (*7*) and pMIN164 (*8*). Plasmid DNA isolated by alkaline lysis or the boiling method can be efficiently transformed into electrocompetent *S. aureus* (*3*). DNA preparations free of contaminants yield the highest transformation efficiencies (*see* Note 2).

2.3. Solutions and Reagents for Electroporation

1. B2 medium (*see* Note 3): 10 g casein hydrolysate, 25 g yeast extract, 5 g glucose, 25 g NaCl, 1 g K_2HPO_4 dissolved in 1 L of water. Adjust pH to 7.5. Autoclave.
2. 0.2M sodium hydrogen maleate: 11.6 g maleic anhydride, 4 g NaOH dissolved in 500 mL water. Autoclave.

Table 1
Staphylococcal Plasmids Commonly Used for Transformation

Plasmid	Size, kbp	Selectable marker	Antibiotic concentration for induction, µg/mL	Antibiotic concentration for selection, µg/mL	Reference
pC194	2.9	Chloramphenicol	0.2	10	10
pE194	3.7	Erythromycin	0.2	10	11
pI258	28	Erythromycin	0.2	10	12
pLI50	5.4	Chloramphenicol	0.3	15	7
pMIN164	8.6	Erythromycin	0.2	10	8
pT181	4.4	Tetracycline	0.1	5	13

3. 4X Penassay broth: 17.5 g penassay medium (antibiotic medium 3; Difco Corporation, Detroit, MI) dissolved in 250 mL water. Autoclave.
4. 2X SMM: 25 mL 0.2M sodium hydrogen maleate, 40 mL 0.1M NaOH. Adjust pH to 6.5. Add 5 mL 1M MgCl$_2$ and 42.7 g sucrose. Dissolve and bring final volume to 125 mL with water. Filter-sterilize.
5. SMMP: 55 mL 2X SMM, pH 6.8, 40 mL 4X penassay broth, 5 mL 10% (w/v) bovine serum albumin. Adjust pH to 7.0 and filter-sterilize.
6. 10% (v/v) Glycerol in water. Filter-sterilize.
7. Trypticase soy agar (Becton Dickinson, Cockeysville, MD): 40 g powder in 1 L of water. Autoclave.

2.4. Antibiotic Selection

Selection of transformants carrying the plasmid or chromosomal marker of interest is based on resistance to the action of heavy metals or antibiotics included in selective culture medium. Antibiotic resistance in staphylococci is often the result of inducible enzymes that inactivate the antibiotic, e.g., chloramphenicol acetyl transferase or a methylase that acts on 23S ribosomal RNA (associated with erythromycin resistance). The antibiotic concentration used for enzyme induction is generally 1/50 the concentration used for selection. Commonly used staphylococcal plasmids, their resistance markers, and antibiotic concentrations used for induction and selection are shown in Table 1.

3. Methods

3.1. Preparation of Electrocompetent Cells

1. Inoculate a single colony of staphylococci into 3 mL of B2 broth. Incubate overnight at 37°C with constant aeration.

Table 2
Reported Plasmid Transformation Efficiencies for *Staphylococcus spp.*

Species	Strain	Plasmid[a]	Size, kb	Transformants/ µg of DNA	Reference
S. epidermidis	ATCC 12228	pSK265[b]	2.9	5.0×10^3	*4*
S. epidermidis	Tu3298	pC194	2.9	3.0×10^5	*2*
S. epidermidis	115	pWN101	10.5	1.6×10^2	*5*
S. carnosus	TM300	pLM6	2.8	1.0×10^2	*14*
S. carnosus	TM300	pC194	2.9	1.0×10^5	*2*
S. carnosus	TM300	pFP80	4.6	3.8×10^2	*14*
S. staphylolyticus	ATCC 1362	pC194	2.9	5.0×10^3	*2*
S. aureus	3A	pLM6	2.8	1.3×10^2	*14*
S. aureus	3A	pC194	2.9	3.0×10^5	*2*
S. aureus	3A	pFP80	4.6	1.3×10^2	*14*
S. aureus	RN4220	pSK265	2.9	5.7×10^5	This study
S. aureus	RN4220	pSK265	2.9	4.0×10^8	*4*
S. aureus	RN4220	pE194	4.0	1.7×10^3	*3*
S. aureus	RN4220	pUB110	4.5	1.2×10^4	*3*
S. aureus	RN4220	pLI50[c]	5.4	2.0×10^5	This study
S. aureus	RN4220	pJCL24[c]	14.8	6.4×10^3	This study
S. aureus	RN4220	pI258	28.2	1.7×10^1	*3*
S. aureus	ATCC 29213	pSK265	2.9	6.1×10^5	*4*

[a]Plasmid DNA was prepared from *S. aureus*, unless indicated otherwise.
[b]pSK265 was derived by insertion of the pUC19 polylinker into the single *Hind*III site of pC194 *(15)*.
[c]Plasmid DNA was prepared from *E. coli*.

2. Add 1.5 mL of the overnight culture to 150 mL of fresh B2 broth in a 1-L flask. Incubate the cells with constant aeration (~250 rpm) at 37°C until the OD at 650 nm reaches 0.5 (*see* Note 3).

3. Chill the culture on ice for 15 min to stop growth.

4. Harvest the cells by centrifugation at 12,000*g* for 15 min at 4°C.

5. Wash the cells three times in an equal volume of sterile water. Wash once with 30 mL of 10% glycerol.

6. After centrifugation, resuspend the bacterial pellet in 15 mL of 10% glycerol, and incubate at 20°C for 15 min.

7. Centrifuge the cell suspension as described above, and suspend the cells in 5 mL of 10% glycerol. The final cell concentration should be $\geq 1 \times 10^{10}$ CFU/mL.

8. Prepare 250-µL aliquots of the electrocompetent cells in sterile 1.5-mL tubes. Use the cells fresh, or flash-freeze aliquots in dry ice. Electrocompetent staphylococci are stable for several months at –70°C (*see* Note 4).

3.2. Electroporation

1. Add DNA samples (5 ng to 2 μg) for electroporation to sterile 1.5-mL tubes (*see* Notes 2, 5, and 6). The volume should not exceed 3 μL. A tube containing no DNA should be included as a control.
2. Thaw the competent bacterial cells at room temperature for several minutes. Add 50 μL of cells to each sample tube, and mix by pipeting.
3. Incubate the samples at ambient temperature for 30 min.
4. Transfer the transformation mixture to a cuvet with a 0.2-cm electrode gap (*see* Note 7). Place the cuvet in the sample chamber and pulse once at 2.5 kV, capacitor at 25 μF, and parallel resistor at 100 Ω, yielding a field strength of 12.5 kV and a time constant of ~2.5 ms (*see* Note 8).
5. Immediately resuspend the cells in 950 μL of SMMP (*see* Note 4). Transfer the mixture to a sterile culture tube, and incubate at 37°C for 1 h in the presence of subinhibitory concentrations of antibiotics (*see* Table 1) to allow for expression of genes encoding antibiotic resistance.
6. Plate aliquots of the electroporation mixtures on trypticase soy agar or other suitable medium containing antibiotics at concentrations suitable for selection (*see* Table 1). Incubate plates up to 48 h at 37°C.
7. Verify transformants by restriction enzyme analysis of isolated plasmid DNA.

4. Notes

1. Except for restriction-negative mutants, such as *S. aureus* RN4220 and SA113, staphylococcal strains and species will not accept DNA propagated in *E. coli*. It is therefore essential that recombinant DNA molecules prepared in *E. coli* be passaged first in one of these mutant strains. After the plasmid is reisolated from *S. aureus,* it can be transferred to other staphylococcal strains and species. Using the methods described above to electroporate *S. aureus* strain RN4220, we have achieved transformation efficiencies of 2×10^5 transformants/μg of pLI50 DNA prepared from *E. coli* and 6×10^5 transformants/μg of pSK265 DNA prepared from *S. aureus.*
2. Although large plasmids generally transform less efficiently than small plasmids, this difference can be partially explained by differences in their molar concentrations. The data in Table 2 indicate that there is considerable variation among laboratories in the transformation efficiencies obtained by electroporation. Numerous factors influence electrotransformation, including the method of preparation of the competent cells, the purity of the DNA, the electroporation medium, the amplitude and duration of the electrical pulse, and the bacterial host strain. Our experience with clinical *S. aureus* isolates shows that their efficiency of transformation may be 10- to 50-fold lower than that of strain RN4220. Increases in

the DNA concentration also affect the frequency of transformation: The number of transformants increases linearly with DNA concentrations ranging from ~1 ng to ~1 μg. Addition of excess DNA to electroporation mixtures does not further increase the number of transformants *(2–4)*.

3. During the preparation of electrocompetent cells from clinical isolates of *S. aureus* of phage type 94/96, we noted that the bacterial cells clumped when the B2 culture reached the exponential phase of growth. As a result, we were unable to obtain accurate optical density readings. When competent cells from the same strains were prepared by the method of Augustin and Gotz *(2)* with use of basic medium, no bacterial clumping occurred. (Basic medium is composed of 1 g peptone, 5 g yeast extract, 1 g glucose, 5 g NaCl, and 1 g K_2HPO_4/L of water; the pH is adjusted to 7.4.)

4. The results of bacterial plate counts performed on the transformation mixture before and after electroporation indicate whether the frozen electrocompetent bacterial cells retain viability after a freeze/thaw cycle and whether the appropriate survival rates are achieved after electroporation. Only ~10% of staphylococci survive exposure to a field strength of 12.5 kV/cm. On one occasion, we failed to electrotransform *S. aureus* despite the display of appropriate time constant and voltage read-out values by the electroporator. The results of viable plate counts performed on a control sample for that experiment revealed that all of the staphylococci had survived electroporation. On inspection, we found that the electrode contacts in the base of the sample chamber had broken; consequently, no electrical pulse was delivered across the cuvet chamber.

5. Kraemer and Iandolo *(3)* reported that they were able to electrotransform *S. aureus* RN4220 directly with a 1.1-kb DNA fragment ligated to linearized pLI50. Although only 10 transformants were obtained from the 1-μg DNA sample, this experiment demonstrates the feasibility of cloning directly within the staphylococcal genetic background. This process eliminates the additional steps of cloning of DNA fragments into shuttle vectors, amplification of the DNA in *E. coli,* and transformation back into the staphylococcus.

6. Although most applications of electroporation involve recombinant plasmid DNA molecules, it is also possible to electroporate the *Staphylococcus* with chromosomal DNA. In one application of this technique, a ~2 μg sample of genomic DNA prepared from a mutant *S. aureus* strain (containing a single Tn*918* insertion) was mixed with competent cells of the parental strain that lacked the transposon. After electroporation, transformants that acquired the transposon were selected on medium containing tetracycline. This selection yielded "back-transformants" of the parental strain that underwent a double homologous recombination event with sequences

flanking the transposon, replacing a portion of the wild-type chromosomal DNA with DNA insertionally inactivated by Tn*918 (9)*.

7. Although electroporation cuvets are disposable, they may be reused a few times if they are rinsed three times with sterile water and then twice with 70% ethanol. Residual ethanol should be removed from the cuvet by vigorous shaking or drying.

8. If the electrical pulse from the electroporator "arcs," the ionic strength of the sample is probably too high. Care should be taken to prepare the DNA in water or in buffer with a low-salt content, e.g., 10 m*M* Tris-HCl, pH 8.0, or 5 m*M* EDTA, pH 8.0. If the volume of DNA added is small (1 μL) relative to the volume of the bacterial cells (50 μL), DNA with a slightly higher salt content may be used. Likewise, electrocompetent cells should be suspended in a medium of low ionic strength, e.g., 10% glycerol *(2)* or 0.5*M* sucrose *(3)*.

References

1. Novick, R. P. (1990) The staphylococcus as a molecular genetic system, in *Molecular Biology of the Staphylococci* (Novick, R. P., ed.), VCH, New York, pp. 1–37.
2. Augustin, J. and Gotz, F. (1990) Transformation of *Staphylococcus epidermidis* and other staphylococcal species with plasmid DNA by electroporation. *FEMS Microbiol. Lett.* **66,** 203–208.
3. Kraemer, G. R. and Iandolo, J. J. (1990) High-frequency transformation of *Staphylococcus aureus* by electroporation. *Curr. Microbiol.* **21,** 373–376.
4. Schenk, S. and Laddaga, R. A. (1992) Improved method for electroporation of *Staphylococcus aureus. FEMS Microbiol. Lett.* **94,** 133–138.
5. Whitehead, S. S., Leavitt, R. W., and Jensen, M. M. (1993) Staphylococcosis of turkeys. 6. Development of penicillin resistance in an interfering strain of *Staphylococcus epidermidis. Avian Dis.* **37,** 536–541.
6. Asheshov, E. H., Coe, A. W., and Porthouse, A. (1977) Properties of strains of *Staphylococcus aureus* in the 94, 96 complex. *J. Med. Microbiol.* **10,** 171–178.
7. Lee, C. Y., Buranen, S. L., and Ye, Z.-H. (1991) Construction of single-copy integration vectors for *Staphylococcus aureus. Gene* **103,** 101–105.
8. Hovde, C. J., Hackett, S. P., and Bohach, G. A. (1990) Nucleotide sequence of the staphylococcal enterotoxin C3 gene: sequence comparison of all 3 type C staphylococcal enterotoxins. *Mol. Gen. Genet.* **220,** 329–333.
9. Albus, A., Arbeit, R. D., and Lee, J. C. (1991) Virulence of *Staphylococcus aureus* mutants altered in type 5 capsule production. *Infect. Immun.* **59,** 1008–1014.
10. Horinouchi, S. and Weisblum, B. (1982) Nucleotide sequence and functional map of pC194, a plasmid that specifies inducible chloramphenicol resistance. *J. Bacteriol.* **150,** 815–825.
11. Horinouchi, S. and Weisblum, B. (1982) Nucleotide sequence and functional map of pE194, a plasmid that specifies inducible resistance to macrolide, lincosamide, and streptogramin type B antibiotics. *J. Bacteriol.* **150,** 804–814.

12. Novick, R. P., Murphy, E., Gryczan, T. J., Baron, E., and Edelman, I. (1979) Penicillinase plasmids of *Staphylococcus aureus:* restriction-deletion maps. *Plasmid.* **2,** 109–129.
13. Iordanescu, S. (1976) Three distinct plasmids originating in the same *Staphylococcus aureus* strain. *Arch. Roum. Pathol. Exp. Microbiol.* **35,** 111–118.
14. De Rossi, E., Brigidi, P., Rossi, M., Matteuzzi, D., and Riccardi, G. (1991) Characterization of Gram-positive broad host-range plasmids carrying a thermophilic replicon. *Res. Microbiol.* **142,** 389–396.
15. Jones, C. and Khan, S. (1986) Nucleotide sequence of the enterotoxin B genes from *Staphylococcus aureus. J. Bacteriol.* **166,** 29–33.

CHAPTER 22

Electroporation and Efficient Transformation of *Enterococcus faecalis* Grown in High Concentrations of Glycine

Brett D. Shepard and Michael S. Gilmore

1. Introduction

Enterococci are the focus of increasing academic and clinical research because of their importance as agents in nosocomial infections that are frequently refractory to many commonly used antimicrobial agents *(1–3)*. To facilitate studies on the pathogenic and drug resistance mechanisms associated with enterococcal infection, techniques for the efficient introduction of exogenous DNA have been developed. This chapter provides a description of a protocol used routinely to transform *Enterococcus faecalis* by electroporation. The efficient transformation achieved by this method results from the combined use of an agent to weaken the cell wall during the production of cells competent for electroporation and an osmotic stabilizer to preserve the integrity of the cell throughout the process.

As in other gram-positive bacteria, the enterococcal cell wall presents a formidable physical barrier. Bacterial cell walls (both gram-positive and negative) gain much of their strength from a rigid layer of peptidoglycan. Proteins and crosslinked polysaccharides contribute to the rigidity within individual layers of the polymer *(4)*. Approximately 40 layers of peptidoglycan are found in the cell walls of gram-positive bacteria, accounting for approx 50% of the total cell wall thickness. In contrast, gram-negative bacteria generally contain a maximum of two to three layers of peptidoglycan, accounting for 5–10% of the total cell-

From: *Methods in Molecular Biology, Vol. 47: Electroporation Protocols for Microorganisms*
Edited by: J. A. Nickoloff Humana Press Inc., Totowa, NJ

wall thickness *(5)*. Thus, although gram-negative bacteria contain an additional outer membrane, the cell wall of gram-positive species is generally thicker and more resistant to physical stress. Methods designed to temporarily penetrate this physical barrier have been developed in order to optimize electroporation techniques.

Initially, methods for the introduction of plasmid DNA into nonnaturally transformable gram-positive species utilized polyethylene glycol to transform bacterial protoplasts *(6)*. Although relatively efficient, this technique is comparatively labor-intensive and technically demanding because of the intrinsic fragility of bacterial protoplasts *(7–9)*. The technical refinement required for successful polyethylene-glycol-mediated transformation of bacterial protoplasts was cited as a potential source of variability when comparing results from different laboratories *(8,10)*. The development of protocols for the electroporation of bacteria provided the promise of an efficient and practical alternative.

Initial applications of electroporation for the transformation of gram-positive bacterial species achieved limited transformation efficiencies *(7,11)*. However, Powell et al. *(7)* found that transformation efficiencies increased when cells of gram-positive *Lactococcus lactis* subsp. *lactis* were treated with lysozyme prior to electroporation, indicating that the intact gram-positive cell wall and/or glycocalyx provided a physical barrier to electroporation and uptake of plasmid DNA. This was supported by experiments showing that cells of *Streptococcus cremoris* cultured with threonine to weaken the cell wall were transformed with low efficiency *(12)*. With other parameters held constant, cells that had not been treated with threonine were not transformed. Threonine treatment has also been observed to increase transformation efficiencies following electroporation of other species of gram-positive bacteria *(13,14)*. However, increases in transformation efficiencies owing to threonine treatment have been observed to be less than those associated with other agents that weaken the cell wall *(9,13,14)*.

Early protocols for the electroporation and transformation of gram-positive species using lysozyme to degrade and weaken regions of the cell wall presented many of the same technical difficulties that were encountered with PEG-mediated protoplast transformation. Extensive treatment with lysozyme led to cell fragility and lysis on application of the electroporating pulse *(7)*. Thus, it appeared that cells required some minimal level of cell wall to survive electroporation and facilitate cell-wall regeneration. Moreover, the protracted phase for regeneration of a functional cell wall by bacteria extensively treated with lysozyme extended the time required before the cells could be placed in hypotonic selective growth media. Finally, gram-positive species vary in susceptibility to cell-wall-hydrolyzing enzymes *(15)*.

In an effort to optimize electroporation procedures for a variety of gram-positive species, the incorporation of alternative agents to weaken cell walls was investigated. A protocol was developed for electroporation of *L. lactis* subsp. *cremoris* that involved culturing the cells in the presence of high concentrations of glycine as an agent to weaken the cell wall *(15)*. The growth media was osmotically stabilized with 0.5*M* sucrose to enhance stability of the weakened cells. This protocol allowed Holo and Nes *(15)* to electroporate and transform, at high efficiencies, several strains of *L. lactis* subsp. *cremoris* that previously had not transformed following treatment with lysozyme. Additionally, no cell lysis was observed on application of the electric pulse. Thus, the use of glycine in osmotically stabilized media allowed for sufficient balance between the need for membrane access by transforming DNA and subsequent viability following electroporation. Moreover, cells treated with glycine were able to regenerate a functional cell wall within a matter of hours.

The use of glycine to weaken the gram-positive bacterial cell wall has been incorporated into many electroporation protocols, resulting in higher efficiencies of transformation than previously obtained with cells either not treated with glycine or treated with lysozyme *(9,10,13,16)*. The procedure outlined in this chapter is based on those previously described *(9,10)* and is used routinely to transform *E. faecalis* by electroporation.

2. Materials

1. *E. faecalis* strains: Originally, Cruz-Rodz and Gilmore *(10)* utilized *E. faecalis* plasmid-free strain JH2-2 *(17)* and the recombination-deficient strain UV202 *(18)*. Similar growth kinetics and survival rates were observed for each strain, but the transformation efficiencies for strain UV202 were observed to be lower by approximately three orders of magnitude. The *E. faecalis* plasmid-free strain FA2-2 *(19)* has been observed to be transformed at rates similar to those reported for JH2-2 *(17;* unpublished results). Dunny et al. *(9)* used a similar procedure to transform *E. faecalis* strains OG1SSp, OG1RF, JH2-2, and UV202.
2. Plasmid DNA: Transformation of *E. faecalis* has been accomplished using plasmid DNA isolated by a variety of methods *(8–10)*. Regardless of the

method of isolation, purification of the plasmid DNA by CsCl density gradient centrifugation in the presence of ethidium bromide is a general standard *(8–10)*. Although minipreparations of plasmid DNA have been used to transform *E. faecalis,* electroporation with DNA not purified by density-gradient centrifugation has been observed to generate significantly lower transformation efficiencies *(9)* *(see* Notes 1–3).

3. TE buffer: 10 mM Tris-HCl and 1 mM EDTA, pH 7.5 *(see* Note 3).

4. M17 media: 37.25 g/L M17 broth (Difco, Detroit, MI) and distilled water to bring the solution to the final volume. Adjust the pH to 6.8–7.0 and autoclave. Once the medium has cooled to approx 60°C, antibiotic(s) appropriate for the particular strain of *E. faecalis* that is to be cultured may be added. Store at 25°C for up to 1 mo. Storage times can be increased by omitting antibiotics from the stock solution.

5. SGM17 media: 37.25 g/L M17 broth, 0.5M sucrose, 8% glycine, and distilled water to bring the solution to the final volume. Adjust the pH to 6.8–7.0 and autoclave. After cooling, add the appropriate antibiotic(s). The medium is generally made fresh as required *(see* Notes 4 and 5).

6. SR plates: (per liter) 10 g tryptone, 200 g sucrose, 5 g yeast extract, 25 g gelatin, 10 g glucose, 15 g agar, 2.5 mM MgCl$_2$, and 2.5 mM CaCl$_2$. Add distilled water to bring the solution to the final volume. Adjust the pH to 6.8 and autoclave. Once the solution has cooled to approx 60°C, add antibiotic(s) appropriate for selection. Gently swirl the medium to distribute the antibiotic(s) evenly, avoiding the introduction of air bubbles that will be transferred to the plates. Pour the molten agar into sterile Petri dishes, and allow the plates to solidify before storage. Invert the plates and store in sealed plastic bags at 4°C. Storage periods should not exceed 1 mo.

7. SGM17MC media: 37.25 g/L M17 broth, 0.5M sucrose, 8% glycine, 10 mM MgCl$_2$, 10 mM CaCl$_2$, and distilled water to bring the solution to the final volume. Adjust the pH to 6.8–7.0 and autoclave. Because the medium will be added to *E. faecalis* cells immediately following electroporation, no antibiotics are added. Store at 4°C *(see* Notes 4–6 and 13).

8. Electroporation buffer: 0.5M sucrose and 10% glycerol. Adjust the pH to 7.0. The buffer will be used for storage of cells and will be present in the incubation mixture following electroporation. Thus, it should be autoclaved for sterility and stored at 4°C. The buffer can be stored for several months.

3. Methods

3.1. Production of Electrocompetent Cells

1. Inoculate 5 mL of M17 medium (supplemented with the appropriate antibiotics) with cells from a single colony of *E. faecalis*. Allow the culture to incubate overnight (approx 12–15 h) at 37°C without aeration.

2. Dispense 100 mL of fresh SGM17 (supplemented with appropriate antibiotics) into a sterile 500-mL flask. Inoculate the medium with 1 mL of the overnight *E. faecalis* culture. Incubate for 18–24 h at 37°C without aeration (*see* Note 7).

3. Collect cells at 1000*g* for 10 min. Centrifugation should be performed within the temperature range of 4–25°C. When removing and transporting the tube from the centrifuge, be careful not to dislodge the pellet of cells.

4. Pour the supernatant from the tube, and wash the cells with 1 vol of ice-cold electroporation buffer.

5. Because cells may have been dislodged from the pellet during the wash, centrifuge the cells at 1000*g* for 10 min at 4–25°C.

6. Repeat steps 4 and 5.

7. Resuspend the cells in electroporation buffer using a minimum volume to recover the cells. Typically, a volume equal to 1/100 of the original culture is used. The resuspended cells should be kept on ice (*see* Note 11).

8. Divide the cells into 40-µL aliquots in sterile microcentrifuge tubes. Once each aliquot has been dispensed, it should be placed on ice immediately. After all aliquots have been collected, store them at –70°C (*see* Note 8).

3.2. Electroporation

1. Thaw a 40-µL aliquot of frozen *E. faecalis* cells on ice. At the same time, chill the plasmid DNA preparation on ice. Place a sterile electroporation cuvet (gap width, 0.2 cm) on ice or in a freezer, and allow it to chill for approx 5 min (*see* Notes 9 and 10).

2. Add 1 µL or less of the DNA preparation (at a concentration of 0.1–1.0 µg/µL) to the aliquot of cells. Mix the cells and the DNA by tapping the base of the tube.

3. Transfer the suspension to the prechilled electroporation cuvet. Gently tap the cuvet so that the suspension settles evenly to the bottom.

4. Wipe any moisture from the cuvet with a Kimwipe. This simple step will help limit uncontrolled discharging or arcing that may occur on application of the pulse.

5. Pulse with 2.5-kV (12.5 kV/cm), 25-µF capacitance, and 200-Ω resistance (*see* Notes 18–20).

6. Immediately add 1 mL of ice-cold SGM17MC to the cells, and place the cuvet on ice for at least 5 min. Immediate placement on ice is important because the closure of the membrane pores produced by the pulse appears to be delayed at 0°C *(6)*.

7. Incubate the electrotransformed cells for 2 h at 37°C without aeration (*see* Notes 12 and 13).

8. Make appropriate dilutions in microcentrifuge tubes using SGM17MC medium. Base the dilutions on final volumes of 100 µL. Spread the 100-µL

aliquots of diluted transformed cells on SR plates (supplemented with the appropriate antibiotics). Incubate the cells at 37°C for 48 h (*see* Notes 14–17).

9. For each series of electroporations, two controls—one lacking the DNA and the other lacking the pulse—should be performed in parallel.

4. Notes
4.1. Plasmid DNA

1. The size of the required plasmid will depend on the experimental design and the size of the DNA fragment that will be carried. For *E. faecalis*, it has been observed that higher transformation efficiencies were associated with the use of smaller plasmids; however, the differences in efficiency between small and large plasmids were minimal when plasmid sizes were below approx 20–30 kbp *(9,10)*.

2. For *E. faecalis* strains, Dunny et al. *(9)* observed that 300 ng of DNA were sufficient for transformation. Luchansky et al. *(8)* used concentrations between 150 and 1000 ng to transform *E. faecalis*. Cruz-Rodz and Gilmore *(10)* transformed *E. faecalis* with only 70 ng of DNA by using a smaller volume of cells. In general, <1 µg of DNA is sufficient for efficient transformation of *E. faecalis (9,11)*. Transformation frequencies have been observed to decrease when the added DNA volume was greater than or equal to one-tenth of the volume of electrocompetent cells *(6)*.

3. Plasmid DNA used for electroporation is typically dissolved in TE buffer in order to limit the concentration of ions added to the electroporation solution. Moderate to high ionic strength DNA solutions have been observed to result in arcing on application of the electrical pulse. The addition of DNA to the cells should not increase the total salt concentration of the electroporation solution by more than 1 mM *(6)*. When DNA concentrations are sufficiently high, 0.5–1.0 µL of DNA directly from a ligation reaction may be added directly to 40 µL of electrocompetent cells for transformation. If larger volumes of a ligation reaction are required to obtain desired numbers of transformants, the DNA is typically precipitated with ethanol and resuspended in a reduced volume of TE prior to addition to the electrocompetent cell preparation.

4.2. Media

4. The percentage of glycine indicated for SGM17 and SGM17MC media should serve as a guide. Higher glycine concentrations generally yield higher transformation efficiencies *(9,10)*. However, 8% glycine may inhibit the growth of some *E. faecalis* strains. Thus, adjustments may be required to allow for sufficient cell growth. Adequate cell growth for electroporation was observed in the presence of 8% glycine for strains JH2-2

and UV202 *(10)*. However, a 4–6% range of glycine was found to be useful for strains JH2-2 and UV202 *(9)*. When first using a strain of *E. faecalis*, it may be useful to set up a series of small-scale overnight cultures with graded percentages of glycine to determine the highest concentration that still allows for adequate growth (*see* Note 7). The concentration of glycine in both media should be equal so as not to change dramatically the osmotic or nutritional environment of the cells between steps of the procedure.

5. Both solid glycine and stock solutions of glycine have been used in SGM17 and SGM17MC media with no difference in transformation frequencies. If a stock solution is used (typically 25% w/v), it should be adjusted to pH 6.8–7.0 and sterilized. For sterilization of stock glycine, vacuum filtration into a sterile container is preferable to autoclaving. Stock solutions of 25% w/v or greater should be kept at 37°C to prevent glycine crystallization.

6. The storage and handling of SGM17MC are a common source of problems. Because the medium lacks antibiotics, it should be handled with extra attention to sterility. The medium is added directly to cold cells following electroporation. Thus, it is typically stored at 4°C. Unlike the rapid and observable growth of contaminants in media stored at room temperature, any contaminants that enter SGM17MC may grow at a slow and initially undetectable rate.

4.3. Cell Growth and Handling

7. For the suggested incubation time (18–24 h) of the primary cultures grown in SGM17, typical optical density values obtained at 560 nm were between 0.3 and 0.7. It has been reported that growth of some *E. faecalis* strains in medium containing 4–6% glycine for periods longer than 12–15 h decreases transformation efficiencies *(9)*. Because the procedure provided in this chapter is based on a higher percentage of glycine in the media, incubation times below 16 h generally do not provide an adequate concentration of cells.

8. Aliquots of cells may be stored at –70°C for up to 6 mo (*see* Note 20). *E. faecalis* cells stored at –70°C have been observed to transform at efficiencies equal to those obtained from cells that were electroporated before freezing *(9,10)*.

9. Repeated freezing and thawing of cells may lead to decreased transformation efficiencies *(16)*.

10. During preparation and following electroporation, cells should be kept on ice as much as possible *(6)*.

11. The growth of cells in the presence of glycine weakens the cell wall. Thus, the cells are fragile. To avoid reduction in viability and a potential decrease

in transformation efficiency, the cells should be handled in a manner that limits the amount of physical stress. When resuspending the cells following centrifugation, avoid the use of a pipet or a vortex apparatus. Allow the pellet to disperse gently in the electroporation buffer. Once the buffer has permeated the pellet, the cells can be resuspended by tapping the centrifuge tube.

12. The cells should incubate at 37°C without aeration for a minimum of 2 h following electroporation. Incubation for at least 2 h before spreading the cells on selective medium is a standard procedure in the electroporation of many gram-positive bacterial species *(9–11,13,15,16)*. The increased growth of colonies associated with the incubation period may result from recovery of injured cells, cell division, or a combination of each *(11)*.

13. The 2-h incubation period following electroporation is primarily for cell recovery. However, this period also allows for the expression of antibiotic resistance genes carried by the introduced plasmid. For this reason, antibiotics are not added to SGM17MC. Addition of antibiotics to the medium may significantly reduce the growth of transformants.

14. Typically, 100 μL of a 1:10 or 1:100 dilution of the transformation mixture are plated on SR agar. The remaining cells may be collected by centrifugation and resuspended in 100 μL of SGM17MC for recovery of transformants when low yields are expected.

15. On an individual SR plate, variation in size among different *E. faecalis* colonies is frequently observed. The variation should not be assumed to be an indicator of contamination by other bacteria until directly tested. The heterogeneity appears to result from different rates of growth among colonies originating from cells in different stages of recovery and metabolic activity following electroporation. If contamination is suspected, crude verification can be efficiently achieved by performing a gram stain with different colonies randomly selected from the plate. When using a gram stain, be sure to include suitable controls of the intended recipient.

16. Depending on the antibiotic and its concentration in the SR plate, there may be growth of satellite colonies at antibiotic levels where such growth is not observed on other media. This may result from the high osmotic strength of SR agar. Satellite growth can be eliminated by increasing the concentration of antibiotic.

17. Electroporated cells may be grown on BHI plates rather than SR plates. However, transformation efficiencies are reduced by several orders of magnitude when BHI plates are substituted for SR plates (unpublished results).

4.4. Electroporation Parameters

18. For *E. faecalis* strains, higher voltages generally produce higher transformation efficiencies *(9,10)*. However, optimal transformation of different

E. faecalis strains may require adjustment of the recommended voltage. The maximum field strength of 12.5 kV/cm produced high transformation efficiencies in strains JH2-2 and UV-202 (10^6 and 10^3 transformants/μg DNA, respectively) *(10)*. We exclusively use a 12.5 kV/cm field strength to transform strain FA2-2. However, Dunny et al. *(9)* recommend a field strength between 8.75 and 10.0 kV/cm, since field strengths above 10.0 kV/cm often resulted in arcing. It should be noted that the electroporation buffer used by Dunny et al. *(9)* contained additional ions that are not included in the buffer described in this chapter. This may have contributed to the arcing observed at higher voltages (*see* Note 20). Using a buffer of equal or similar composition to that outlined in this chapter, high transformation efficiencies (e.g., 10^3–10^7 transformants/μg DNA) for other gram-positive species were observed at a field strength of 12.5 kV/cm *(15,16)*.

19. Transformation has been observed to be efficient with time constants of 4.5–4.7 ms *(9,10)*. The recommended settings of 25 μF and 200 Ω yield time constants within this range. Time constants outside of the range may necessitate adjustments in the resistance of the DNA preparation or the cell aliquot *(6)*. If adjustments in either the settings or preparations are necessary, the following equation may be used as a guideline:

$$\text{Time constant } (\tau) = \text{capacitance } (C) \times \text{resistance } (R) \qquad (16)$$

20. When arcing occurs, it is often the result of moisture on the sides of the electroporation cuvet. Thus, it is essential to dry the cuvet with a Kimwipe prior to electroporation. Another common source of arcing is in the cell preparation used for electroporation. If there are many ions in the preparation, arcing may occur. Thus, it is important to wash and store cells adequately with a nonionic electroporation buffer. Arcing may also be the result of the accumulation of deposits on the electrodes found along the sides of the electroporation chamber. The deposits are often easily removed by gentle cleaning with a small wire brush. Occasionally, cracks develop in the solder connecting the electrodes to the chamber. In such situations, the electrodes should be repaired or replaced, because arcing will tend to occur with each electric pulse. Additionally, when using cells stored for more than 6 mo, we have observed an increased incidence of arcing on application of the electrical pulse.

References

1. Schaberg, D., Culver, D. H., and Gaynes, R. P. (1991) Major trends in the microbial etiology of nosocomial infection. *Am. J. Med.* **91(Suppl. 3B),** 72S–75S.
2. Moellering, R. C., Jr. (1992) Emergence of *Enterococcus* as a significant pathogen. *Clin. Infect. Dis.* **14,** 1173–1178.

3. Leclercq, R., Dutka-Malen, S., Brissom-Noël, A., Molinas, C., Derlot, E., Arthur, M., Duval, J., and Courvalin, P. (1992) Resistance of *Enterococci* to aminoglycosides and glycopeptides. *Clin. Infect. Dis.* **15,** 495–501.
4. Tsien, H. C., Shockman, G. D., and Higgins, M. L. (1978) Structural arrangement of polymers within the wall of *Streptococcus faecalis. J. Bacteriol.* **133,** 372–386.
5. Beveridge, T. J. (1981) Ultrastructure, chemistry, and function of the bacterial wall. *Int. Rev. Cytol.* **72,** 229–317.
6. Potter, H. (1993) Application of electroporation in recombinant DNA technology. *Methods Enzymol.* **217,** 461–478.
7. Powell, I. B., Achen, M. G., Hillier, A. J., and Davidson, B. E. (1988) A simple and rapid method for genetic transformation of lactic streptococci by electroporation. *Appl. Environ. Microbiol.* **54,** 655–660.
8. Luchansky, J. B., Muriana, P. M., and Klaenhammer, T. R. (1988) Application of electroporation for transfer of plasmid DNA to *Lactobacillus, Lactococcus, Leuconostoc, Listeria, Pediococcus, Bacillus, Staphylococcus, Enterococcus,* and *Propionibacterium. Mol. Microbiol.* **2,** 637–646.
9. Dunny, G. M., Lee, L. N., and LeBlanc, D. J. (1991) Improved electroporation and cloning vector system for Gram-positive bacteria. *Appl. Environ. Microbiol.* **57,** 1194–1201.
10. Cruz-Rodz, A. L. and Gilmore, M. S. (1990) High efficiency introduction of plasmid DNA into glycine treated *Enterococcus faecalis* by electroporation. *Mol. Gen. Genet.* **224,** 152–154.
11. McIntyre, D. A. and Harlander, S. K. (1989) Genetic transformation of intact *Lactococcus lactis* subsp. *lactis* by high-voltage electroporation. *Appl. Environ. Microbiol.* **55,** 604–610.
12. van der Lelie, D. J., van der Vossen, J. M. B. M., and Venema, G. (1988) Effect of plasmid incompatibility on DNA transfer to *Streptococcus cremoris. Appl. Environ. Microbiol.* **54,** 865–871.
13. Bhowmik, T. and Steele, J. L. (1993) Development of an electroporation procedure for gene disruption in *Lactobacillus helveticus* CNRZ 32. *J. Gen. Microbiol.* **139,** 1433–1439.
14. Park, S. F. and Stewart, G. S. A. B. (1990) High-efficiency transformation of *Listeria monocytogenes* by electroporation of penicillin-treated cells. *Gene* **94,** 129–132.
15. Holo, H. and Nes, I. F. (1989) High-frequency transformation, by electroporation, of *Lactococcus lactis* subsp. *cremoris* grown with glycine in osmotically stabilized media. *Appl. Environ. Microbiol.* **55,** 3119–3123.
16. Bringel, F. and Hubert, J.-C. (1990) Optimized transformation by electroporation of *Lactobacillus plantarum* strains with plasmid vectors. *Appl. Microbiol. Biotechnol.* **33,** 664–670.
17. Jacob, A. E. and Hobbs, S. J. (1974) Conjugal transfer of plasmid-borne multiple antibiotic resistance in *Streptococcus faecalis* var. *zymogenes. J. Bacteriol.* **117,** 360–372.
18. Yagi, Y. and Clewell, D. B. (1980) Recombinant-deficient mutant of *Streptococcus faecalis. J. Bacteriol.* **143,** 966–970.
19. Clewell, D. B., Tomich, P. K., Gawron-Burke, G., Franke, A. E., Yagi, Y., and An, F. Y. (1982) Mapping of *Streptococcus faecalis* plasmids pAD1 and pAD2 and studies relating to transposition of Tn917. *J. Bacteriol.* **152,** 1220–1230.

CHAPTER 23

Introduction of Recombinant DNA into *Clostridium* spp.

Mary K. Phillips-Jones

1. Introduction

The importance of the Clostridia both in terms of their pathogenicity toward humans and their potential role in biotechnological processes has prompted many studies of the genetic manipulation of these organisms. Obviously, a prerequisite for such studies is a means of introducing recombinant DNA into individual viable cells, and the first reports of successful plasmid transformation involved use of protoplasts or L-forms *(1–4)*, or conjugal cointegrate transfer techniques *(5,6)*. However, these methods proved tedious or time-consuming to perform, and sometimes gave inconsistent results. Some workers suggested that it was the presence of DNases produced by many species of *Clostridium* that contributed to the low or inconsistent transformation efficiencies obtained by these methods *(7,8)*. Even so, protoplast transformation is still used routinely and successfully for the introduction of shuttle plasmids into *Clostridium acetobutylicum* (e.g., *9*).

The most widely adopted method for introducing plasmid DNA into *Clostridium* species is electroporation, and a number of different protocols have been reported. In these reports, the species studied were *Clostridium perfringens (10–17)*, *Clostridium botulinum (18)*, or *C. acetobutylicum (19–24)*. Some aspects of the electroporation methods devised are common to all three species, so it should be possible to derive similar suitable methods for other species of *Clostridium*.

From: *Methods in Molecular Biology, Vol. 47: Electroporation Protocols for Microorganisms*
Edited by: J. A. Nickoloff Humana Press Inc., Totowa, NJ

The choice of transformation strain may be of importance. Some workers have suggested that only 10% of cultured strains are electroporatable (R.Titball, personal communication), and it is clear that some workers prefer to use specific transformation strains. In the case of *C. acetobutylicum*, it has again been suggested that production of DNases or specific restriction endonucleases contributes to lower transformation efficiencies *(20,25)*, although others report no such problems using the same strains *(24)*. For *C. perfringens*, it has been suggested that nucleases do not affect efficiency of plasmid transformation. Even so, different types (A–C) and strains of this species vary in their ability to be transformed, and in order to optimize efficiencies, some adjustments to electroporation protocols are often beneficial *(14)*.

A wide variety of plasmid vectors have been developed for introducing recombinant DNA into *Clostridium*. For stable establishment within Clostridial cells, the most important features of plasmid vectors are (1) an origin of replication (or replicon) that is functional in the *Clostridium* species concerned and (2) an antibiotic resistance gene that is expressed and functional, for selection of plasmid-containing cells. In addition, in order to aid vector construction and plasmid isolation, most vectors (shuttle vectors) also carry replicons and selection markers for plasmid selection in *Escherichia coli*. Thus, most vectors previously used in electroporation-mediated transformation are based on the insertion of cryptic Clostridial plasmids (possessing Clostridial replicons) and Clostridial or gram-positive antibiotic resistance genes, into well-characterized *E. coli* plasmid vectors. Alternatively, some constructs, e.g., pFNK1, are based on *Bacillus subtilis* plasmids (rather than *E. coli*). Such plasmids are termed *Bacillus subtilis-C. acetobutylicum* shuttle vectors and typically possess very little DNA derived from *E. coli (20)*. In addition, previous studies have made use of pAMβ1 from *Streptococcus faecalis*, which has been shown to replicate efficiently in *C. acetobutylicum* and *C. perfringens (5,10)*. The electroporation protocol described below is based on the methods described by Phillips-Jones *(13,16)*.

2. Materials

2.1. Plasmid Vectors and Strains

Table 1 lists many of the plasmid vectors that have been successfully introduced into various Clostridial species and strains using electroporation (*see* Note 1). Earlier vectors designed for protoplast or L-form trans-

Table 1

Plasmid Vectors and Transformation Strains Used in Electroporation of *Clostridium* spp.

Plasmid	Plasmid characteristics	Transformation strain	Reference
C. perfringens			
pAMβ1	From *S. faecalis* OG1-X	ATCC 3624A	*10,14,28*
pHR106	*C. perfringens* Cm^R	ATCC 3624A	*10,14,29*
pHR106		Strain 13	*11,29*
pAK201	*C. perfringens* Cm^R	ATCC 3624A	*12,14*
pSB92A2	*C. perfringens* Cm^R	P90.2.2	*13*
pPS14	pSB92A2 plus *luxAB*	P90.2.2	*16*
pJIR418	MCS and *lacZ'* of pUC18 *C. perfringens* Cm^R, Em^R	Strain 13 derivatives	*15*
pJIR750	As pJIR418; Cm^R only	Strain 13 derivatives	*17*
pJIR751	As pJIR418; Em^R only	Strain 13 derivatives	*17*
C. acetobutylicum			
pMTL500E	Em^R gene of pAMβ1	ATCC 8052	*19*
pSYL2	Em^R gene of pAMβ1 *C. butyricum* pCBU2 replicon	ATCC 824 NCIMB 8052	*21,30*
pSYL7	*C. perfringens* pJU122 replicon	NCIMB 8052	*21*
pSYL9	pAMβ1 replicon	NCIMB 8052	*21*
pSYL14	*B. subtilis* pI1M13 replicon	NCIMB 8052	*21*
pFNK1	*B. subtilis/C. acetobutylicum* shuttle vector; *B. subtilis* replicon	ATCC 824	*20*
pAK301	pMTL500E plus *engB*	ATCC 824	*24*
pCAK1	Phagemid from which single-stranded DNA can be prepared	ATCC 824	*23*
C. botulinum			
pGK12	Em^R Cm^R, replicates in *E. coli, B. subtilis,* and *S. lactis*	Type A	*18,31*

formation, but that may also be suitable for electroporation-induced transformation, have appeared in earlier reviews and are not included here *(26–29).*

2.2. Culture Media

1. Cooked meat medium is supplied by Oxoid, Unipath Ltd., Basingstoke, Hampshire, UK.
2. TGY broth: 30 g Trypticase Peptone (BBL; Becton Dickinson Microbial Systems, Cockeysville, MD), 20 g glucose, 1 g L-cysteine, 10 g yeast

extract, and distilled water to 1 L. Adjust pH to 7.8 (pH falls to approx 7.1 during autoclaving). Autoclave at 121°C for 15 min.

3. BHI agar: 12.5 g calf brain infusion solids, 5.0 g beef heart infusion solids, 10.0 g protease peptone, 5.0 g NaCl, 2.0 g dextrose, 2.5 g disodium phosphate (anhydrous), 15 g agar, and distilled water to 1 L. Adjust pH to 7.4. Autoclave at 121°C for 15 min. Alternatively, use 37 g BHI powder (Oxoid) with 15 g agar and 1 L distilled water.

4. For selection of pSB92A2-containing cells, which acquire resistance to chloramphenicol owing to the presence of the *C. perfringens catP* gene, add chloramphenicol to a final concentration of 5 µg/mL to cooled (50°C) sterile BHI agar prior to pouring plates. Prepare a stock solution of 10 mg/mL in ethanol, filter-sterilize using a 0.22-µm membrane, and store at −20°C for up to 3 mo. **Caution: Chloramphenicol is toxic; wear gloves and avoid exposure to the powdered form.**

2.3. Solutions and Reagents for Electrotransformation

1. Electroporation buffer: 15% (v/v) glycerol. Autoclave at 121°C for 15 min. Store anaerobically at 4°C.

2. Alternative electroporation buffer (SMP): 92.4 g sucrose, 0.2 g MgCl$_2$ · 6H$_2$O, 70 mL 0.1M sodium phosphate, pH 7.4, and distilled water to 1 L. The sodium phosphate solution is sterilized separately by autoclaving at 121°C for 15 min and added aseptically to cooled, autoclaved (121°C/15 min) sucrose-MgCl$_2$ solution. Store anaerobically at 4°C (*see* Note 2).

3. Plasmid DNA: CsCl-purified plasmid DNA at a concentration of approx 1 mg/mL in TE buffer (10 mM Tris-HCl, 1 mM EDTA, pH 8.0). Store at −20°C.

3. Methods

3.1. Preparation of Cells for Electroporation

1. Prepare stock cultures of *C. perfringens* by inoculating cooked meat medium in metal screw-capped universal bottles and incubating anaerobically at 37°C for 2–4 d or until good growth is observed. Screw down the caps until tight, and store anaerobically at room temperature (*see* Note 3).

2. Inoculate 10 mL of prewarmed TGY broth with 0.1 mL of stock culture, and incubate anaerobically overnight at 37°C.

3. Inoculate 100 mL of fresh TGY broth (which has been preincubated anaerobically at 37°C), with 3 mL of the fresh overnight culture (*see* Note 4). Incubate anaerobically at 37°C with gentle shaking to ensure thorough suspension of the cells, until cells have just entered the stationary phase of growth (approx 6 h for *C. perfringens* P90.2.2) (*see* Notes 1, 5, and 6).

4. Transfer the culture to sterile, precooled centrifuge tubes (*see* Note 7). Harvest the cells by centrifugation at 4°C for 20 min at 3000*g*, ensuring that the rotor is precooled prior to use. Pour off the supernatant.
5. Resuspend the cells in 10 mL of cold electroporation buffer (*see* Note 2). Wash cells gently using a pipet tip, and avoid vortexing. Harvest the cells once more by centrifuging for 20 min at 3000*g*. Resuspend the cell pellets gently in 5 mL of cooled electroporation buffer, and place on ice for 20 min.

3.2. Electroporation Procedure

1. Pipet 0.8-mL samples of the cells prepared above into 0.4-cm electroporation cuvets precooled on ice. Incubate on ice for 10 min.
2. Add approx 1–10 µg plasmid DNA/cuvet, invert to mix, and incubate for 10 min on ice (*see* Notes 6 and 8).
3. Immediately deliver an electric pulse of 2.5 kV (6.25 kV/cm), 25 µF (*see* Notes 9–11). Incubate the cuvet on ice for 10 min.
4. Add 4.8 mL of TGY broth (preincubated anaerobically at room temperature) (*see* Note 12). Incubate anaerobically for 3 h at 37°C to allow gene expression (*see* Note 13).

3.3. Selection of Transformants

1. Plate 0.1-mL cell samples on BHI-selective plates that have been pre-incubated anaerobically at 37°C for 20 h. In the case of pSB92A2 and other *cat*-possessing plasmids, use BHI containing 5 µg/mL chloramphenicol. Alternatively, harvest 1 mL of cells by microcentrifugation (12,000*g*) for 2 min, remove the supernatant TGY broth, and resuspend in 0.1 mL fresh TGY. Plate the 0.1 mL on BHI antibiotic selective plates as usual (*see* Note 12).
2. Incubate inoculated plates anaerobically for 1–3 d at 37°C before counting transformants.

4. Notes

1. Table 1 lists many of the *C. perfringens, C. acetobutylicum,* and *C. botulinum* strains that have been successfully electrotransformed to date. Most electroporation protocols use whole cells, although one study pretreated cells of *C. perfringens* strain 13 with lysostaphin, a peptidase that weakens the peptidoglycan layer of the cell wall, prior to electroporation *(11)*. The protocols differ in the phase of growth at which cells are harvested prior to electroporation. Some workers have found that exponential phase cells give optimal transformation efficiencies *(10,18,19,21);* others have used late exponential cells *(14,23)* or early stationary cells *(12,13,16,20)*. The differences in transformation efficiencies may reflect strain differences. In the case of *C. acetobutylicum* DSM792, it has been shown that production of DNases is

optimal in stationary phase cells, highlighting the possibility of reduced transformation efficiencies during this phase of growth for this strain *(25)*. It is not known whether the same is generally true for all strains of this species.

2. Choice of electroporation buffer may be important. In a comparison of two electroporation buffers (15% glycerol and SMP), for electrotransforming *C. perfringens* P90.2.2, use of 15% glycerol consistently resulted in higher numbers of transformants than SMP *(13)*. However, SMP buffer is preferred by some workers for different *Clostridium* species and strains *(10,11,13,18,19)* and is described in Section 2.3. Alternative electroporation buffers are SP (270 mM sucrose, 5 mM NaH$_2$PO$_4$, pH 7.4) *(20,30)* and 10% (w/v) polyethylene glycol (PEG) 8000 in distilled water *(18,23,24)*.

3. Different species of Clostridia vary with respect to their oxygen tolerance. *C. perfringens* is a relatively oxygen-tolerant species. For other species, it may be necessary to maintain anaerobic conditions for culture storage. Also, it may be necessary to perform all electroporation steps, apart from the pulse delivery, in an anaerobic cabinet.

4. A high level of inoculation appears to be required; 3% appears to be adequate.

5. Assessment of growth phase is most conveniently performed by taking 1-mL samples at regular intervals and measuring the culture absorbance at 600 nm.

6. The methods described above were designed for electroporation of *C. perfringens* P90.2.2. A notable feature of Clostridial electroporation is the variety of conditions required for different species and strains. When using a strain for which no information is available regarding electroporation conditions, it is advisable to try a number of different electroporation buffers and DNA concentrations, as well as to examine the effects of growth phase when cells are harvested, growth medium and pH, and composition of selection media. A useful preliminary experiment is to determine the proportion of cells killed during the electroporation procedure; aim for a kill rate of 90–99%. Two thorough studies of factors affecting the efficiency of *C. perfringens* electroporation-mediated transformation were reported by Allen and Blaschek *(14)* and Scott and Rood *(11)*.

7. Once the cell culture has been harvested, work through the procedure as quickly as possible. Keep the cells cool (4°C) at all times where low temperatures are indicated in the protocol.

8. The effect of plasmid DNA concentration on electrotransformation efficiencies can be marked *(14)*. High concentrations can result in very low efficiencies. The DNA concentrations used previously range from approx 16 ng to 10 µg/mL *(13,20,23)*. The most commonly used concentration is 1–5 µg/mL *(10–12,18,21,23)*.

9. The time constant should be approx 8.3 ms.
10. Although the Gene Pulser manufacturers (Bio-Rad Laboratories, Richmond, CA) strongly recommend the use of the pulse-controller unit at field strengths >6.0 kV/cm (in order to avoid serious damage to the Gene Pulser apparatus should any arcing occur), optimal transformation was obtained when the pulse controller was omitted. In order to minimize the chances of arcing, care should be taken to ensure that samples contain low ionic strength, i.e., ensure minimal carryover of culture medium and use high resistance electroporation buffers, such as 15% glycerol. Alternatively, the pulse controller can be set to the ∞ resistance setting, which means that an open circuit exists in parallel with the sample, but again, the chances of arcing are increased.
11. In the majority of studies, the voltage used is the maximum 2.5 kV, and the field strength is 6.25 kV/cm. Other studies have used 2.0 kV (e.g., *20*).
12. One report has suggested that a reduction in pH of the expression and selective media to 6.4 is beneficial for optimum transformation efficiency *(14)*.
13. Times required for gene expression and cell recovery following electroporation can vary, even using the same strain, plasmid, and conditions. The 3-h gene expression time indicated in the protocol is a minimum time, and can be up to 5 or 6 h.

References

1. Heefner, D. L., Squires, C. H., Evans, R. J., Kopp, B. J., and Yarus, M. J. (1984) Transformation of *Clostridium perfringens. J. Bacteriol.* **159,** 460–464.
2. Soutschek-Bauer, E., Hartl, L., and Staudenbauer, W. L. (1985) Transformation of *Clostridium thermohydrosulfuricum* DSM 568 with plasmid DNA. *Biotech. Lett.* **7,** 705–710.
3. Mahony, D. E., Mader, J. A., and Dubel, J. R. (1988) Transformation of *Clostridium perfringens* L-forms with shuttle plasmid DNA. *Appl. Environ. Microbiol.* **54,** 264–267.
4. Reysset, G., Hubert, J., Podvin, L., and Sebald, M. (1988) Transfection and transformation of *Clostridium acetobutylicum* strain N1-4081 protoplasts. *Biotechnol. Tech.* **2,** 199–204.
5. Oultram, J. D., Davies, A., and Young, M. (1987) Conjugal transfer of a small plasmid from *Bacillus subtilis* to *Clostridium acetobutylicum* by cointegrate formation with plasmid pAMβ1. *FEMS Microbiol. Lett.* **42,** 113–119.
6. Minton, N. P., Brehm, J. K., Oultram, J. D., Swinfield, T. J., and Thompson, D. E. (1988) Construction of plasmid vector systems for gene transfer in *Clostridium acetobutylicum,* in *Anaerobes Today* (Hardie, J. M. and Borriello, S. P., eds.), Wiley, Chichester, pp. 125–134.
7. Blaschek, H. P. and Klacik, M. A. (1984) Role of DNase in recovery of plasmid DNA from *Clostridium perfringens. Appl. Environ. Microbiol.* **48,** 178–181.

8. Lin, Y.-L. and Blaschek, H. P. (1984) Transformation of heat-treated *Clostridium acetobutylicum* protoplasts with pUB110 plasmid DNA. *Appl. Environ. Microbiol.* **48,** 737–742.

9. Truffaut, N., Hubert, J., and Reysset, G. (1989) Construction of shuttle vectors useful for transforming *Clostridium acetobutylicum. FEMS Microbiol. Lett.* **58,** 15–20.

10. Allen, S. P. and Blaschek, H. P. (1988) Electroporation-induced transformation of intact cells of *Clostridium perfringens. Appl. Environ. Microbiol.* **54,** 2322–2324.

11. Scott, P. T. and Rood, J. I. (1989) Electroporation-mediated transformation of lysostaphin-treated *Clostridium perfringens. Gene* **82,** 327–333.

12. Kim, A. Y. and Blaschek, H. P. (1989) Construction of an *Escherichia coli–Clostridium perfringens* shuttle vector and plasmid transformation of *Clostridium perfringens. Appl. Environ. Microbiol.* **55,** 360–365.

13. Phillips-Jones, M. K. (1990) Plasmid transformation of *Clostridium perfringens* by electroporation methods. *FEMS Microbiol. Lett.* **66,** 221–226.

14. Allen, S. P. and Blaschek, H. P. (1990) Factors involved in the electroporation-induced transformation of *Clostridium perfringens. FEMS Microbiol. Lett.* **70,** 217–220.

15. Sloan, J., Warner, T. A., Scott, P. T., Bannam, T. L., Berryman, D. I., and Rood, J. I. (1992) Construction of a sequenced *Clostridium perfringens–Escherichia coli* shuttle plasmid. *Plasmid* **27,** 207–219.

16. Phillips-Jones, M. K. (1993) Bioluminescence *(lux)* expression in the anaerobe *Clostridium perfringens. FEMS Microbiol. Lett.* **106,** 265–270.

17. Bannam, T. L. and Rood, J. I. (1993) *Clostridium perfringens–Escherichia coli* shuttle vectors that carry single antibiotic resistance determinants. *Plasmid* **29,** 233–235.

18. Zhou, Y. and Johnson, E. A. (1993) Genetic transformation of *Clostridium botulinum* Hall A by electroporation. *Biotech. Lett.* **15,** 121–126.

19. Oultram, J. D., Loughlin, M., Swinfield, T.-J., Brehm, J. K., Thompson, D. E., and Minton, N. P. (1988) Introduction of plasmids into whole cells of *Clostridium acetobutylicum* by electroporation. *FEMS Microbiol. Lett.* **56,** 83–88.

20. Mermelstein, L. D., Welker, N. E., Bennett, G. N., and Papoutsakis, E. T. (1992) Expression of cloned homologous fermentative genes in *Clostridium acetobutylicum* ATCC 824. *Biotechnology* **10,** 190–195.

21. Lee, S. Y., Bennett, G. N., and Papoutsakis, E. T. (1992) Construction of *Escherichia coli–Clostridium acetobutylicum* shuttle vectors and transformation of *Clostridium acetobutylicum* strains. *Biotech. Lett.* **14,** 427–432.

22. Mermelstein, L. D. and Papoutsakis, E. T. (1993) *In vivo* methylation in *Escherichia coli* by the *Bacillus subtilis* phage 3T I methyltransferase to protect plasmids from restriction upon transformation of *Clostridium acetobutylicum* ATCC 824. *Appl. Environ. Microbiol.* **59,** 1077–1081.

23. Kim, A. Y. and Blaschek, H. P. (1993) Construction and characterization of a phage-plasmid hybrid (Phagemid), pCAK1, containing the replicative form of viruslike particle CAK1 isolated from *Clostridium acetobutylicum* NCIB 6444. *J. Bacteriol.* **175,** 3838–3843.

24. Kim, A. Y., Attwood, G. T., Holt, S. M., White, B. A., and Blaschek, H. P. (1994) Heterologous expression of endo-β-1,4-D-glucanase from *Clostridium cel-*

lulovorans in *Clostridium acetobutylicum* ATCC 824 following transformation of the *eng*B gene. *Appl. Environ. Microbiol.* **60**, 337–340.

25. Burchhardt, G. and Durre, P. (1990) Isolation and characterization of DNase-deficient mutants of *Clostridium acetobutylicum. Current Microbiol.* **21**, 307–311.
26. Rood, J. I. and Cole, S. T. (1991) Molecular genetics and pathogenesis of *Clostridium perfringens. Microbiol. Rev.* **55**, 621–648.
27. Young, M., Minton, N. P., and Staudenbauer, W. L. (1989) Recent advances in the genetics of the clostridia. *FEMS Microbiol. Rev.* **63**, 301–326.
28. Young, M., Staudenbauer, W. L., and Minton, N. P. (1989) Genetics of *Clostridium.* in *Clostridia,* Biotechnology Handbooks, vol. 3 (Minton, N. P. and Clarke, D. J., eds.), Plenum, London, pp. 63–103.
29. Roberts, I., Holmes, W. M., and Hylemon, P. B. (1988) Development of a new shuttle plasmid system for *Escherichia coli* and *Clostridium perfringens. Appl. Environ. Microbiol.* **54**, 268–270.
30. Lee, S. Y., Mermelstein, L. D., Bennett, G. N., and Papoutsakis, E. T. (1992) Vector construction, transformation and gene amplification in *Clostridium acetobutylicum* ATCC 824. *Annals NY Acad. Sci.* **665**, 39–51.
31. Kok, J., van der Vossen, J. M. B. M., and Venema, G. (1984) Construction of plasmid cloning vectors for lactic streptococci which also replicate in *Bacillus subtilis* and *Escherichia coli. Appl. Environ. Microbiol.* **48**, 726–731.

CHAPTER 24

Electroporation of Mycobacteria

T. Parish and N. G. Stoker

1. Introduction

The genus *Mycobacterium* is usually divided into the fast-growing and slow-growing species, and among the latter group are several important pathogens of humans. *Mycobacterium tuberculosis,* the causative agent of tuberculosis, is responsible for the greatest number of deaths caused by bacterial infections worldwide. In addition, leprosy (caused by *Mycobacterium leprae*) afflicts millions of people, causing deformity and death. In recent years, the incidence of tuberculosis has risen in the Western world, and infection of immunocompromised patients by members of the *Mycobacterium avium* complex has become a serious problem. Many other mycobacterial species are able to cause infections, not only in humans, but also in animals (e.g., *Mycobacterium bovis* in cows and badgers) and birds (e.g., *Mycobacterium peregrinum, Mycobacterium avium*). The study of the pathogenicity of these organisms is therefore of great importance.

The study of mycobacterial genomes has exploded during the last 10 years. Initially, no systems were available for the direct manipulation of mycobacterial genes in mycobacteria, so *Escherichia coli* was used as the primary cloning host. Several genomic libraries were created *(1–5)* in *E. coli.* Although these proved useful for the identification of many protein antigens *(6),* the use of *E. coli* as a cloning host has several limitations. It is now known that many mycobacterial promoters do not function at all in *E. coli.* Therefore, it is difficult to study the expression and control of mycobacterial genes in such a host. In addition, certain

From: *Methods in Molecular Biology, Vol. 47: Electroporation Protocols for Microorganisms*
Edited by: J. A. Nickoloff Humana Press Inc., Totowa, NJ

posttranslational modifications of proteins do not take place in *E. coli,* and therefore, the antigenicity and properties of proteins expressed in *E. coli* may differ *(7–9).*

The development of electroporation techniques for use with both fast- and slow-growing mycobacterial species has provided a means for the genetic manipulation of mycobacteria directly. Many types of DNA have been introduced into various mycobacterial species using this technology, including plasmids, cosmids, integrating vectors, and transposon delivery vectors. The use of high-efficiency electroporation in conjunction with these vector systems has facilitated many studies, for example, investigations into mycobacterial gene expression *(10),* the production of recombinant BCG vaccines *(11),* and the production of transposon mutant libraries *(12).*

The fast-growing *Mycobacterium smegmatis* is now commonly used as a cloning host for the study of genes from pathogenic mycobacteria. Although *M. smegmatis* mc^{2}6, taken from ATCC607, showed poor transformation efficiencies by electroporation, a mutant strain designated mc^{2}155 was isolated that showed transformation efficiencies of 10^4–10^5/μg of DNA *(13).* Using plasmid and phage DNAs, evidence has also been provided that *M. smegmatis* mc^{2}155 does not possess a restriction and modification system *(13).* This is an obvious advantage, since it means that exogenous DNAs will not be degraded on entry into the cell.

Although *M. smegmatis* has been the most widely used species, many other species of mycobacteria have been transformed using electroporation, including *M. bovis* BCG *(14–17), Mycobacterium vaccae (18–20), Mycobacterium phlei (21), Mycobacterium w (19), Mycobacterium fortuitum (17,21), Mycobacterium aurum (22,23),* and *Mycobacterium parafortuitum (24).* The efficiencies for electroporation vary among different mycobacterial species (Table 1). Other factors that can influence the efficiency are the DNA used and the selectable marker carried on the transforming DNA. The differences observed in efficiencies found for different markers may reflect the stability of the gene products in the mycobacterial cell.

Several vector systems have been developed for use in mycobacteria, including plasmids, integrating vectors, and transposon delivery systems. Table 2 shows some of the vector systems that have been successfully used in mycobacteria. Vectors for use in mycobacteria must carry suitable selectable markers; the choice of selection markers is critical and is dependent on the particular species of mycobacteria being used. Table 3

Table 1
Reported Electroporation Efficiencies for Mycobacterial Species

Species	Efficiency[a]	Selection	Reference
Fast growers			
M. aurum	100	Kanamycin	*22*
M. parafortuitum	300	Kanamycin	*24*
	30	Streptomycin	*24*
M. smegmatis	10	Kanamycin	*14,25*
M. smegmatis mc^2155	10^5–10^6	Kanamycin	*13,18,40,41*
M. smegmatis 1–2c	2×10^2	Kanamycin	*19*
	5×10^3	Hygromycin	*19*
M. vaccae	10^3–10^5	Hygromycin	*19*
Slow growers			
M. bovis BCG	10^3–10^5	Kanamycin	*18,25,27,28*
M. tuberculosis	10^4	Kanamycin	*40*
M. w	10^3–10^5	Hygromycin	*19*

[a]Number of transformants/µg DNA.

Table 2
Vectors Used in Mycobacteria

Vector	Type	Based on[a]	Selection	Reference
pJRD215	Cosmid	RSF1010	Kanamycin/streptomycin	*22*
pMSC1	Cosmid	pAL5000	Kanamycin	*42*
pRR3	Plasmid	pAL5000	Kanamycin	*28*
pYT937	Plasmid	pMSC262	Kanamycin	*21*
pEP3	Plasmid	pNG2	Hygromycin	*16*
pBL525	Plasmid	D29 phage	Kanamycin	*41*
pUS903	Integrating vector	IS900	Kanamycin	*18*
pMV361	Integrating vector	L5 phage	Kanamycin	*11*

[a]Extrachromosomal plasmid replicon, mycobacteriophage integration system, or insertion element.

shows selection markers that have been successfully used for various species. For *M. smegmatis,* kanamycin resistance has been widely used *(25),* since the commonly used mc^2155 strain is sensitive to kanamycin at low concentrations and the rate of spontaneous mutation to kanamycin resistance for mc^2155 is 10^{-7}–10^{-9} *(25,26).* Kanamycin resistance has also been used as a selectable marker in *M. aurum, M. parafortuitum, M. tuberculosis,* and *M. bovis* BCG *(15,22,24,27,28).* An alternative to kanamycin resistance is hygromycin resistance, which has been used in

Table 3
The Use of Antibiotic Resistance Genes as Selectable Markers in Mycobacteria

Species	Antibiotic selection	Reference
Fast growers		
M. aurum	Streptomycin	22
	Kanamycin	22
M. fortuitum	Kanamycin	21
M. parafortuitum	Kanamycin	24
	Streptomycin	24
M. phlei	Kanamycin	21
M. smegmatis	Kanamycin	14,25,28,40,41
	Hygromycin	16,19
	Chloramphenicol[a]	13
	Tetracycline	29
	Streptomycin	22
	Sulfonamide	30
	Gentamicin	30
M. vaccae	Hygromycin	19
M. bovis BCG	Kanamycin	14,15,27,28,40
	Hygromycin	16
	Chloramphenicol[b]	21
M. tuberculosis	Kanamycin	27,28
M. w	Hygromycin	19

[a]Used for screening, not for direct selection.
[b]In conjunction with kanamycin.

both *M. smegmatis* and BCG *(16)*. Vectors carrying kanamycin resistance have been unable to transform at least two species that are potential vaccine candidates, *M. w* and *M. vaccae (19)*. Hygromycin resistance has been successfully used as a selectable marker in these species *(19)*.

It has been reported that tetracycline can be used with *M. smegmatis (29)*. However, this antibiotic is not suitable for slow-growing mycobacteria, since tetracycline is unstable over the time required for culture (3–6 wk). Ampicillin resistance is not suitable for use in mycobacteria, since they are naturally resistant to β-lactams. Chloramphenicol resistance cannot be used for direct selection owing to the high rate of spontaneous mutations (10^{-4}–10^{-5}) *(13)*, although it has been used in conjunction with other antibiotic resistance genes *(13,21)*. Genes conferring resistance to sulfonamides *(30)*, gentamicin *(30)*, or streptomycin *(24)* may also be used for selection, although it may not be advisable to introduce such antibiotic resistances into pathogenic mycobacteria because of biohazard considerations, nor would they be appropriate for use in live recombinant vaccines owing to the possibility of transfer to other pathogenic bacteria. (Streptomycin, sulfonamides, and kanamycin may all be used in treatment of infections caused by pathogenic mycobacteria.) Therefore, nonantibiotic resistance markers are better suited, such as the L5 phage superimmunity gene, which confers resistance to mycobacteriophage L5 infection *(31)*.

2. Materials
2.1. Media and Solutions for Growth of Mycobacteria and E. coli

1. Lowenstein Jensen (LJ) slopes (Difco, West Molsey, Surrey, UK): Store at 4°C.
2. Middlebrook 7H9 broth (Difco): Dissolve in deionized water at 4.7 g/900 mL and autoclave.
3. Middlebrook ADC enrichment (Difco), containing bovine albumin fraction V, dextrose, catalase and NaCl: Store at 4°C *(see* Note 3).
4. Middlebrook OADC enrichment (as ADC with oleic acid; *see* Note 3) (Difco): Store at 4°C.
5. Tween 80 (Sigma, Poole, Dorset, UK): Prepare as a 20% v/v stock, filter-sterilize through a 0.2-µm membrane, and store at 4°C *(see* Note 5).
6. For fast-growing species, such as *M. smegmatis,* media should be prepared using 7H9 broth supplemented with 10% v/v ADC and 0.05% v/v Tween 80.
7. For slow-growing species, such as *M. tuberculosis* or *M. bovis* BCG, media should be prepared using 7H9 broth supplemented with 10% v/v OADC and 0.05% v/v Tween 80.
8. 2*M* glycine (Analar-grade, Sigma): Autoclave *(see* Note 7).
9. LB (Difco) for culture of *E. coli* should be prepared at 25 g/L and autoclaved. Agar for plates should be added at 15 g/L prior to autoclaving.

2.2. Solutions and Equipment for Preparation of Electrocompetent Cells

1. 10% v/v Glycerol: Sterilize by autoclaving.
2. Electroporation cuvets: 0.2-cm gap electrodes *(see* Note 12).
3. Electroporation apparatus with pulse controller *(see* Notes 11 and 12).
4. DNA in solution *(see* Notes 10 and 13): This should be free from salts, enzymes, and other substances. To purify DNA, it can be ethanol-precipitated and thoroughly washed with 70% ethanol (this will also remove excess salts). The concentration of DNA should be approx 0.2–1 mg/mL.
5. Lemco broth: 10 g/L peptone, 5 g/L Lemco powder, 5 g/L NaCl; autoclave and supplement with 0.05% v/v final concentration Tween 80. For plates, Tween 80 should be omitted, and agar should be added to 1.5% final concentration prior to autoclaving.

Table 4
Antibiotic Selection for Mycobacteria

Antibiotic	Stock solution	Working concentration
Chloramphenicol	34 mg/mL in ethanol	40 µg/mL
Gentamicin	50 mg/mL in water	20 µg/mL
Hygromycin	50 mg/mL in PBS[a]	100 µg/mL[b]
Kanamycin	50 mg/mL in water	10–50 µg/mL
Streptomycin	20 mg/mL in water	30 µg/mL

[a]Phosphate-buffered saline.
[b]200 µg/mL for *E. coli.*

2.3. Media and Reagents for Selection of Transformants

1. Middlebrook 7H10 agar: 19 g agar base/900 mL; autoclave.
2. Middlebrook ADC or OADC enrichment (*see* Section 2.1., and Note 3).
3. Cycloheximide (Sigma): Stock solution of 20 mg/mL in water, store at –20°C (*see* Note 16).
4. For plates, Middlebrook 7H10 agar should be supplemented with 10% v/v ADC enrichment (fast-growing species) or 10% v/v OADC enrichment (slow-growing species) plus selection antibiotic. Plates for slow-growing species must be poured to a depth of 6–8 mm, i.e., 90-mm diameter plates with 40 mL agar, to prevent drying out during the long incubation period required for growth. Cycloheximide should be added to a final concentration of 100 µg/mL (*see* Note 16).
5. Kanamycin sulfate (Sigma): 50 mg/mL stock (filter-sterilize); store at –20°C.
6. Hygromycin B (Boehringer Mannheim, Lewes, East Sussex, UK): 50 mg/mL stock in phosphate-buffered saline; store at 4°C in the dark.
7. Selection plates: 10–30 µg/mL kanamycin or 50–100 µg/mL hygromycin (*see* Table 4 for other antibiotics).

2.4. Solutions for Plasmid Preparation

1. GET buffer: 50 m*M* glucose, 10 m*M* EDTA, pH 8.0, 25 m*M* Tris-HCl, pH 8.0; filter-sterilize through 0.2-µm membrane. EDTA should be prepared as a 0.5*M* solution and buffered to pH 8.0 with NaOH (EDTA only dissolves fully at pH 8.0) before diluting to final concentration. Lysozyme (Sigma) should be added to 10 mg/mL before use.
2. Alkaline SDS solution; 0.2*M* NaOH, 1% sodium dodecyl sulfate (SDS). SDS should be prepared as a 20% solution and autoclaved.
3. 3*M* sodium acetate: Adjust to pH 4.8 with glacial acetic acid and sterilize by autoclaving.

2.5. Solutions and Equipment
for Preparation of Electrocompetent E. coli

1. 10 mM Tris-HCl, pH 7.5, 1 mM MgCl$_2$: Tris buffer should be prepared as a 1M stock, buffered with HCl, and autoclaved, MgCl$_2$ should be prepared as a 1M stock; filter-sterilize through a 0.2-μm membrane. Dilute with sterile distilled water to achieve final concentration.
2. LB and LB agar plates containing suitable antibiotic for selection of transformants (*see* Section 2.1.).

3. Methods

Caution: Since some mycobacterial species are pathogenic to humans, appropriate containment facilities for each species should be used for all procedures (*see* Note 1).

3.1. Mycobacterial Culture

Mycobacteria should be maintained in the laboratory by regular subculture on LJ slopes. Inocula for overnight or small cultures can be taken directly off the slope (*see* Note 2).

3.1.1. Fast-Growing Species

1. Inoculate 5 mL of Lemco broth with a loopful of mycobacteria; the cells can be dispersed using a vortex (*see* Notes 2 and 5). Incubate at 37°C with shaking (100 rpm) overnight.
2. Inoculate a large-scale culture (100–500 mL in 250–1000 mL conical flask) with a 1/100 dilution of the overnight culture and continue incubation at 37°C with shaking until OD$_{600}$ = 0.8–1.0 (usually between 16 and 24 h; *see* Note 5).

3.1.2. Slow-Growing Species

1. Inoculate 5 mL of 7H9 broth (containing OADC and Tween 80) with a loopful of mycobacteria, vortex to disperse cells, and incubate at 37°C with shaking (100 rpm) for 10–15 d (*see* Note 2).
2. Inoculate a large-scale culture (100–500 mL) with a 1/100 dilution of the overnight culture and continue incubation at 37°C with shaking until OD$_{600}$ is 0.5–1.0, usually between 14 and 28 d (*see* Notes 5 and 6).
3. Optional: add 0.1 vol 2M glycine (final concentration is 1.5% w/v) 24 h before harvesting cells (*see* Note 7).

3.2. Preparation of Electrocompetent Cells

1. Incubate cells on ice for 1.5 h (*see* Notes 8 and 13) before harvesting by centrifugation at 3000g for 10 min. This incubation improves transformation efficiencies by fourfold.

Table 5
Growth Conditions for Mycobacterial Transformants

Species	Growth temperature	Length of incubation
Fast growers		
M. aurum	37°C	3–5 d
M. phlei	37°C	3–5 d
M. smegmatis	37°C	3–5 d
M. vaccae	30°C	3–7 d
Slow growers		
M. tuberculosis	37°C	3–4 wk
M. bovis BCG	37°C	3–4 wk
M. w	37°C	10–14 d

2. Wash cells three times in ice-cold 10% glycerol. Reduce the volume each time, e.g., for 100 mL, wash 1 = 25 mL, wash 2 = 10 mL, and wash 3 = 5 mL. Finally, resuspend in 1/100–1/500 original culture volume of ice-cold 10% glycerol.
3. At this stage, cells may be frozen in a dry ice/ ethanol bath and stored in aliquots at –70°C for future use. Cells frozen in this way should be thawed on ice and used as required.

3.3. Electroporation

1. Add approx 1 µg salt-free DNA (no more than 5 µL; *see* Notes 10 and 13) to 0.4 mL mycobacterial suspension, and leave on ice for 10 min (*see* Note 9). Transfer to a 0.2-cm electrode gap electroporation cuvet (*see* Note 9). The cuvet should be chilled on ice before use (*see* Note 13).
2. Place cuvet in electroporation chamber and subject to one pulse of 2.5 kV, 25 µF, with the resistance set at 1000 Ω (*see* Notes 9, 11–13).
3. Incubate cuvet on ice for 10 min, transfer cell suspension to a sterile universal bottle, add 5 mL of Lemco broth, and incubate at 37°C for 2 h for fast growers (*see* Note 14); add 5 mL of 7H9 (plus OADC and Tween), and incubate at 37°C for 3 h for slow growers. This step allows expression of any antibiotic resistance gene carried on the DNA.
4. Harvest cells by centrifugation at 3000*g* for 10 min, and plate out suitable dilutions (to give 30–300 colonies/plate) on 7H10 agar plus ADC or OADC enrichment and appropriate antibiotic (*see* Notes 15 and 16).
5. Incubate plates at 37°C until colonies become visible; this will take 3–4 d for fast growers and 2–4 wk for slow growers (*see* Table 5 for species requirements). When using slow-growing species, the plates must be sealed with parafilm to prevent them from drying out (*see* Note 16).

3.4. Growth of Mycobacterial Transformants

3.4.1. Fast-Growing Species

1. Inoculate 5 mL of Lemco broth plus selection antibiotic with transformant colonies.
2. Incubate at 37°C for 2–3 d.

3.4.2. Slow-Growing Species

1. Inoculate 5 mL of 7H9 (plus OADC and Tween) containing appropriate selection antibiotic with transformant colonies.
2. Incubate at 37°C with slow shaking (100 rpm) for 2 wk.

3.5. Plasmid Preparation from Mycobacteria

1. Harvest a 1.5-mL culture, resuspend in 100 μL of GET containing 10 mg/mL lysozyme, and incubate at 37°C for 24 h.
2. Add 200 μL of alkaline SDS, and incubate at 37°C for 1 h.
3. Add 150 μL of 3M Na acetate pH 4.8, and incubate on ice for 1 h.
4. Centrifuge for 15 min at 10,000g, and recover supernatant.
5. Precipitate plasmid DNA with 800 μL of absolute ethanol overnight at –20°C, recover by 15 min of centrifugation at 10,000g, and redissolve pellet in 40 μL of TE.
6. Plasmid DNA can be used to retransform *E. coli* for larger-scale plasmid preparation; for direct visualization of plasmid DNA, 20 μL can be electrophoresed on a 0.8% agarose gel (*see* Note 17).

3.6. Electroduction Between Mycobacteria and E. coli

3.6.1. Preparation of Electrocompetent E. coli

1. Inoculate 100 mL of LB with a loopful of *E. coli,* and incubate at 37°C with shaking (250 rpm) until $OD_{660} = 0.6$.
2. Incubate cells on ice for 10 min.
3. Harvest cells at 3000g for 10 min, and wash in 10 mM Tris-HCl, pH 7.5, and 1 mM MgCl$_2$.
4. Resuspend cells in same buffer at 10^9 cells/mL ($OD_{660} = 20$), roughly 1/30 of original volume.

3.6.2. Electroduction of E. coli

1. Pick a transformed mycobacterial colony from selection plate, and mix with 20 μL of ice-cold 10% glycerol.
2. Vortex and leave on ice for 10 min.
3. Add to 0.4 mL of *E. coli* electrocompetent cells, and leave on ice for 10 min.
4. Place in electroporation chamber, and give a single pulse of 2.5 kV, 45 μF (∞ Ω resistance).

5. Transfer cells to a sterile tube, add 4 mL of LB, and incubate for 1 h at 37°C.
6. Plate on LB agar containing appropriate antibiotic for 12–16 h.
7. Slow-growing species, such as *M. bovis* BCG, should yield 10–100 *E. coli* colonies. *M. smegmatis* should yield about 10^4 *E. coli* transformants (*see* Note 17).

4. Notes

1. Pathogenic mycobacteria represent a serious biohazard. Therefore, all culture and genetic manipulation must be carried out in appropriate containment facilities inside a Class I safety cabinet. In most countries, genetic manipulation involving pathogenic mycobacteria or their DNA must be met with approval by the relevant authorities. In any case, risk assessment must form the first part of any experiment with pathogenic mycobacteria. A list of mycobacterial species and the type of containment required should be consulted prior to use.
2. Mycobacteria are relatively slow-growing organisms; the fast-growing species have a generation time of 2–3 h, and the slow-growing species of around 20 h. This often leads to a problem with contamination of cultures, since many common contaminants have a much quicker doubling time and will rapidly outgrow mycobacteria. It is extremely important to maintain a good aseptic technique, especially with slow growers. It is often wise to set up duplicate cultures in case one becomes contaminated. Cultures should be checked for purity using acid-fast staining at all stages *(32)*.
3. Both ADC and OADC supplements are extremely heat-labile, and should only be added to 7H10 or 7H9 media after cooling. *M. bovis* BCG can be grown in medium supplemented with ADC rather than OADC, but growth is slower. Growth of mycobacterial cultures is enhanced by the provision of up to 10% CO_2 in the air above the medium.
4. Mycobacteria have chemically resistant cell walls that are difficult to lyse. Thus, they are able to survive high voltages even when pulses have long time constants. Several factors, outlined below, affect the efficiency of transformation; these include the growth phase of cells when harvested, electroporation media, and the field strength and time constant of the delivered pulse.
5. Growth of cells: Mycobacterial cells, particularly *M. tuberculosis,* have a tendency to clump together in culture; this is because of the thick waxy nature of the mycobacterial coat. The addition of Tween 80, a nonionic detergent, to media reduces the amount of clumping and provides a more homogenous suspension of cells. The medium used for growth of mycobacteria for electroporation is not important, and a variety of different recipes are used, the most common being Middlebrook 7H10. In general,

mycobacterial cultures should be removed from the incubator when in the logarithmic phase of growth. In contrast to *E. coli,* the cells can be harvested at any point from the early to late log phase (A_{600} of 0.2–1.0).

6. Since mycobacteria have the ability to remain dormant for long periods of time, cultures may contain many cells that are not in an active phase of growth. The maintenance of slow-growing species in a midlog phase of growth by regular dilution of cultures over several months improves the efficiency of transformation (probably by increasing the number of actively growing cells) *(33).*

7. For slow-growing species, such as BCG, the addition of glycine (to a final concentration of 1.5%) to young growing cultures can improve transformation efficiencies *(33,34).* Glycine affects the cell wall of mycobacteria and presumably makes DNA entry easier. Ideally, glycine should be added 1–2 d prior to harvesting. The best efficiencies are obtained with younger cultures of 4–7 d *(15).*

8. Preincubation on ice: Once cultures have reached the required stage of growth, they should be removed from the incubator and incubated on ice for 1.5 h prior to harvesting. This results in a fourfold increase in transformation efficiency *(35).* Longer incubations on ice result in reduced efficiency probably because of increased cell lysis. This may also increase the possibility of arcing during pulse delivery.

9. Pulse delivery: It is important to have an even cell suspension for electroporation, since any clumping of cells will lead to arcing and a reduced transformation efficiency. During the incubation on ice prior to pulse delivery, the cells may settle in the tube, and it is necessary to redistribute them using a pipet or a vortex immediately prior to the high-voltage pulse. This step serves to resuspend the cells and to ensure thorough mixing of the DNA. In addition, care must be taken to ensure that no bubbles are introduced between the two electrodes of the cuvet.

10. DNA type and concentration: The efficiency of electroporation depends on the type of DNA used for transformation; some vectors have been unable to transform particular mycobacterial species, and the efficiency often depends on the choice of selectable marker (*see* Table 3). The efficiency of transformation is not affected by the amount of DNA added; addition of 0.5–500 ng DNA produces the same efficiency, and as much as 5 µg can be used. The volume of DNA used is, however, critical; for small volumes of cell suspensions, the addition of a large amount of DNA in water will alter the conductivity of the suspension. Therefore, it is important that not more than 5 µL of DNA solution are added to the cell suspension.

11. Pulse conditions: The use of a pulse controller in addition to the actual electroporation apparatus allows control over the parallel resistance and,

therefore, the time constant, with higher parallel resistance producing a longer time constant. Recent observations have shown that the optimum time constant is 15–25 ms (1000 Ω resistance) *(35)*. The use of 0.2-cm electrode gap cuvets, as opposed to the 0.4-cm gap cuvets originally used, results in a higher field strength *(35)*. The electroporation medium also has an effect on the time constant. Use of diluted glycerol provides a high resistance medium, allowing longer time constants to be achieved.

12. There are many different electroporation devices available commercially; we use the Bio-Rad (Hempstead, Hertfordshire, UK) Gene Pulser, but any apparatus that can deliver high-voltage pulses can be used. There are also different makes of cuvets available; although the gap or path length may be the same, the maximum volume of the cell suspension can vary from 50–400 µL. The volume of cell suspension used does not seem to affect the efficiency *(36)*, so it does not matter which cuvets (and therefore what volume of cells) are used.

13. Arcing: The use of the pulse-controller apparatus serves to reduce the probability of arcing when using high voltages applied to high-resistance media, although it may still occur. Factors that cause arcing include the presence of lysed cells in the sample, salts in the DNA solution, and electroporation with cuvets that have not been chilled on ice. These factors can be minimized by ensuring that during preparation of electrocompetent cells, the preincubation on ice is no longer than 1.5 h. Also, ensure that the cuvets are chilled on ice before use; the cuvet slide may also be chilled. Always ensure that the DNA for transformation is free from salts and other contaminants; ethanol precipitation and washing with 70% ethanol can be used to purify DNA, which should preferably be dissolved in sterile deionized, distilled water. The settings for the pulse are important as well. Increasing the parallel resistance to ∞ Ω increases the possibility of arcing; therefore, a setting of 1000 Ω produces more consistent results. In some cases, arcing may be violent enough to blow the lid off the electroporation cuvet, dispersing the cell suspension over the inside of the electroporation chamber (thereby creating aerosols). When working with pathogenic organisms, it is imperative that the pulse be delivered with the electroporation chamber placed inside a Class I safety cabinet and that appropriate disinfectants (freshly diluted 1% Hycolin) are at hand. For nonpathogenic species, the pulse can be delivered on the bench, but Hycolin should be available to deal with any spills.

14. The dilution of cells immediately after the pulse is important. Cells should be diluted at least 10-fold and incubated for several hours prior to plating. Omission of this step leads to greatly reduced efficiencies *(36)*. Presumably, the dilution allows better recovery from the pulse and, therefore, greater survival of transformants.

15. The problem of clumping is also important when plating cells after electroporation. It is important to ensure that resistant colonies have arisen from single cells, so the cells must be thoroughly resuspended before plating. For particularly "sticky" cells, the suspension may be passed through a 23-gage needle several times prior to plating. **Caution:** This must not be attempted with pathogenic mycobacteria owing to the risk of needle-stick injuries. Appropriate dilutions may also help to alleviate this problem by thinning out the suspension. It is important to dilute cells prior to plating, since the cell suspensions used for electroporation are very concentrated. If the cells are not diluted, it may be very difficult to visualize true resistant colonies against a background lawn of sensitive cells. This is owing to aggregation, which protects some cells from the effects of the antibiotic.

16. Since slow-growing organisms take up to 4 wk to form colonies from single cells, it is important to pour plates to a depth of 6–8 mm and to wrap them securely in parafilm to prevent drying out during the incubation period. Cycloheximide can be added to plates at 100 µg/mL to prevent fungal contamination, which is possible during such long incubation times. If fungal contamination is recurrent, it may be better to use Lemco agar plates with cycloheximide, since 7H10 agar appears to interfere slightly with the action of cycloheximide (L. Brooks, personal communication). The long incubation period also means that antibiotic-containing plates should be freshly poured for each experiment, since this will minimize the loss of antibiotic activity.

17. Although several methods exist for the preparation of plasmid DNA from mycobacterial species *(37,38)*, none of them yield sufficiently large quantities for restriction enzyme analysis, partly because of the low copy number of most mycobacterial vectors, and often the DNA isolated is of poor quality. Therefore, it is advisable to retransform plasmids isolated from mycobacteria into *E. coli* (using a *rec*A deletion strain, such as TG2 or JM109) and prepare plasmid DNA from the more amenable species. Restriction enzyme analysis can then be used to confirm the identity of the plasmids thus isolated. Since all mycobacterial species used in electroporation are wild type with respect to *rec*A, recombination and rearrangements of plasmid DNA can and do occur. Therefore, it is important to check the identity of all DNAs thus introduced into mycobacteria. An alternative method is to use direct electroduction from mycobacteria to *E. coli (39)*; this involves taking a mycobacterial colony, resuspending it in a small amount of glycerol, and mixing with electrocompetent *E. coli*. The mixture is then pulsed using conditions suitable for *E. coli* electroporation, and *E. coli* transformants are selected on media containing the relevant antibiotic. *E. coli* transformants appear after 16 h, which is much sooner than the appearance of the mycobacterial colonies (2–3 d). This

method is much quicker and may be more convenient for slow-growing species, where the growth of mycobacterial transformants for plasmid preparation may take 2–4 wk.

References

1. Clark-Curtiss, J. E., Jacobs, W. R., Docherty, M. A., Richie, L. R., and Curtiss, R. W., III (1985) The molecular analysis of DNA and the construction of genomic libraries of *M. leprae. J. Bacteriol.* **161,** 1093–1102.
2. Thole, J. E. R., Dauwerse, H. G., Das, P. K., Groothius, D. G., Schouls, L. M., and van Embden, J. D. A. (1985) Cloning of *Mycobacterium bovis* BCG DNA and expression of antigens in *Escherichia coli. Infect. Immun.* **50,** 800–806.
3. Young, R. A., Mehra, V., Sweetser, D., Buchanan, T., Clark-Curtiss, J., Davis, R. W., and Bloom, B. R. (1985) Genes for the major protein antigens of the leprosy parasite *M. leprae. Nature* **316,** 450–452.
4. Young, R. A., Bloom, B. R., Grosskinsky, C. M., Ivanyi, J., Thomas, D., and Davis, R. W. (1985) Dissection of *M. tuberculosis* antigens using recombinant DNA. *Proc. Natl. Acad. Sci. USA* **82,** 2583–2587.
5. Jacobs, W. R., Jr., Docherty, M. A., Curtiss, R., III, and Clark-Curtiss, J. E. (1986) Expression of *Mycobacterium leprae* genes from a *Streptococcus mutans* promoter in *Escherichia coli* K-12. *Proc. Natl. Acad. Sci. USA* **83,** 1926–1930.
6. Young, D. B, Kaufmann, S. H. E., Hermans, P. W. M., and Thole, J. E. R. (1992) Mycobacterial protein antigens: a compilation. *Mol. Microbiol.* **6,** 133–145.
7. Garbe, T., Harris, D., Vordermeier, M., Lathigra, R., Ivanyi, J., and Young, D. (1993) Expression of the *Mycobacterium tuberculosis* 19-kilodalton antigen in *Mycobacterium smegmatis*: immunological analysis and evidence of glycosylation. *Infect. Immun.* **61,** 260–267.
8. Zhang, Y., Lathigra, R., Garbe, T., Catty, D., and Young, D. (1991) Genetic analysis of superoxide dismutase, the 23 kilodalton antigen of *Mycobacterium tuberculosis. Mol. Microbiol.* **5,** 381–391.
9. Thangaraj, H. S., Lamb, F. I., Davis, E. O., Jenner, P. J., Jeyakumar, L. H., and Colston, M. J. (1990) Identification, sequencing and expression of *M. leprae* superoxide dismutase, a major antigen. *Infect. Immun.* **58,** 1937–1942.
10. Das Gupta, S. K., Bashyam, M. D., and Tyagi, A. K. (1993) Cloning and assessment of mycobacterial promoters by using a plasmid shuttle vector. *J. Bacteriol.* **175,** 5186–5192.
11. Stover, C. K., de la Cruz, V. F., Fuerst, T. R., Burlein, J. E., Benson, L. A., Bennett, L. T., Bansal, G. P., Young, J. F., Lee, M. H., Hatfull, G. F., Snapper, S. B., Barletta, R. G., Jacobs, W. R., Jr., and Bloom, B. R. (1991) New use of BCG for recombinant vaccines. *Nature* **351,** 456–460.
12. Guilhot, C., Otal, I., Rompaey, I. V., Martin, C., and Gicquel, B. (1994) Efficient transposition in mycobacteria: construction of *Mycobacterium smegmatis* insertional mutant libraries. *J. Bacteriol.* **176,** 535–539.
13. Snapper, S. B., Melton, R. E., Mustafa, S., Kieser, T., and Jacobs, W. R., Jr. (1990) Isolation and characterization of efficient plasmid transformation mutants of *Mycobacterium smegmatis. Mol. Microbiol.* **4,** 1911–1919.

14. Matsuo, K., Yamaguchi, R., Yamazaki, A., Tasaka, H., Terasaka, K., Totsuka, M., Kobayashi, K., Yukitake, H., and Yamada, T. (1990) Establishment of a foreign antigen secretion system in mycobacteria. *Infect. Immun.* **58,** 4049–4054.

15. Goto, Y., Taniguchi, H., Udou, T., Mizuguchi, Y., and Tokunaga, T. (1991) Development of a new host vector system in mycobacteria. *FEMS Microbiol. Lett.* **83,** 277–282.

16. Radford, A. J. and Hodgson, A. L. M. (1991) Construction and characterization of a mycobacterium-*Escherichia coli* shuttle vector. *Plasmid* **25,** 149–153.

17. Villar, C. A. and Benitez, J. (1992) Functional analysis of pAL5000 plasmid in *Mycobacterium fortuitum. Plasmid* **28,** 166–169.

18. Dellagostin, O. A., Wall, S., Norman, E., O'Shaughnessy, T., Dale, J. W., and McFadden, J. (1993) Construction and use of integrative vectors to express foreign genes in mycobacteria. *Mol. Microbiol.* **10,** 983–993.

19. Garbe, T. R., Barathi, J., Barnini, S., Zhang, Y., Abou-Zeid, C., Tang, D., Mukherjee, R., and Young, D. B. (1994) Transformation of mycobacterial species using hygromycin resistance as selectable marker. *Microbiology* **140,** 133–138.

20. Houssaini-Iraqui, M., Clavel-Seres, S., Rastogi, N., and David, H. L. (1992) The expression of the *Mycobacterium aurum* carotenogenesis operon is not repressed by the repressor of *Mycobacterium vaccae* photoinducible carotenogenesis. *FEMS Microbiol. Lett.* **99,** 233–236.

21. Qin, M., Taniguchi, H., and Mizuguchi, Y. (1994) Analysis of the replication region of a mycobacterial plasmid, pMSC262. *J. Bacteriol.* **176,** 419–425.

22. Hermans, J., Martin, C., Huijberts, G. N. M., Goosen, T., and de Bont, J. A. M. (1991) Transformation of *Mycobacterium aurum* and *Mycobacterium smegmatis* with the broad host range Gram-negative cosmid vector pJRD215. *Mol. Microbiol.* **5,** 1561–1566.

23. Houssaini-Iraqui, M., Lazraq, R., Clavel-Seres, S., Rastogi, N., and David, H. L. (1992) Cloning and expression of *Mycobacterium aurum* carotenogenesis genes in *Mycobacterium smegmatis. FEMS Microbiol. Lett.* **90,** 239–244.

24. Hermans, J., Suy, I. M. L., and de Bont, J. A. M. (1993) Transformation of Gram-positive micro-organisms with the Gram-negative broad-host-range cosmid vector pJRD215. *FEMS Microbiol. Lett.* **108,** 201–204.

25. Snapper, S. B., Lugosi, L., Jekkel, A., Melton, R., Kieser, T., Bloom, B. R., and Jacobs, W. R., Jr. (1988) Lysogeny and transformation in mycobacteria: stable expression of foreign genes. *Proc. Natl. Acad. Sci. USA* **85,** 6987–6991.

26. Woodley, C. L. and David, H. L. (1976) Effects of temperature on the rate of the transparent to opaque colony type transition in *Mycobacterium avium. Antimicrobial Ag Chemother.* **9,** 113–119.

27. Lugosi, L., Jacobs, W. R., Jr., and Bloom, B. R. (1989) Genetic transformation of BCG. *Tubercle* **70,** 159–170.

28. Ranes, M. G., Rauzier, J., Lagranderie, M., Gheorghiu, M., and Gicquel, B. (1990) Functional analysis of pAL5000, a plasmid from *Mycobacterium fortuitum:* construction of a "mini" mycobacterium-*Escherichia coli* shuttle vector. *J. Bacteriol.* **172,** 2793–2797.

29. Hatfull, G. F. (1993) Genetic transformation of mycobacteria. *Trends Microbiol.* **1,** 310–314.

30. Gormley, E. P. and Davies, J. (1991) Transfer of plasmid RSF1010 by conjugation from *Escherichia coli* to *Streptomyces lividans* and *Mycobacterium smegmatis*. *J. Bacteriol.* **173,** 6705–6708.
31. Donnelly-Wu, M., Jacobs, W. R., Jr., and Hatfull, G. R. (1993) Superinfection immunity of mycobacteriophage L5: applications for genetic transformation of mycobacteria. *Mol. Microbiol.* **7,** 407–417.
32. Cruickshank, R. (1965) *Medical Microbiology: A Guide to the Laboratory Diagnosis and Control of Infection,* 11th ed. E. & S. Livingstone Limited, London, pp. 652–655.
33. Aldovini, A., Husson, R. N., and Young, R. A. (1993) The *uraA* locus and homologous recombination in *Mycobacterium bovis* BCG. *J. Bacteriol.* **175,** 7282–7289.
34. Husson, R. N., James, B. E., and Young, R. A. (1990) Gene replacement and expression of foreign DNA in mycobacteria. *J. Bacteriol.* **172,** 519–524.
35. Cirillo, J. D., Weisbrod, T. R., and Jacobs, W. R., Jr. (1994) Efficient electrotransformation of *Mycobacterium smegmatis*. Bio-Rad Laboratories Bulletin 1360 US/EG.
36. Hinshelwood, S. T. (1992) Molecular biology of histidine biosynthesis in *Mycobacterium smegmatis*. Ph.D. thesis, University of London.
37. Crawford, J. T. and Bates, J. H. (1979) Isolation of plasmids from mycobacteria. *Infect. Immun.* **24,** 979–981.
38. Jacobs, W. R., Jr., Kalpana, G. V., Cirillo, J. D., Pascopella, L., Snapper, S. B., Udani, R. A., Jones, W., Barletta, R. G., and Bloom, B. R. (1991) Genetic systems for mycobacteria. *Methods Enzymol.* **204,** 537–555.
39. Baulard, A., Jourdan, C., Mercenier, A., and Locht, C. (1992) Rapid mycobacterial plasmid analysis by electroduction between *Mycobacterium* spp. and *Escherichia coli. Nucleic Acids Res.* **20,** 4105.
40. Kalpana, G. V., Bloom, B. R., and Jacobs, W. R., Jr. (1991) Insertional mutagenesis and illegitimate recombination in mycobacteria. *Proc. Natl. Acad. Sci. USA* **88,** 5433–5437.
41. David, M., Lubinsky-Mink, S., Ben-Zvi, A., Ulitzer, S., Kuhn, J., and Suissa, M. (1992) A stable *Escherichia coli-Mycobacterium smegmatis* plasmid shuttle vector containing the mycobacteriophage D29 origin. *Plasmid* **28,** 267–271.
42. Hinshelwood, S. and Stoker, N. G. (1992) An *Escherichia coli–Mycobacterium* shuttle cosmid vector, pMSC1. *Gene* **110,** 115–118.

CHAPTER 25

Electrotransformation
of the Spirochete *Borrelia burgdorferi*

D. Scott Samuels

1. Introduction

Borrelia burgdorferi is an etiologic agent of Lyme disease, the most common arthropod-borne disease in the United States *(1,2)*. The bacterium, a member of the spirochete phylum, has a genome predominantly composed of linear DNA molecules *(3,4)*. Formulating a medium in which *B. burgdorferi* grows in vitro was the first step toward a genetic understanding of the physiology and pathogenesis of the organism *(5,6)*. The growth of *B. burgdorferi* as single colonies in solid medium *(7–9)* has facilitated mutant isolation by selection *(10,11)*, although a defined medium for selection of auxotrophs is not currently available. The transformation system described in this chapter will be useful for manipulating the spirochete on a molecular genetic level.

Electroporation is the use of an electric pulse to permeabilize cell membranes reversibly *(12)* and is an extremely efficient method of genetically transforming bacteria *(13)*. Electrotransformation has been used to disrupt a hemolysin gene in the spirochete *Serpulina hyodysenteriae,* an etiologic agent of swine dysentery, by homologous recombination *(14)*. The effect of electroporation buffers and capacitance on the survival of *B. burgdorferi* has been reported *(15)*, and electrotransformation has been employed to insert point mutations conferring antibiotic resistance into the *gyrB* gene by homologous recombination *(16)*.

This chapter provides detailed methods for introducing DNA into *B. burgdorferi* by electroporation and for the selection of transformants (or

From: *Methods in Molecular Biology, Vol. 47: Electroporation Protocols for Microorganisms*
Edited by: J. A. Nickoloff Humana Press Inc., Totowa, NJ

spontaneous mutants) on solid medium. We typically obtain transformation efficiencies of 10^3 transformants/µg of linear DNA. The protocol may work for other species of the genus *Borrelia*. The most critical parameter appears to be the growth phase or cell density of the culture when collected. The cells are washed extensively, twice in cold phosphate-buffered saline and three times in a cold osmotically buffered low-ionic-strength solution, and concentrated to about 10^{10} cells/mL. They are genetically transformed by adding DNA in a low-ionic-strength solution and pulsing with a short-duration, high-intensity exponential decay electric field. Transformants are selected in solid medium containing an antibiotic (or an antibody). Unfortunately, there are both biosafety and physiological limitations on the use of many antibiotics with *B. burgdorferi* species.

2. Materials

1. *B. burgdorferi* strain B31 is available from American Type Culture Collection (Rockville, MD).
2. Barbour-Stoenner-Kelly (BSK) II medium (without gelatin): 8% (v/v) 10X CMRL-1066 (without L-glutamine and sodium bicarbonate; Life Technologies, Gaithersburg, MD), 4 g/L Neopeptone (Difco, Detroit, MI), 40 g/L bovine serum albumin (BSA; fraction V, Pentex; Miles, Kankakee, IL), 1.6 g/L Yeastolate (TC; Difco), 4.8 g/L N-2-hydroxyethylpiperazine-N'-2-ethanesulfonic acid (HEPES), 4 g/L glucose, 0.56 g/L sodium citrate, 0.64 g/L sodium pyruvate, 0.32 g/L N-acetyl-D-glucosamine, 1.76 g/L sodium bicarbonate, and 6.6% rabbit serum (trace hemolyzed; Pel-Freez, Rogers, AR). Adjust to pH 7.6 with 1N NaOH, stir slowly for 2–3 h, and sterilize by filtration (successively through a prefilter, a 1.2-µm filter, a 0.45-µm filter, and a 0.22-µm filter). Store at 4°C for up to 2 mo (*see* Note 1).
3. Dulbecco's phosphate-buffered saline (dPBS): 8 g/L NaCl, 0.2 g/L KCl, 1.15 g/L Na_2HPO_4, and 0.2 g/L KH_2PO_4. Sterilize by filtration and store at 4°C.
4. dPBS with divalent cations (dPBS^{2+}): Add 0.1 g/L $CaCl_2$ and 0.2 g/L $MgCl_2 \cdot 6H_2O$ to dPBS.
5. Electroporation solution (EPS): 93 g/L sucrose and 15% (v/v) glycerol. Sterilize by filtration and store at 4°C.
6. Plating-BSK (P-BSK) medium: Add 83 g BSA, 8.3 g Neopeptone, 10 g HEPES, 1.2 g sodium citrate, 8.3 g glucose, 1.3 g sodium pyruvate, 0.7 g N-acetyl-D-glucosamine, 3.7 g sodium bicarbonate, and 4.2 g Yeastolate to 1 L of water (18 MΩcm). Adjust to pH 7.5 with 1N NaOH, stir slowly for 2–3 h, and sterilize by filtration (successively through a prefilter and a 0.22-µm filter). Store at 4°C for up to 6 mo.

7. 1.7% agarose: Use high-strength, analytical-grade.
8. Antibiotic solution (for selection of transformants): 50 mg/mL of coumermycin A_1 in dimethyl sulfoxide (*see* Note 2). Store at –20°C for up to 6 mo. 25 mg/mL Novobiocin (a less expensive coumarin antibiotic, can be used in place of coumermycin A_1 (Sigma, St. Louis, MO). Make fresh or store in small fractions at –20°C in the dark for up to 6 mo.
9. 5% sodium bicarbonate: prepared fresh and filter-sterilized.

3. Methods
3.1. Preparation of Competent Cells

1. Inoculate 500 mL of BSK II medium in a 500-mL screw-top bottle with 1 mL of a late-log-phase culture (*see* Note 3). Incubate at 32–34°C (without agitation) until the culture reaches a density of about 5×10^7 cells/mL (*see* Note 4). This requires 36–96 h.
2. Transfer culture to two sterile 250-mL screw-top centrifuge bottles and cap.
3. Centrifuge at 4000g for 20 min at 4°C. Decant the supernatant fraction and resuspend each cell pellet in 30 mL of cold dPBS (*see* Note 5).
4. Transfer cells to two sterile 50-mL screw-top centrifuge tubes and cap.
5. Centrifuge at 3000g for 10 min at 4°C. Decant the supernatant fraction, and resuspend each cell pellet in 30 mL of cold dPBS.
6. Centrifuge at 3000g for 10 min at 4°C. Decant the supernatant fraction, and resuspend each cell pellet in 10 mL of cold EPS.
7. Transfer cells to two sterile 14-mL polypropylene tubes and cap.
8. Centrifuge at 2000g for 10 min at 4°C. Decant the supernatant fraction, and resuspend each cell pellet in 10 mL of cold EPS. Repeat.
9. Centrifuge at 2000g for 10 min at 4°C. Decant the supernatant fraction, and pool the cell pellets in 0.6 mL of cold EPS (*see* Note 6).
10. Distribute 50-µL aliquot fractions of the cell suspension into sterile 1.7-mL tubes on ice (*see* Notes 7 and 8).

3.2. Electroporation

1. Cool electroporation cuvets (0.2-cm electrode gap) to 4°C.
2. Transfer 1–5 µL of a solution containing 0.3–1 µg of DNA in water (*see* Notes 9 and 10) to the cell suspension, mix gently, and incubate on ice for about 1 min.
3. Transfer the cell/DNA mixture to a chilled electroporation cuvet. Cap the cuvet, and shake the cell/DNA mixture to the bottom of the cuvet so that it spans the two electrodes.
4. Place the cuvet in the pulse generator, and deliver a single exponential decay pulse of 2.5 kV, 25 µF, and 200 Ω. This should produce a time constant of 4–5 ms (*see* Notes 10 and 11).

5. Immediately (within 1 min), add 1 mL of BSK II medium (at room temperature) without antibiotics, and mix the cell suspension by pipeting up and down.

6. Transfer the entire mixture to a sterile 14-mL tube that contains an additional 9 mL of BSK II medium (at room temperature) and incubate (without agitation) at 32–34°C for 20 h.

3.3. Selection of Transformants

1. Mix 240 mL of P-BSK medium, 38 mL of 10X CMRL-1066, and 12 mL of rabbit serum. Equilibrate the mixture at 55°C in a water bath. Autoclave 200 mL of 1.7% agarose, equilibrate to 55°C, and combine with the medium mixture. Add 20 mL of fresh 5% sodium bicarbonate with antibiotics (the final volume is 510 mL) (*see* Notes 2 and 12).

2. Transfer 15 mL of the molten medium into 12–14 100-mm dishes, and allow to solidify at room temperature. Equilibrate the remainder of the molten medium at 42°C.

3. Transfer 0.1 mL of BSK II medium containing the electroporated cells to a 50-mL tube. Add 20 mL of the molten medium (at 42°C), and mix by pipeting up and down once. Transfer the mixture to the plates containing the solidified bottom agarose medium and allow to solidify at room temperature.

4. Centrifuge the remaining 9.9 mL of culture at 8000*g* for 5 min, resuspend in 1 mL of supernatant fraction, and plate as above.

5. Incubate the plates at 32–34°C in a humidified 5% CO_2 atmosphere. Colonies will appear in about 14 d.

6. Isolate single colonies by picking with a plugged 15-cm Pasteur pipet (with bulb). Transfer to 10 mL of BSK II in the presence of antibiotics. Cultures will reach late-log phase in 6–9 d.

4. Notes

1. The quality of BSA varies by source and lot. We have found Miles to be a reliable source. However, we reserve 5- or 10-kg batches and test samples for the ability to support the growth of *B. burgdorferi*. Pretested BSK II medium without gelatin can be purchased from Sigma (BSK-H), but it is expensive. We have stored BSK-H medium (without serum) at 4°C for up to 2 yr and found that it can support the growth of highly passaged strain B31 on the addition of fresh serum.

2. The only antibiotic that is not clinically useful and has been shown to be effective for selection of resistant mutants is coumermycin A_1 *(11,17)*. However, a preliminary report suggests that *B. burgdorferi* can be electrotransformed with a gene that confers chloramphenicol resistance

(D. Persing and D. Podzorski, personal communication), and the spirochete *S. hyodysenteriae* has been electrotransformed with a gene that confers kanamycin resistance *(14)*. In addition, bactericidal antibodies have been used to select for mutants of *B. burgdorferi (10,18,19)*.

3. *B. burgdorferi* is a class 2 human pathogen and therefore should be handled in a class II biological safety cabinet (laminar flow hood). In addition, BSK II medium is rich, and all procedures should be performed aseptically. Introduction of recombinant DNA into a class 2 pathogen requires permission from the Institutional Biosafety Committee before initiation of the experiments according to Section IIIB of the Guidelines for Research Involving Recombinant DNA Molecules (Federal Register).

4. The cell density (or growth phase) is a significant factor for successful electrotransformation, as is the case with other bacteria *(12,20)*. The cells will not transform efficiently if the cell density is too high (when the color of the medium changes). We have had success electrotransforming cultures harvested at $1-7 \times 10^7$ cells/mL, although a low cell density ($1-2 \times 10^7$ cells/mL) requires pelleting the cells at a higher g force (up to $5000g$) and adjusting the final volume of the cell suspension (*see* Note 6). Cell density should be determined using a Petroff Hausser Counting Chamber (Hausser Scientific Partnership, Horsham, PA). Dilute 0.1 mL of the culture with 0.9 mL of cold dPBS^{2+} and place in the counting chamber. Count cells over all 25 groups of 16 small squares in all planes using a dark-field microscope. Multiply the number of cells counted by 5×10^5 to calculate cells/mL. Alternatively, cell density can be determined by spectrophotometry *(17)*. Centrifuge 10 mL of the culture at $5000g$ for 10 min. Decant the supernatant fraction, and resuspend the cell pellet in 1 mL of dPBS^{2+}. Centrifuge at $8000g$ for 5 min. Decant the supernatant fraction, resuspend the cell pellet in 1 mL of dPBS^{2+}, and measure the A_{600}. Multiply the A_{600} by 1.4×10^8 to calculate cells/mL in the culture.

5. Thorough washing is important to remove components of the medium (*see* Note 10). Cell pellets are resuspended in both dPBS and EPS by pipeting followed by vortex mixing. These treatments do not appear to affect cell viability.

6. The final cell concentration should be $1-5 \times 10^{10}$ cells/mL (with a final volume of about 0.9 mL). The volume of EPS used to resuspend the final cell pellet may have to be adjusted to account for initial cell number and efficiency of decanting.

7. We find that use of presterilized aerosol-resistant pipet tips (with aerosol barriers) helps to maintain sterility when handling small volumes of liquid.

8. We have not examined the effect of temperature on transformation efficiency, but maintaining the competent cells at 4°C is generally considered to yield optimal efficiencies *(12,20)*. As with other bacterial species *(20)*,

competent cells can be stored at –70°C without a significant loss of trans-formation efficiency.

9. We routinely obtain 1000 or more transformants/μg of DNA with strain B31, although we have only used linear DNA generated by PCR as an electrotransformation substrate *(16)*. Linear molecules are 1000-fold less efficient in electrotransformation of *Escherichia coli* than circular molecules *(12)*, and we are currently constructing circular replicons for use in *B. burgdorferi*. We have recently shown that *B. burgdorferi* can be transformed with oligonucleotides.

10. Electroporation in the presence of high-ionic-strength solutions causes arcing (and a lower time constant). Two arcs will kill all of the *B. burgdorferi* cells. We use the Wizard DNA purification system (Promega, Madison, WI) and elute the DNA at a high concentration in water. Trans-formation efficiency generally increases with DNA concentration *(12,20)*.

11. Preliminary studies suggest that one pulse effected higher transformation efficiencies than multiple pulses and that varying the resistance from 100–400 Ω affected the time constant, but did not significantly alter the trans-formation efficiency.

12. An antibiotic concentration that inhibits bacterial growth in liquid culture by 80–90% relative to growth in the absence of antibiotics has been used to select for spontaneous mutants and transformants in solid medium *(11,16,17)*. We currently use 0.2 μg/mL coumermycin A_1 or 5 μg/mL novobiocin for selection.

Acknowledgments

I thank Claude Garon for support and Kit Tilly for manuscript review.

References

1. Steere, A. C. (1989) Medical progress: Lyme disease. *N. Engl. J. Med.* **321,** 586–596.
2. Barbour, A. G. and Fish, D. (1993) The biological and social phenomenon of Lyme disease. *Science* **260,** 1610–1616.
3. Hinnebusch, J. and Tilly, K. (1993) Linear plasmids and chromosomes in bacteria. *Mol. Microbiol.* **10,** 917–922.
4. Saint Girons, I., Old, I. G., and Davidson, B. E. (1994) Molecular biology of the *Borrelia,* bacteria with linear replicons. *Microbiology* **140,** 1803–1816.
5. Burgdorfer, W., Barbour, A. G., Hayes, S. F., Benach, J. L., Grunwaldt, E., and Davis, J. P. (1982) Lyme disease—a tick-borne spirochetosis? *Science* **216,** 1317–1319.
6. Barbour, A. G. (1984) Isolation and cultivation of Lyme disease spirochetes. *Yale J. Biol. Med.* **57,** 521–525.
7. Kurtti, T. J., Munderloh, U. G., Johnson, R. C., and Ahlstrand, G. G. (1987) Colony formation and morphology in *Borrelia burgdorferi. J. Clin. Microbiol.* **25,** 2054–2058.
8. Bundoc, V. G. and Barbour, A. G. (1989) Clonal polymorphisms of outer mem-brane protein OspB of *Borrelia burgdorferi. Infect. Immun.* **57,** 2733–2741.

9. Rosa, P. A. and Hogan, D. M. (1992) Colony formation by *Borrelia burgdorferi* in solid medium: clonal analysis of *osp* locus variants, in *First International Conference on Tick-Borne Pathogens at the Host-Vector Interface: An Agenda for Research* (Munderloh, U. G. and Kurtti, T. J., eds.), University of Minnesota, St. Paul, pp. 95–103.

10. Šadžiene, A., Rosa, P. A., Thompson, P. A., Hogan, D. M., and Barbour, A. G. (1992) Antibody-resistant mutants of *Borrelia burgdorferi:* in vitro selection and characterization. *J. Exp. Med.* **176**, 799–809.

11. Samuels, D. S., Marconi, R. T., Huang, W. M., and Garon, C. F. (1994) *gyrB* mutations in coumermycin A_1-resistant *Borrelia burgdorferi. J. Bacteriol.* **176**, 3072–3075.

12. Shigekawa, K. and Dower, W. J. (1988) Electroporation of eukaryotes and prokaryotes: a general approach to the introduction of macromolecules into cells. *BioTechniques* **6**, 742–751.

13. Trevors, J. T., Chassy, B. M., Dower, W. J., and Blaschek, H. P. (1992) Electrotransformation of bacteria by plasmid DNA, in *Guide to Electroporation and Electrofusion* (Chang, D. C., Chassy, B. M., Saunders, J. A., and Sowers, A. E., eds.), Academic, San Diego, pp. 265–290.

14. ter Huurne, A. A. H. M., van Houten, M., Muir, S., Kusters, J. G., van der Zeijst, B. A. M., and Gaastra, W. (1992) Inactivation of a *Serpula (Treponema) hyodysenteriae* hemolysin gene by homologous recombination: importance of this hemolysin in pathogenesis of *S. hyodysenteriae* in mice. *FEMS Microbiol. Lett.* **92**, 109–114.

15. Sambri, V. and Lovett, M. A. (1990) Survival of *Borrelia burgdorferi* in different electroporation buffers. *Microbiologica* **13**, 79–83.

16. Samuels, D. S., Mach, K. E., and Garon, C. F. (1994) Genetic transformation of the Lyme disease agent *Borrelia burgdorferi* with coumarin-resistant *gyrB. J. Bacteriol.* **176**, 6045–6049.

17. Samuels, D. S. and Garon, C. F. (1993) Coumermycin A_1 inhibits growth and induces relaxation of supercoiled plasmids in *Borrelia burgdorferi,* the Lyme disease agent. *Antimicrob. Agents Chemother.* **37**, 46–50.

18. Cinco, M. (1992) Selection of a *Borrelia burgdorferi* antigenic variant by cultivation in the presence of increasing amounts of homologous immune serum. *FEMS Microbiol. Lett.* **92**, 15–18.

19. Coleman, J. L., Rogers, R. C., and Benach, J. L. (1992) Selection of an escape variant of *Borrelia burgdorferi* by use of bactericidal monoclonal antibodies to OspB. *Infect. Immun.* **60**, 3098–3104.

20. Dower, W. J., Chassy, B. M., Trevors, J. T., and Blaschek, H. P. (1992) Protocols for the transformation of bacteria by electroporation, in *Guide to Electroporation and Electrofusion* (Chang, D. C., Chassy, B. M., Saunders, J. A., and Sowers, A. E., eds.), Academic, San Diego, pp. 485–499.

CHAPTER 26

Yeast Transformation and the Preparation of Frozen Spheroplasts for Electroporation

Lisa Stowers, James Gautsch, Richard Dana, and Merl F. Hoekstra

1. Introduction

Transformation methods for the baker's yeast *Saccharomyces cerevisiae* have allowed for rapid advances in molecular genetic studies with this model microorganism, for the development of alternative expression systems to prepare recombinant proteins, and for the development of novel cloning strategies to identify and characterize heterologous genes. Developments in yeast transformation began with experiments that involved the enzymatic digestion and removal of cell-wall material to produce spheroplasts that were competent for taking up DNA *(1)*. This method, although more difficult to master than other methods for routine work, was efficient and allowed for the screening of yeast genomic libraries. As the number of yeast researchers increased, so did the variety of transformation methods, and in the early 1980s a lithium salt-mediated transformation method was developed as an alternative to spheroplast methods *(2)*. Lithium acetate-mediated transformation is widely used in the yeast community for strain manipulation and for gene-cloning experiments.

In the mid to late 1980s additional alternative approaches for transformation were introduced. One of these options involved a physical method

From: *Methods in Molecular Biology, Vol. 47: Electroporation Protocols for Microorganisms*
Edited by: J. A. Nickoloff Humana Press Inc., Totowa, NJ

for yeast transformation in which cells were resuspended with DNA and glass beads, and the mixture was agitated prior to selection for transformants. A second approach was the extension of mammalian and prokaryotic electroporation methods to yeast. Because of their size and growth conditions, the electroporation of yeast uses conditions similar to those for bacterial electroporation. Early electroporation methods were relatively inefficient. However, recent developments led to increased transformation efficiencies, allowing the transformation of colonies stored on Petri dishes, and the preparation and storage of frozen electrocompetent cells.

One of the most significant advances in yeast transformation experiments came with the development of high-efficiency yeast transformation methods by Burgers and Percival *(3)*. Driven by the need to use transformation methods for mutant identification, Burgers and Percival systematically evaluated a variety of transformation parameters with the goal of obtaining the highest transformation efficiencies possible. The spheroplast transformation modifications described by these workers facilitated YAC cloning methods and extensive screening of heterologous cDNA libraries constructed in yeast expression plasmids.

High-efficiency transformation of intact yeast cells and spheroplasts is a complex procedure, and improved methods are now available *(3–5)*. However, inconsistent results are often a problem and can result from the use of impure reagents, subtle variations in technique, unique properties of the individual yeast strain, or the physiological state of the cells. In an attempt to simplify and improve high-efficiency transformation procedures, this chapter first describes yeast electroporation methods and then describes a method for the preparation of frozen spheroplasts, which can be readily used for both electroporation and nonelectroporation transformation protocols. The latter methods are particularly useful when small amounts of DNA are available, when screening genomic or cDNA plasmid libraries, and for two-hybrid system screens *(6,7)*. Frozen spheroplasts prepared with high-quality reagents and stabilized as described below can provide 10^7 transformants/µg of plasmid DNA. Frozen cells that transform with high efficiency offer convenience, can save considerable time, and provide consistent results. The frozen spheroplasts offer flexibility in transformation, since they can be used for electroporation or for standard spheroplast methods.

2. Materials

High-quality reagents are essential for high-efficiency electroporation. We examined numerous sources of reagents for optimal transformation efficiency. The largest variations in this method are found using different lots of polyethylene glycol, of agar for plating, and of sorbitol. We recommend side-by-side batch comparisons to optimize efficiency. Careful selection of reagents can increase transformation efficiencies 10- to 100-fold. A yeast spheroplast transformation kit is available (BIO 101, Inc., San Diego, CA) that includes selected reagents as listed below.

2.1. Solutions

1. $1M$ sorbitol (transformation-grade).
2. SCEM: $1M$ sorbitol, $0.1M$ sodium citrate, pH 5.8, 10 mM EDTA, 30 mM β-mercaptoethanol (β-ME). SCEM is prepared as a filter-sterilized solution without β-ME and stored frozen in 50-mL aliquots. β-ME is added just before use.
3. PEG: 20% polytheylene glycol 8000, 10 mM Tris-HCl, pH 7.5, 10 mM $CaCl_2$. PEG is filter-sterilized and stored in 10-mL aliquots at –20°C.
4. CaST solution: 10 mM $CaCl_2$, $1M$ sorbitol, 10 mM Tris-HCl, pH 7.5. Filter-sterilize and store in frozen aliquots.
5. Frozen cell storage solution: 40% glycerol, 14% dimethylsulfoxide, $0.2M$ mannitol, $0.32M$ sucrose, $0.1M$ sorbitol, $0.2M$ trehalose. Filter-sterilize and store frozen.
6. Electroporation buffer: 0.3 mM Na_2HPO_4, 0.02 mM KH_2PO_4, 10% glycerol. Autoclave to sterilize.
7. Lyticase stock solution: Dissolve lyticase (Sigma, St. Louis, MO) in 50 mM Tris-HCl, pH 7.5, 5 mM β-ME, 20% glycerol at a final concentration of 25,000 U/mL. Flash freeze the stock immediately after preparation in 100-μL aliquots in liquid N_2. Store the frozen stock at –70°C.
8. Carrier DNA: 10 mg/mL calf thymus or salmon sperm DNA, sonicated to reduce viscosity.

2.2. Media

1. YPD: 1% bacto-yeast extract, 2% bacto-peptone, 2% glucose. Media may be solidified with 2% bacto-agar.
2. Synthetic media: Prepare particular yeast selection media depending on the auxotrophic markers being complemented during transformation (3). For selection of transformants, media is supplemented with $1M$ sorbitol.
3. SOS: $1M$ sorbitol, 6.5 mM $CaCl_2$, 0.25% bacto-yeast extract, 0.5% bacto-peptone, 0.5% glucose.

3. Methods

3.1. Electroporation of Intact Yeast Cells

1. Grow yeast cells in 500 mL of YPD with vigorous shaking to a cell concentration of $5–10 \times 10^7$ cells/mL (*see* Note 1).
2. Harvest cells by centrifugation in a cold rotor at $4000g$ for 5 min at 4°C. Be careful to discard as much medium as possible.
3. Resuspend cells in 500 mL of ice-cold sterile H_2O, and repeat step 2.
4. Repeat step 3 using 250 mL of ice-cold sterile water.
5. Resuspend the cell pellet in 10 mL of ice-cold, sterile 1M sorbitol, wash by centrifugation, and remove the supernatant.
6. Resuspend the cell pellet in 0.5 mL of ice-cold 1M sorbitol, and store the cells on ice until needed.
7. Add 20 µL of concentrated yeast and 5 µL or less transforming DNA to a 1.5-mL tube. This mixture can be stored briefly on ice.
8. Transfer to an ice-cold, electroporation cuvet with a 0.15-cm electrode gap.
9. Pulse once with 400 V, 10 µF, and low resistance (e.g., with a BRL Cell-Porator).
10. Transfer the electroporated cells to a 1.5-mL tube containing 0.5 mL of ice-cold 1M sorbitol.
11. Spread aliquots onto sorbitol-containing selective media, and incubate at 30°C until colonies form (usually 2–3 d; *see* Note 2).

3.2. Electroporation of Frozen Spheroplasts

1. Inoculate a single yeast colony into 50 mL of YPD and grow with agitation to a density of 2×10^7 cells/mL. The use of higher-density cultures often results in fewer transformants/µg of DNA.
2. Pellet cells in a 50-mL conical tube by centrifuging for 5 min at $400g$.
3. Resuspend cells in 20 mL of sterile H_2O, centrifuge at $400g$ for 5 min to pellet cells, and discard supernatant.
4. Gently resuspend the cell pellet in 10 mL of 1M sorbitol. Centrifuge at $400g$ for 5 min, and discard the supernatant.
5. Gently resuspend the cells in 10 mL of SCEM. Add 40 µL of freshly thawed lyticase solution, gently mix, and incubate at 30°C. Check for spheroplasts at 15-min intervals by placing 2 µL of cells in 20 µL of 1% sodium dodecyl sulfate and observing under a microscope. Spheroplasts should lyse, and either be invisible or appear as ghost cells. Compare the detergent-treated cells to 2 µL of cells in 20 µL of 1M sorbitol, where cells should be clearly visible. Continue to incubate with lyticase until 90–95% of the cells have formed spheroplasts (*see* Note 3).
6. Centrifuge gently at $300g$ for 5 min in round-bottom (e.g., Falcon 2051) 17 × 100 mm plastic tubes to pellet cells. Gently discard the supernatant. The

cells are extremely fragile at this point. Because of the gentle centrifugation, often not all of the cells will pellet. Carefully aspirate supernatant without disturbing the pellet.

7. Gently resuspend in spheroplasts in 10 mL of 1*M* sorbitol by gently tapping the tube. Centrifuge at 300*g* for 5 min, discard supernatant, and repeat this wash to eliminate lyticase thoroughly.

8. Gently resuspend cells in 2 mL of CaST solution. Add 2 mL of cell-storage solution. Aliquot cells in 2-mL sterile cryovials, and store at –70°C.

9. As needed, thaw cells quickly in a 37°C water bath, and promptly wash three times in round-bottom tubes with 3 mL of 1*M* sorbitol.

10. After the final wash, resuspend spheroplasts in 1.9 mL of 1*M* sorbitol with 0.1 mL of electroporation buffer. Incubate on ice for 5 min. Pellet cells, decant supernatant, resuspend in 2 mL of 1*M* sorbitol, and add 10 μL of electroporation buffer.

11. In a 12-mL round-bottom tube, combine 1 μL of carrier DNA with 0.1–10 μg of nonreplicating or linear plasmid DNA. For autonomously replicating plasmid DNA, add 100–400 ng of DNA without carrier DNA *(8)*. Add 100 μL of cells, and swirl to mix. Incubate at 30°C for 20 min.

12. Transfer 30 μL of the cell/DNA mixture to an electroporation cuvet with 0.15-mm electrode gap, and pulse once with 400 V, 10-μF capacitance, and low resistance. Dilute cells with 100 μL of cold 1*M* sorbitol. Because these cells are spheroplasted, top agar is required for regeneration: Add 5 mL of top agar (selective media containing 1*M* sorbitol, no warmer than 50°C), and plate directly on selective medium with 1*M* sorbitol *(see* Notes 4–7).

3.3. Transformation of Spheroplasts Using PEG

1. Gently add 1 mL of PEG by pipeting the solution along the side of a tube containing 100 μL of spheroplasts plus DNA (Section 3.2., step 11). Gently swirl to mix, and incubate at room temperature for 10 min.

2. Centrifuge at 200–300*g* for 2 min to pellet the cells. This gentle spin will form a loose pellet that is frequently difficult to visualize. For optimum efficiency, we recommend round-bottom tubes in this stage of the procedure. Quickly discard supernatant by decanting or aspirating. Although the pellet is loose at this stage, it is better to discard some of the cells than to achieve a tight pellet by longer centrifugation or greater *g* forces.

3. Gently resuspend the cells in 150 μL of SOS, and incubate at 30°C for 40 min without agitation. Add 5 mL of top agar, and pour onto selective plates to regenerate spheroplasts as described above. Incubate at 30°C until transformant colonies appear.

4. Notes

1. We do not recommend using OD_{600} measurements to determine cell concentrations, since significant strain variation exists in the conversion between OD_{600} and growth phase. If working with a well-characterized strain, standard curves can be used to convert OD_{600} values to cell concentrations.

2. It has been reported that lithium acetate, dithiothreitol, or β-ME treatment increase the efficiency of electrotransformation of intact yeast cells, and that the sorbitol cell suspension can be stored at –70°C if glycerol is added to a final concentration of 15% *(8–10)*. We have not experimented with these variations and, for routine transformation experiments, find that the method as described yields sufficient transformants for most applications. Furthermore, *Schizosaccharomyces pombe* also can be efficiently electroporated using procedures similar to those outlined above *(11)*.

3. The efficiency of spheroplasting can be evaluated by spreading 200 μL of the cells onto nonselective, nonisotonic media (YPD). The spheroplasts will not grow in this nonisotonic environment, and after 24 h, only a few hundred colonies should appear. If a lawn of colonies appears, it indicates that spheroplasting was inefficient and that digestion time should be increased. Alternatively, a loss of lyticase activity owing to improper storage conditions may have occurred and a fresh enzyme preparation may be required.

4. Frozen spheroplasts offer convenience and high transformation efficiency. The ability to obtain high efficiencies from frozen cells is apparently the result of the combination of glycerol, DMSO, and the complex sugars. The use of freeze/thawing, addition of DMSO during the heat-shock thawing step, and the use of carrier DNA are factors that increase transformation efficiency *(3–5)*. Although consistent results for both electroporation and spheroplast transformation for strain AB1380 *(3)* were obtained using the procedures described here, other yeast strains may require alterations in the time of lyticase exposure for the preparation of spheroplasts to optimize rates of transformation, regeneration, and growth of transformants.

5. To determine the efficiency of transformation, wash cells after thawing as described in Section 3.2., step 9, and add 0.1 μg of a control *ARS* plasmid. Electroporate and plate on selective media containing sorbitol. After 3–5 d at 30°C, there should be several thousand colonies on these test plates. When transforming with a large plasmid or YAC, transformation frequencies will be lower than with the test plasmid.

6. Nonselective media can be used to check the viability of spheroplasts. Mix 200 μL of cells with 5 mL of YPD top agar containing 1*M* sorbitol, and pour onto a YPD agar plate containing 1*M* sorbitol. Under these conditions, all cells should grow, and a lawn of cells should appear after 24 h.

Absence of a lawn indicates low cell viability at the time of transformation. If this occurs, decrease the time of incubation with the enzyme, use slower centrifugation speeds, and use minimal agitation during all resuspension steps.

7. Spheroplasts are transformed with an efficiency between 10^6 and 10^7 transformants/µg of plasmid DNA. Freezing and immediately thawing spheroplasts in the presence of cell storage solution significantly increases transformation efficiency over that obtained with spheroplasts that are not frozen and thawed. Competent spheroplasts are stable at $-70°C$ for >6 mo.

References

1. Hinnen, A., Hicks, J. B., and Fink, G. R. (1978) Transformation of yeast. *Proc. Natl. Acad. Sci. USA* **75**, 1929–1934.
2. Ito, H., Fukuda, Y., Murata, K., and Kimura, A. (1983) Transformation of intact yeast cells treated with alkali cations. *J. Bacteriol.* **153**, 163–168.
3. Burgers, P. M. J. and Percival, A. (1987) Transformation of yeast spheroplasts without cell fusion. *Anal. Biochem.* **163**, 391–401.
4. Gallego, C., Casas, C., and Herrero, E. (1993) Increased transformation levels in intact cells of *Saccharomyces cerevisiae* aculeacin A-resistant mutants. *Yeast* **9**, 523–526.
5. Geitz, R. D., Jean, A. S., Woods, R. A., and Schistl, R. H. (1992) Improved method for high efficiency transformation of intact yeast cells. *Nucleic Acids Res.* **20**, 1425–1434.
6. Fields, S. and Song, O. (1989) A novel genetic system to detect protein-protein interactions. *Nature* **340**, 245,246.
7. Harper, J. W., Adami, G. R., Wei, N., Keyomarski, K., and Elledge, S. J. (1993) The p21 Cdk-Interacting protein Cip1 is a potent inhibitor of G1 cyclin-dependent kinases. *Cell* **75**, 805–816.
8. Simon, J. R. (1993) Transformation of intact yeast cells by electroporation. *Methods Enzymol.* **217**, 478–483.
9. Ju, Q. and Warner, J. R. (1991) Competent *Saccharomyces cerevisiae* cells can be frozen and used for transforming with high frequency. *Trends Genet.* **7**, 242.
10. Hill, J., Ian, K. A., Donald, G., and Griffiths, D. E. (1991) DMSO-enhanced whole cell yeast transformation. *Nucleic Acids Res.* **19**, 5791.
11. Zhao, Y., Hopkins, K. M., and Lieberman, H. B. (1993) A method for the preparation and storage of frozen, competent *Schizosaccharomyces pombe* spheroplasts. *Biotechniques* **15**, 238,239.

CHAPTER 27

Ten-Minute Electrotransformation of *Saccharomyces cerevisiae*

Martin Grey and Martin Brendel

1. Introduction

The introduction of exogenous DNA into prokaryotic and eukaryotic cells is one of the most frequently used procedures in the daily laboratory work of a molecular geneticist. In recent years, methods based on high-voltage electric shocks have been established for various species (reviewed in this book). In the case of *Saccharomyces cerevisiae,* several procedures have been described that emphasize maximal transformation efficiency *(1–5).* In this case, many parameters had to be optimized that require accurate, and consequently time-consuming, controls.

Since the most favored organism for screening and amplification of plasmids is *E. coli* because of its several advantages (e.g., short generation time and high transformation efficiency), the transformation of yeast usually involves the introduction of a defined plasmid (with the exception of screening a gene bank) and it is, therefore, not the prime objective of such an experiment to obtain thousands of identical transformants. With this in mind, we have developed an extremely simple method that is reduced to basic steps and emphasizes rapidity in execution rather than maximal transformation efficiency. It is possible to transform a yeast strain without any preparation of cell cultures with defined growth phase, just by scraping the cells off the agar surface of Petri dish cultures and subjecting them to transformation by electroporation.

This 10-min protocol should contribute to flexibility in the laboratory routine of molecular geneticists working with *S. cerevisiae.*

From: *Methods in Molecular Biology, Vol. 47: Electroporation Protocols for Microorganisms*
Edited by: J. A. Nickoloff Humana Press Inc., Totowa, NJ

2. Materials

1. Electroporation buffer: 18.2 g sorbitol and distilled H_2O to 100 mL. Sterilize by autoclaving. Add 2 mL of sterile HEPES buffer ($1M$, pH 7.3, Gibco, Grand Island, NY) to a final concentration of 20 mM (*see* Note 2). Store at room temperature.
2. Sorbitol plates: Standard selective plates containing $1M$ sorbitol. Add before autoclaving medium (*see* Note 6).
3. Plasmid DNA: Miniprep DNA *(6)* dissolved in either sterile distilled water or TE buffer (*see* Note 3).

3. Methods

1. Collect cells of the desired yeast strain from a YEPD plate, normally kept in the refrigerator at 4°C, using a sterile transfer loop, and suspend them in a 1.5-mL tube containing approx 1 mL of electroporation buffer. Vortex briefly to ensure that clumps of cells have dispersed (*see* Notes 1 and 2).
2. Centrifuge the cells at room temperature for 2 s at 9000g.
3. Decant the supernatant, and resuspend the cells in the remaining liquid by vortexing. Transfer 40 µL of the suspension into a sterile electroporation cuvet.
4. Add 200–1000 ng of plasmid DNA dissolved in up to 5 µL TE buffer or distilled H_2O and mix gently (*see* Note 3).
5. Pulse at 1.4 kV, 200 Ω, 25 µF using cuvets with a 0.2-cm electrode gap (e.g., Gene Pulser, Bio-Rad, Richmond, CA). Monitor the pulse duration if possible (*see* Notes 4 and 5).
6. Immediately add approx 1 mL of electroporation buffer, and mix the suspension carefully by gently pipeting up and down.
7. Using the same pipet, spread aliquots of the transformed cells directly onto appropriate selective agar plates containing $1M$ sorbitol (*see* Note 6) without any further treatment. Incubate at 28–30°C until colonies appear (3–5 d).

4. Notes

1. The yeast strains used in this protocol have been stored on solid complete medium (YEPD) in a refrigerator before transformation. Fresh-grown cells achieve maximum transformation efficiency, whereas the yield of transformants will decrease with extended storage time by a factor of approximately two after 6 wk in the refrigerator (Fig. 1).
2. The presence of 20 mM HEPES in the electroporation buffer increases the transformation efficiency by a factor of about 2.5 irrespective of the age of the yeast cells (Fig. 1), but HEPES is not essential.
3. Generally, an optimal transformation efficiency is achieved if DNA of high purity is used. Since this protocol emphasizes simplicity, all parameters

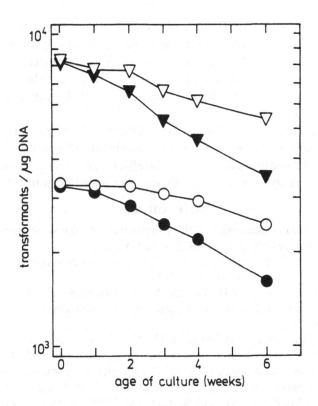

Fig. 1. Transformation efficiency of haploid yeast strain MG 5123-6B (*Mat*a, *ura3-52, his5-2, ade2-1*) with YEp24 (New England Biolabs, Beverly, MA) in relation to storage time of the culture on solid medium at 4°C. At zero weeks of age, the cells had been growing at 30°C for 3 d. Filled triangles and circles, number of transformants in presence and absence of HEPES, respectively. Open symbols, transformation efficiency corrected for cell survival ([total number of subjected cells]/[number of living cells]) before electroporation.

and data concerning efficiency have been developed using crude plasmid DNA obtained by a rapid isolation technique *(6)*.

4. A critical parameter is the duration of the current pulse, which should be approx 4.5 ms with the conditions described in our protocol. If the resistance between the two electrodes in the cuvet is too low owing to a high salt concentration in the sample, the pulse duration will not exceed 1 ms and the transformation efficiency will decrease significantly. In this case, the transformation should be repeated with a new sample and reduced voltage (e.g., 1.35 instead of 1.4 kV).

5. Among the abovementioned pulse parameters, only the voltage should be varied, whereas the other parameters should remain constant. The value given in the protocol is suitable for all yeast strains tested in our laboratory, but may not be optimal for all strains. If a particular strain is used frequently for transformations, it may be useful to determine the optimum voltage for that strain.

6. The plates that are used to select transformed cells should contain $1M$ sorbitol. If such plates are not available, standard selective plates without sorbitol may be substituted. Although the efficiency is reduced in this case by 2- to 10-fold, the method normally still yields sufficient transformants.

References

1. Becker, D. M. and Guarente, L. (1991) High-efficiency transformation of yeast by electroporation. *Methods Enzymol.* **194,** 182–187.
2. Delorme, E. (1989) Transformation of *Saccharomyces cerevisiae* by electroporation. *Appl. Environ. Microbiol.* **55,** 2242–2246.
3. Hashimoto, H., Morikawa, H., Yamada, Y., and Kimura, A. (1985) A novel method for transformation of intact yeast cells by electroinjection. *Appl. Microbiol. Biotechnol.* **21,** 336–339.
4. Karube, I., Tamiya, E., and Matsuoka, H. (1985) Transformation of *Saccharomyces cerevisiae* spheroplasts by high electric pulse. *FEBS Lett.* **182,** 90–94.
5. Manivasakam, P. and Schiestl, H. (1993) High efficiency transformation of *Saccharomyces cerevisiae* by electroporation. *Nucleic Acids Res.* **21,** 4414,4415.
6. Del Sal, G., Manfioletti, G., and Schneider, C. (1988) A one-tube plasmid DNA mini-preparation suitable for sequencing. *Nucleic Acids Res.* **16,** 9878.

CHAPTER 28

Electroporation
of *Schizosaccharomyces pombe*

Mark T. Hood and C. S. Stachow

1. Introduction

Many of the techniques that have been developed for the manipulation of the budding yeast *Saccharomyces cerevisiae* have now been adapted to be used on the alternative host, *Schizosaccharomyces pombe*. One particularly important technique is the introduction of exogenous DNA into the yeast cell. One of the earlier methods requires generating protoplasts with a cell-wall-degrading enzyme prior to the introduction of DNA, generally giving transformation efficiencies of $2–3 \times 10^4$ transformants/µg DNA *(1)*. When combined with the cationic liposome-forming reagent Lipofectin, the protoplast method can generate transformation efficiencies of 7.0×10^5 transformants/µg DNA *(2)*. Protocols for transforming intact yeast have been developed that involve treatment of cells with monovalent cations, polyethylene glycol (PEG), and a 25-min heat pulse *(3,4)*. Although the transformation efficiencies by these methods are lower than those of the standard protoplast method, they are not as tedious or time-consuming. Electroporation, subjecting cells to a controlled electrical pulse, is a transformation technique that has recently gained popularity. The main advantage of electroporation is the ease and time required to generate transformants. In addition, because of its biophysical nature, electroporation works well with a wide variety of cell types *(5–8)*. It has also been used to incorporate a number of different molecules into cells *(9–12)*. The procedure for electroporation presented

From: *Methods in Molecular Biology, Vol. 47: Electroporation Protocols for Microorganisms*
Edited by: J. A. Nickoloff Humana Press Inc., Totowa, NJ

below was developed as an easier and less time-consuming alternative for transformation of *S. pombe*.

2. Materials

2.1. Cells and Plasmid Vectors

As with most transformation methods, transformation efficiency gained by electroporation will vary depending on the type of vector and strain of cell used *(13–15)*. However, with *S. cerevisiae,* vectors of various size and construction have been successfully transformed into cells by electroporation *(16)*. The variations seemed to be owing more to intrinsic genetic properties of the vectors and the host than to physical properties, such as size (although there probably exists a size limitation; *see* Note 5). The protocol presented here was developed using *S. pombe* h- ura 4-294 with the cloning vector pFL20, which contains *S. pombe* autonomously replicating sequences, and the *S. cerevisiae URA3* gene, which complements the *S. pombe ura4-294* mutation *(17)*.

2.2. Growth Media and Solutions

1. YDP: 1% yeast extract, 2% dextrose, and 2% bacto-peptone.
2. YNB selection plates: 0.67% yeast nitrogen base without amino acids, 2% glucose, and 1.5% agarose.
3. TE buffer: 10 mM Tris-HCl and 1 mM EDTA, pH 8.0.
4. Polyethylene glycol 4000 (PEG): 60% PEG 4000 (Sigma, St. Louis, MO) in TE buffer, filter-sterilized.

3. Methods

3.1. Preparation of Cells

1. Inoculate 50 mL of YDP with a single colony from a YDP agar plate, and incubate with moderate aeration through shaking at 30°C until late-log phase (OD$_{595}$ ~2.1 or approx 20 h) *(see* Note 1).
2. Harvest cells by centrifugation at approx 3000g for 5 min at 4°C, and resuspend pellet in 1 mL of TE. Repeat centrifugation, and resuspend in 1 mL of TE. For each transformation, mix 50 µL of cells with 50 µL of 60% PEG 4000.
3. Add between 1 ng and 1 µg of plasmid DNA *(see* Note 2) to the cell/PEG suspension, and mix thoroughly by tapping the side of the tube. Incubate suspension on ice until ready for electroporation *(see* Note 3).

3.2. Electroporation

1. Place the electrocuvets (one per transformation) and holder in the freezer (–20°C) 20 or 30 min prior to electroporation.

2. Remove a cold electrocuvet and holder from the freezer, and transfer the 100 μL cell/DNA suspension into the cuvet. Tap the cuvet on the bench top to bring all of the solution to the bottom. Deliver a pulse of 1.7 kV, 1 μF, 600 Ω (*see* Note 4).

3. Dilute the suspension directly in the electrocuvet with 300 μL of TE, and transfer the entire 400 μL to a microcentrifuge tube. Plate 10 and 100 μL aliquots on selection agar plates and incubate at 30°C. Transformed colonies should appear in 3–6 d. We have found that allowing the electroporated cells to remain in PEG for up to 40 min prior to dilution and plating increases the number of transformants (*see* Note 5). However, if maximal transformation efficiency is not a requirement, this incubation can be omitted.

4. Notes

1. When subjected to an electrical pulse, younger cell cultures are more fragile than cultures in late-log or stationary phase, as shown through viability studies. This undoubtedly explains in part the lower transformation efficiencies obtained with younger cultures. Although stationary phase cell cultures display high viability after electroporation, transformation efficiencies are low.

2. With pFL20, we have found that the saturating amount of DNA per 100 μL of cells is approx 1 μg and that higher transformation efficiencies can be obtained with less DNA. This may vary depending on the DNA vector used *(18)*. Also, the DNA should be in a low-ionic-strength buffer and at a concentration that will allow addition of the desired amount in a volume of <5 μL. This, along with washing cells with TE prior to electroporation, will help to reduce the possibility of arcing during electroporation. Arcing, caused by electroporation of samples with substantial ionic strength, results in loss of sample and generally low transformation frequencies.

3. Some authors have reported that competent cells may be stored frozen at –70°C without loss of transformation efficiency *(18)*. It is advisable to wash thawed cells prior to electroporation.

4. The electroporation conditions used here generally give a decay time of 0.5 ms.

5. The reason for increased transformation efficiencies if electroporated cells are allowed to remain in PEG seems to be related to the pore size generated by the electrical pulse. Through the use of fluorescent macromolecules, it was shown that PEG not only maintains, but increases the size of the pores *(19)*. This is at the cost of viability, but longer incubation in PEG may allow uptake of larger DNA molecules. If DNA size is small and maximal number of transformants is not important, this incubation can be omitted to save time (*see* Fig. 1).

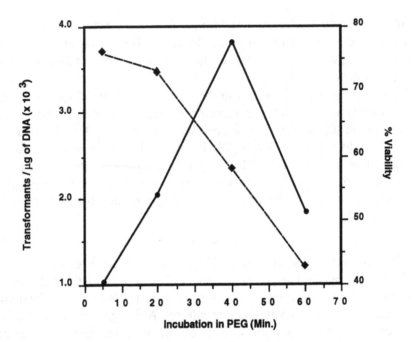

Fig. 1. Effect of PEG incubation on transformation efficiency and viability of electroporated *S. pombe* cells. Plasmid DNA (1 μg) was added to 100 μL suspensions of electroporated *S. pombe* cells after 5, 20, 40, and 60 min of incubation in PEG. Cells were allowed 5 min for DNA uptake followed by dilution with 300 μL of TE buffer. Aliquots of 10 and 100 μL were spread onto duplicate YNB selection plates for determination of transformation efficiency (solid line), and 100-μL aliquots were serially diluted and spread on duplicate YDP agar plates for determination of viability (dashed line).

6. Protocols for electroporation using various additional manipulations have been published for other yeasts. Some of these techniques include generating protoplasts prior to the electrical pulse *(13)*, providing continuous osmotic support with sorbitol *(16)*, using single-stranded carrier DNA at the time of the electrical pulse *(20)*, and pretreatment with dithiothreitol *(21)*. Improvement of transformation efficiency may be obtained if such treatments are applied to the electroporation of *S. pombe*.

7. We have routinely reused electrocuvets without noticeable loss in transformation efficiency. After cleaning with distilled water, the cuvets are autoclaved for 15 min to sterilize. The plastic covers, which are not autoclavable, are kept in ethanol until use. Reused cuvets should be checked for visible damage that occasionally occurs on autoclaving.

Table 1
Current Published Protocols for Transformation of *S. pombe*

Method[a]	Efficiency[b]	Estimated time[c]	Reference
Protoplast	$2-3 \times 10^4$	3 h	*1*
Protoplast/lipofectin	7.0×10^5	2.5 h	*2*
LiAc/minimal media	4.6×10^5	4.5 h	*15*
LiCl	$4-9 \times 10^3$	2 h	*4*
Electroporation	$1 \times 10^5-10^6$	30 min	*18*
Electroporation	8×10^3	10 min	*19*

[a]The titles listed here include only a defining step in the procedure.
[b]Transformation efficiency is defined as transformants/µg DNA. Different yeast strains and transforming vectors were used in each of the protocols.
[c]Estimated time may vary.

8. Table 1 lists the current published protocols for transformation of *S. pombe*. Actual efficiencies may vary depending on the transforming vector and yeast strain.

References

1. Beach, D. and Nurse, P. (1981) High frequency transformation of the fission yeast *Schizosaccharomyces pombe*. *Nature* **290**, 140–142.
2. Allshire, R. C. (1990) Introduction of large linear minichromosomes into *Schizosaccharomyces pombe* by an improved transformation procedure. *Proc. Natl. Acad. Sci. USA* **87**, 4043–4047.
3. Ito, H., Fukuda, Y., Murata, K., and Kimura, A. (1983) Transformation of intact yeast cells treated with alkali cations. *J. Bacteriol.* **153**, 163–168.
4. Bröker, M. (1987) Transformation of intact *Schizosaccharomyces pombe* cells with plasmid DNA. *BioTechniques* **5**, 516–518.
5. Neumann, E., Schaefer-Ridder, M., Wang, Y., and Hofschneider, P. H. (1982) Gene transfer into mouse lyoma cells by electroporation in high electric fields. *EMBO J.* **1**, 841–845.
6. Fromm, M., Taylor, L. P., and Walbot, V. (1985) Expression of genes transferred into monocot and dicot plant cells by electroporation. *Proc. Natl. Acad. Sci. USA* **82**, 5824–5828.
7. Gibson, W. C., White, J. C., Laird, P. W., and Borst, P. (1987) Stable introduction of exogenous DNA into *Trypanosoma brucei*. *EMBO J.* **6**, 2457–2461.
8. Dower, W. J., Miller, J. F., and Ragsdale, C. W. (1988) High efficiency transformation of *E. coli* by high voltage electroporation. *Nucleic Acids Res.* **16**, 6127–6145.
9. Fromm, M., Callis, J., Taylor, L. P., and Walbot, V. (1987) Electroporation of DNA and RNA into plant protoplast. *Methods Enzymol.* **153**, 351–367.
10. Uno, I., Fukami, K., Kato, H., Takenawa, T., and Ishikawa, T. (1988) Essential role for phophatidylinositol 4,5-biphosphate in yeast cell proliferation. *Nature* **333**, 188–190.

11. Weaver, J. C., Harrison, G. I., Bliss, J. G., Mourant, J. R., and Powell, K. T. (1988 Electroporation: high frequency of occurrence of a transient high-permeability stat in erythrocytes and intact yeast. *FEBS Lett.* **229**, 30–34.

12. Yamamoto, T., Moerschell, R. P., Wakem, L. P., Ferguson, D., and Sherman, I (1992) Parameters affecting the frequencies of transformation and co-transformi tion with synthetic oligonucleotides in yeast. *Yeast* **8**, 935–948.

13. Karube, I., Tamiya, E., and Matsuoka, H. (1985) Transformation of *Saccharomy ces cerevisiae* spheroplast by high electrical pulse. *FEBS Lett.* **182**, 90–94.

14. Delorme, E. (1989) Transformation of *Saccharomyces cerevisiae* by electropore tion. *Appl. Environ. Microbiol.* **55**, 2242–2246.

15. Okazaki, K., Okazaki, N., Kume, K., Jinno, S., Tanaka, K., and Okayama, F (1990) High-frequency transformation method and library transducing vectors fo cloning mammalian cDNAs by trans-complementation of *Schizosaccharomyce pombe*. *Nucleic Acids Res.* **18**, 6485–6489.

16. Becker, D. M. and Guarente, L. (1991) High efficiency transformation of yeast b transformation. *Methods Enzymol.* **194**, 182–187.

17. Lossen, R. and Lacroute, F. (1983) Plasmid carrying the yeast OMP decarboxylas structural and regulatory genes: transcription regulation in a foreign environmen *Cell* **32**, 371–377.

18. Prentice, H. L. (1991) High efficiency transformation of *Schizosaccharomyce pombe* by electroporation. *Nucleic Acids Res.* **20**, 621.

19. Hood, M. T. and Stachow, C. S. (1992) Influence of polyethylene glycol on th size of *Schizosaccharomyces pombe* electropores. *Appl. Environ. Microbiol.* **58** 1201–1206.

20. Manivasakam, P. and Schiestl, R. H. (1993) High efficiency transformation of *Sac charomyces cerevisiae* by electroporation. *Nucleic Acids Res.* **21**, 4414,4415.

21. Sánchez, M., Iglesias, F. J., Santamaria, C., and Domínguez, A. (1993) Transfoi mation of *Kluyveromyces lactis* by electroporation. *Appl. Environ. Microbiol.* **59** 2087–2092.

CHAPTER 29

Gene Transfer by Electroporation of Filamentous Fungi

M. Kapoor

1. Introduction

The fungi encompass an enormous array of species, ranging from microscopic uninucleate, unicellular forms to multinucleate, coenocytic, highly differentiated macroscopic morphological forms. The wealth of diversity provided by a wide variety of life cycles, the availability of asexual and sexual modes of reproduction, and the presence of both haploid and diploid phases of the life cycle add further dimensions to this fascinating group of eukaryotic microbes. The tremendous potential of various taxa of filamentous fungi for biotechnological applications has yet to be realized. The filamentous fungi are emerging as a source of suitable hosts for expression of mammalian and other eukaryotic genes to yield products of commercial interest. They are ideally suited to this end by virtue of their normal mode of nutrition, based on the capacity for secretion of degradative enzymes, instrumental in conversion of complex growth substrates into simpler, readily utilizable derivatives that are absorbed by the growing fungal hyphae. Several species of filamentous fungi are known to produce pharmaceuticals, antibiotics, metabolites, phytohormones, and other industrially important products. There is a great deal of interest in the potential for the use of filamentous fungi as biocontrol agents—as antagonists of other fungal phytopathogens, as bioherbicides and bioinsecticides.

Species of Ascomycetes, such as *Neurospora crassa* and *Aspergillus nidulans,* are also of interest as useful model systems for investiga-

From: *Methods in Molecular Biology, Vol. 47: Electroporation Protocols for Microorganisms*
Edited by: J. A. Nickoloff Humana Press Inc., Totowa, NJ

tions of the molecular mechanisms of differentiation and regulation of eukaryotic gene expression. For virtually half a century of genetic research, *N. crassa*—along with the model prokaryote *Escherichia coli* and the eukaryote *Drosophila melanogaster*—has been the organism of choice for investigations of the fundamental principles of genetic recombination, eukaryotic genome organization and biochemical genetics, and regulation of metabolic pathways. The availability of a vast collection of auxotrophic and other mutants, together with the timely development of suitable experimental tools for molecular biological investigations, has rendered these model organisms invaluable for biological research.

Although some members of the Ascomycetes and Basidiomycetes are genetically well characterized, a vast majority of the fungal species, particularly the so-called imperfect fungi, lacking a sexual cycle, are not amenable to direct genetic analysis. The ability to carry out genetic manipulations with the objective of improving the yield of commercially important products of fungal species and to exploit filamentous fungi in biotechnology as host organisms for expression of eukaryotic genes is dependent on success in transferring cloned genes into fungal cells. Therefore, it is essential to have access to a variety of methods for transformation since individual species may require distinctive treatments, commensurate with the chemical composition of the cellular envelope. The use of spheroplasts in transformation experiments with filamentous fungi has been a common practice, ever since a suitable procedure was developed for *N. crassa* spheroplast formation by Case et al. *(1)*. Protocols based on slight modifications of the original method have been adapted successfully for use with several fungal species *(2)*. Although transformation of protoplasts results in high yields of transformants for some fungal species, notably *N. crassa,* the experimental procedure for preparation of protoplasts requires careful standardization of the individual steps. Often problems are encountered owing to the lack of reproducibility when different batches of commercial cell-wall-degrading enzyme preparations, such as Novozyme, are employed. Following transformation, regeneration of the cell wall is essential since protoplasts are extremely fragile. The success of the regeneration step, although vital in this procedure, can be variable. Furthermore, many of the fungal species do not yield protoplast preparations that are suitable for transformation experiments.

Consequently, attempts have been made to develop transformation protocols that do not require protoplasts. One of these procedures involves permeabilization of germinating *N. crassa* conidia by treatment with lithium acetate *(3)* followed by polyethylene glycol, adapted from a method originally developed for *S. cerevisiae* by Ito et al. *(4)*. However, this procedure has not been widely used, because lithium salts are found to be toxic for some fungal species and the overall transformation efficiencies have been reported to be generally low. A transformation system for filamentous fungi, independent of protoplast formation and potentially toxic alkali cations, is furnished by electroporation *(5)*.

2. Materials

2.1. Selectable Markers

The choice of dominant selectable markers depends on the availability of cloned marker genes that can be expressed in the recipient cells, yielding easily scorable phenotypic characteristics. Theoretically any fungal gene with a promoter region that is recognizable by the recipient's transcriptional apparatus can be employed. In this respect, the *N. crassa* gene encoding the benomyl-resistant mutant of β-tubulin *(6)* has proven to be a versatile marker for fungal species that are susceptible to benomyl, and it has been employed in the construction of cosmid vectors for genomic libraries.

A vector constructed with the hygromycin phosphotransferase gene of *E. coli,* pCSN44 *(7),* has proven to be suitable in many instances, the prime requisite being the susceptibility of the target fungus to hygromycin B. Alternatively, appropriate selectable, homologous marker genes can be used in conjunction with auxotrophic mutants as recipients for transforming DNA. A good example of this approach uses the qa-2 gene of *N. crassa,* encoding the catabolic dehydroquinase gene and the recipient strain R-206A, which is a double mutant, deficient in the catabolic as well as the biosynthetic dehydroquinase (*see* Notes 1–3 and Table 2). Table 1 lists the selectable markers and the plasmids employed by us using the protocols described in this chapter.

2.2. Recipient Cell Type and Pretreatment

The electroporation-based transformation procedure, originally developed by us for use with *N. crassa (5,8),* has been found to be readily applicable to other species of filamentous fungi, such as *Aspergillus*

Table 1
Species of Filamentous Fungi, Plasmids
for Electroporation and Selectable Markers Employed

Recipient species/strains	Selectable markers	Plasmids	Reference
Neurospora crassa			
Wild type	*E. coli* hygB[r]	pCSN44	8
(74-OR23-IVA)	*N. crassa* ben[r]	pBEN	
Auxotrophic mutant	*N. crassa* qa-2[+]	Bsqa	5,8
R-206A			
Pencillium urticae	*E. coli* hygB[r]	pCSN44	8
(NRRL 2159A)			
Aspergillus oryzae	*N. crassa* ben[r]	pBEN	8
(ATCC 14895)			
Beauveria bassiana	*N. crassa* ben[r]	pBEN	Unpublished
(ATCC 7159)			

Table 2
Genes and Plasmids Used for Transformation of Germinated Conidia of *N. crassa*

Gene	Cotransformed Selectable marker	Phenotype conferred
N. crassa gdh-1	*N. crassa* ben[r]	Benomyl resistance
(NAD[+]-glutamate	*E. coli* hygB[r]	Hygromycin B resistance
dehydrogenase		
N. crassa hspe-1	*N. crassa* ben[r]	Benomyl resistance
	E. coli hygB[r]	Hygromycin B resistance
N. crassa hsps-1	*N. crassa* qa-2[+]	Dehydroquinase production
Human mtIIA	None	Cadmium chloride
(metallothionein gene		resistance
family)		

oryzae, Leptosphaeria maculans, Penicillium urticae, and *Beauveria bassiana (B. sulfurescens).* With sporulating species, it is feasible to use sexual or asexual spores—macro- or microconidia, ascospores, and pycnidiospores. Although intact spores, with walls of varying thickness and chemical composition, are not suitable unless spheroplasts are formed first, spores during early stages of germination are excellent recipients for exogenous DNA, following brief pretreatment with a suitable enzyme (*see* Notes 4 and 5).

2.3. Solutions and Reagents
for Growth of Fungi and Electroporation

1. Electroporation buffer: 1 mM HEPES buffer, pH 7.5, 50 mM mannitol.
2. β-glucuronidase (Sigma [St. Louis, MO] type H-1) is supplied in powder form. It is a partially purified preparation from *Helix pomatia* containing 300–400 U/mg β-glucuronidase activity and 15–40 U/mg of sulfatase activity (*see* Note 4).
3. Trace element solution: Dissolve successively in 95 mL water: 5.0 g citric acid · 1H$_2$O, 5 g ZnSO$_4$ · 7H$_2$O, 1.0 g Fe(NH$_4$)$_2$(SO$_4$)$_2$ · 6H$_2$O, 0.25 g CuSO$_4$ · 5H$_2$O, 0.05 g MnSO$_4$ · 1H$_2$O, 0.05 g H$_3$BO$_3$ (anhydrous), and 0.05 g Na$_2$MoO$_4$ · 2H$_2$O. Adjust total volume to 100 mL. Store at room temperature with 1 mL chloroform.
4. Biotin stock solution: Dissolve 5.0 mg in 100 mL of 50% ethanol. Store at –20°C in 2.5-mL aliquots.
5. Fries' medium: In a total volume of 500 mL, dissolve 2.5 g (NH$_4$)$_2$ tartrate; 0.5 g NH$_4$NO$_3$, 0.5 g KH$_2$PO$_4$, 0.25 g MgSO$_4$ · 7H$_2$O, 0.05 g CaCl$_2$, 0.05 g NaCl, 0.05 mL biotin solution, 0.05 mL trace element solution. Sucrose (1.5%) may be used as the carbon source.
6. Vogel's minimal medium (for 50X stock): Dissolve successively in 750 mL distilled water: 124 g sodium citrate · 2H$_2$O, 250 g KH$_2$PO$_4$ (anhydrous), 100 g NH$_4$NO$_3$ (anhydrous), 5 g CaCl$_2$ · 2H$_2$O (dissolve separately in 10 mL water and add slowly), 2.5 mL biotin stock solution, 5.0 mL trace element solution. Adjust final volume to 1 L. Store at room temperature with 5 mL chloroform as a preservative. The pH of the 1X strength medium will be close to 5.8; adjustment is not necessary.
7. Czapek-Dox medium is available from Difco, Detroit, MI.
8. Mineral medium: For 1 L of medium, mix 2 g (NH$_4$)$_2$SO$_4$, 1 g K$_2$HPO$_4$, 0.5 g MgSO$_4$, 0.5 g KCl, 0.3 g ZnSO$_4$, 0.01 g FeSO$_4$ · 7H$_2$O dH$_2$O to 900 mL. Autoclave, and then add 100 mL of 30% (w/v) glucose presterilized by filtration.
9. Plasmid DNA: CsCl-purified or relatively crude, "miniprep" quality DNA is satisfactory.

3. Methods

3.1. Electroporation Protocol for Neurospora crassa

1. Germinate conidia from 7-d-old 50-mL cultures of the wild-type strain (74A) or 15-d-old 50-mL cultures of strain R-206A (auxotrophic for aromatic amino acids) in 0.5X Fries' medium *(9)*, containing the appropriate growth supplements, for 2 h at 30°C, while shaking at 150 rpm in a rotary shaker. Culture R-206A in a medium containing a carbon source supple-

mented with 80 μg/mL each of L-phenylalanine, L-tyrosine, and L-tryptophan, 2 μg/mL of *p*-aminobenzoic acid, and 0.2 μg/mL inositol (*see* Note 6).

2. Immediately after the emergence of the germ tubes, add 1 mg/mL solid β-glucuronidase (Sigma: type H1), and continue the treatment for another 2 h under the conditions specified in step 1 (*see* Notes 4 and 7).

3. Wash the germinated conidia free of the growth medium by centrifugation at 3000*g* for 5 min, and transfer to the electroporation buffer. Repeat the washing step three times by centrifugation, and resuspension of the conidial pellet.

4. Suspend the pellet in 0.5–1.0 mL of the electroporation buffer, and to 100-μL conidial suspension, add 1–2 μg of transforming DNA (in 1 or 2 μL). Chill the mixture on ice for 15 min, and subject to electroporation at room temperature. Controls, without added DNA, are treated in exactly the same manner.

5. Pulse once with a field strength of 12.5 kV/cm; capacitance, 25 μF; resistance, 400 Ω. This will yield a time constant of about 5 ms.

6. Immediately following electroporation, add 1 mL of minimal medium (with supplements, if the recipient strain is auxotrophic) to the electroporation cuvet, suspend the contents rapidly, transfer the mixture to a sterile test tube, and incubate for 2–3 h at 30°C, while shaking. This step is important for optimal recovery of transformants. Optimal recovery time should be determined empirically for each species/strain.

7. Finally, select the transformants by plating appropriate dilutions of the spore suspension on a suitable selection medium. The choice of the selection medium will depend on the plasmid/gene employed for transformation (*see* Note 8 *[10]*).

8. Fungal colonies appear within 48 h on benomyl plates and within 72–96 h on hygromycin B plates. Transfer individual colonies to agar slants containing the selection medium (*see* Notes 9 and 10 *[11–13]*).

3.2. Electroporation Protocol for Aspergillus oryzae

1. Prepare a dense spore suspension using a 10-d-old culture in 25 mL of Czapek-Dox liquid medium. Grow the culture at 30°C for 6 h with shaking or until germ tube emergence is evident by microscopic examination (*see* Note 6).

2. Add β-glucuronidase (1 mg/mL), and continue incubation while shaking under the same growth conditions for an additional 2 h.

3. Centrifuge the germinating spore suspension at 5900*g* in sterile glass centrifuge tubes for 10 min.

4. Gently remove the supernatant containing the enzyme using a sterile Pasteur pipet, resuspend the germinated spores in the electroporation buffer, and centrifuge at 5900*g* for 10 min. Resuspend the spores in the

Table 3
Experimental Conditions and Results of Electrotransformation
of Filamentous Fungi with a Dominant Selective Marker

Conditions/results	Fungal species		
	N. crassa	*P. urticae*	*A. oryzae*
Number of conidia ($\times 10^6$)	3.0–6.0	8.0	2.5
Percent viability	54–57	~68	52–55
Pretreatment			
β-glucuronidase	+	+	+
Chitinase	–	–	–
Selectable marker	Hygromycin B resistance	Hygromycin B resistance	Benomyl resistance
Transformation efficiency (stable transformants/ μg DNA)	~1.8×10^3	~2.6×10^3	1–6×10^2

same buffer, and repeat the washing step two more times to ensure complete removal of β-glucuronidase. After the final washing step, suspend the pellet in 500 μL of the electroporation buffer.

5. Transfer 100 μL of spore suspension to a sterile 1.5-mL tube, and add 1 μL of transforming DNA (1–5 μg) (*see* Note 1). Incubate the mixture on ice for 15 min. Prepare control samples without added transforming DNA.
6. Transfer the mixture of the spore suspension and the transforming DNA to an electroporation cuvet, and pulse once with 11–12.5 kV/cm, capacitance, 25 μF. A time constant of 4.6–4.8 ms will result.
7. Add 1 mL of an appropriate growth medium, and incubate the suspension while shaking for 30–60 min. This step is necessary to ensure proper recovery of transformants (*see* Note 11).
8. Finally, plate appropriate dilutions of the spore suspension on selective medium. For average transformation efficiencies *see* Table 3.

3.3. Protocol for Electroporation of Hyphal Fragments

In the case of filamentous fungal species where spore formation is either inefficient or the spores are not suitable for electroporation for other reasons, it is feasible to use hyphal fragments directly in electroporation experiments. An example is the following protocol that we have used with *Beauveria* species.

1. Prepare the initial inoculum by transferring approx 3-mm (approx 50 mg wet wt) of hyphal mass from a slant, using a sterile platinum loop, to 25 mL of mineral medium (*14*) in a 125-mL Erlenmeyer flask.

2. Allow the culture to grow at room temperature while shaking at 100 rpm on a rotary shaker for 48 h. During this time, the hyphal clusters increase in mass at a slow rate.

3. Transfer 100–200-µL vol of the medium with the fungal material to sterile 1.5-mL microcentrifuge tubes, and vortex at a moderate speed (e.g., setting of 4 of 5) for approx 30 s to fragment the hyphae. Pellet the mycelial fragments by centrifugation at 12,000g for 10 min at room temperature.

4. Remove the supernatant, and wash the pelleted fungal cells three times with sterile distilled water to remove the salts in the growth medium. Washing is carried out by resuspension of the cells followed by centrifugation. Following the final wash, suspend the cells in 500 µL of distilled water by vortexing (*see* Note 12).

5. To a sterile electroporation cuvet, add 5 µg plasmid DNA (1 µg/µL), 200 µL of the fungal cell suspension, and 200 µL sterile distilled water, and mix the contents gently. Pulse once with a field strength of 12.5 kV, capacitance, 25 µF, resistance, 400 Ω. This will produce a time constant of about 5 ms.

6. Dilute the electroporated mixture with 1 mL of mineral medium, transfer to a sterile 1.5-mL tube, and incubate at room temperature for 2 h. Plate appropriate volumes of the suspension (50–200 µL) on medium containing 1.5% agar.

7. After the surface has dried, overlay the plates with 5–6 mL of soft agar (0.7% agar + *B. bassiana* liquid medium containing 2 or 5 µg/mL benomyl), and incubate the plates at room temperature. Benomyl-resistant colonies will arise within 3–4 d. No colonies should be observed on controls electroporated in the absence of plasmid DNA and plated subsequently on benomyl-containing medium. Transformation efficiencies are estimated on the basis of the number of benomyl-resistant colonies recovered/µg of input plasmid DNA (*see* Note 13).

4. Notes

1. Cotransformation along with the dominant selectable gene has proven to be an effective mode of introducing target DNA into the recipient cells. The frequency of uptake of target DNA (nonselectable marker) along with the selectable marker is often sufficiently high for most purposes.

2. The overall efficiency of electrotransformation, assessed by determination of the average number of stable transformants recovered/µg of input DNA varies depending on the selectable marker and the host species/strain (Table 2). For instance, the auxotrophic strain R-206A of *N. crassa*, when used as the recipient with qa-2 DNA, resulted in a low yield of transformants (21 stable transformants/µg DNA), whereas the wild-type host strain was transformed with a markedly greater efficiency.

3. The optimal electroporation parameters were found to be comparable for all the filamentous fungal species tested (Table 3). The empirically determined combination of high field strength (12.5 kV/cm; the maximum output of our instrument) and capacitance value of 25 µF resulted in <50% reduction in viability of the treated cells.

4. A gentle treatment of the germinating spores with a suitable hydrolytic enzyme will often suffice to attain the requisite weakening of the cell walls. For this purpose, an appropriate enzyme should be selected in view of the chemical composition of the cell wall of the particular species. Commercial preparations containing cellulase, β-glucanase, β-glucuronidase, and chitinase can be used alone or in combination, depending on the target species. In this procedure, enzymatic treatment is brief to permit a slight weakening of the cell wall without extensive degradation. Protoplast formation is not required for success of electroporation. In fact, protoplasts, being extremely fragile, are susceptible to damage by the electroporation treatment.

5. For asexual spores (conidia), treatment with the mycolytic enzyme β-glucuronidase, at a concentration of 1 mg/mL, was found to be adequate. It is advisable to monitor the effect of the enzyme on cell viability and the rate of germination by frequent microscopic examination. Under the specified conditions, β-glucuronidase treatment was not detrimental for hyphal growth or cell viability, and spheroplast formation was not observed. However, if it is found necessary to use a higher concentration of β-glucuronidase and/or prolonged exposure, a reduction in viability may result. Treatment of pycnidiospores of *Leptosphaeria maculans* with thick, pigmented, chitinous spore walls requires the concomitant use of chitinase and β-glucuronidase *(8)*.

6. The age of the starting material is a critical factor in the yield of transformants. Since wild-type and mutants strains often differ widely in their rate of growth, the most suitable age should be determined empirically for each species and strain.

7. For each strain, the progress of germination should be monitored microscopically. For faster-growing strains, incubation intervals should be adjusted accordingly. Enzymatic treatment should not be excessively long; normally 1–2 h will be sufficient.

8. For *N. crassa,* we use Vogel's minimal medium *(10)* with 1% sorbose, 0.1% glucose, 0.1% fructose, and 1.5% agar (supplemented with 2 µg/mL benomyl for selection of resistant transformants). If hygromycin B-resistance is used as a selectable marker, the *N. crassa* conidial suspension is plated on the sorbose-glucose-fructose-agar medium without hygromycin B. The plates are incubated for 24 h, after which a 0.7% soft-agar overlay containing 200 µL/mL hygromycin B is applied. This two-step procedure

has been found to improve the yield of transformants, since a combination of sorbose and hygromycin B may prove to be lethal for the early stages of hyphal growth.

Alternatively, for selection of hygromycin B-resistant transformants of wild-type *N. crassa,* electroporated cells can be plated on δ Vogel's minimal medium—0.05% yeast extract—0.05% casein hydrolysate—0.01% fructose—1.5% agar, supplemented with 200 μg/mL hygromycin B. Following incubation for 24 h at 30°C, the plates are layered with 5 mL agar containing the same growth medium without hygromycin B.

9. Average values for transformation efficiency are given in Table 3.

10. We have utilized the above procedure for transformation of *N. crassa* germinating conidia using plasmids harboring heterologous genes, such as the human mtIIA *(11),* as well as homologous *N. crassa* genes. Efficient cotransformation was obtained for genes encoding NAD-specific glutamate dehydrogenase, *gdh*-1 *(12),* heat-shock protein 80, *hspe*-1 *(13),* 70-kDa heat-shock-protein, *hsps*-1 *(15),* and qa-2 *(1),* with benomyl-resistant β-tubulin *(6)* and hygromycin phosphotransferase *(7)* genes as selectable markers (Table 2). Southern blot hybridization showed the integration of introduced DNA at ectopic sites in the host genome *(8).*

11. Longer recovery times may be required if the yield of transformants is low. It may be necessary to adjust the recovery conditions for other species/strains of *Aspergillus.*

12. In contrast with germinating spores, treatment with β-glucuronidase is not necessary for successful transformation of hyphal fragments. Following transformation, hyphal fragments can regenerate readily, and a considerable simplification of the procedure is achieved. If hyphal fragments are used, however, care should be exercised in determining the appropriate stage of hyphal growth that yields the best results. Younger hyphae are best suited for electroporation without β-glucuronidase pretreatment.

13. Estimates of hyphal transformation efficiencies are approximate, since accurate counts of hyphal fragments are difficult owing to the variation in fragment size and the possibility of aggregation.

14. These electroporation procedures are simple, convenient, and inexpensive methods for DNA-mediated transformation of filamentous fungi. Neither spheroplast formation nor the use of potentially toxic chemicals is required; germinated spores and hyphal fragments are used directly. Furthermore, relatively crude preparations of plasmid DNA, such as those generated by "minipreps," yield satisfactory results. The yield of transformants is sufficiently high for most genetic manipulations and biotechnological applications.

15. The above protocol can be adapted for use with virtually any filamentous fungal species, provided appropriate selectable markers are available.

Acknowledgments

This work was supported by an operating grant from the Natural Sciences and Engineering Research Council (NSERC) of Canada. The contribution of B. N. Chakraborty, N. A. Patterson, C. Turnnir, and other members of my laboratory in developing and testing the electroporation protocols is greatly appreciated.

References

1. Case, M. E., Schweizer, M., Kushner, S. R., and Giles, N. H. (1979) Efficient transformation of *Neurospora crassa* by utilizing hybrid plasmid DNA. *Proc. Natl. Acad. Sci. USA* **76**, 5259–5263.
2. Fincham, J. R. S. (1989) Transformation in fungi. *Microbiol. Rev.* **53**, 148–170.
3. Dhawale, S. S., Paietta, J., and Marzluf, G. A. (1984) A new, rapid and efficient transformation procedure for *Neurospora*. *Curr. Genet.* **8**, 77–79.
4. Ito, H., Fukuda, Y., Murata, K., and Kimura, A. (1983) Transformation of intact yeast cells treated with alkali cations. *J. Bacteriol.* **153**, 163–168.
5. Chakraborty, B. N. and Kapoor, M. (1990) Transformation of filamentous fungi by electroporation. *Nucleic Acids Res.* **18**, 6737.
6. Orbach, M. J., Porro, E. B., and Yanofsky, C. (1986) Cloning and characterization of the gene for β-tubulin from a benomyl-resistant mutant of *Neurospora crassa* and its use as a dominant selective marker. *Mol. Cell. Biol.* **6**, 2452–2461.
7. Staben, C., Jensen, B., Singer, M., Pollock, J., Schechtman, M., Kinsey, J., and Selker, E. (1989) Use of a bacterial hygromycin B resistance gene as a dominant selectable marker in *Neurospora crassa* transformation. *Fungal Genet. Newslett.* **36**, 79–82.
8. Chakraborty, B. N., Patterson, N. A., and Kapoor, M. (1991) An electroporation-based system for high-efficiency transformation of germinated conidia of filamentous fungi. *Can. J. Microbiol.* **37**, 858–863.
9. Davis, R. N. and de Serres, F. J. (1970) Genetic and microbiological research techniques for *Neurospora crassa*. *Methods Enzymol.* **17**, 79–143.
10. Vogel, H. J. (1956) A convenient growth medium for *Neurospora crassa* (N medium). *Microb. Genet. Bull.* **13**, 42,43.
11. Karin, M. and Richards, R. I. (1982) Human metallothionein genes—primary structure of the metallothionein-II gene and a related processed gene. *Nature* **299**, 797–802.
12. Kapoor, M., Vijayaraghavan, Y., Kadonaga, R., and LaRue, K. E. (1993) NAD$^+$-specific dehydrogenase of *Neurospora crassa*: cloning, complete nucleotide sequence, and gene mapping. *Biochem. Cell Biol.* **71**, 205–219.
13. Roychowdhury, H. A., Wong, D., and Kapoor, M. (1992) *hsp80* of *Neurospora crassa*: cDNA cloning, gene mapping, and studies of mRNA accumulation under stress. *Biochem. Cell Biol.* **70**, 1356–1367.
14. Kergomard, A., Renard, M. F., and Veschambre, H. (1982) Microbial reduction of a,β-unsaturated ketones by *Beauveria sulfurescens*. *J. Org. Chem.* **47**, 792–798.
15. Kapoor, M., Curle, C. A., and Runham, C. (1995) The *hsp* 70 gene family of *Neurospora crassa*: cloning, sequence analysis, expression, and genetic mapping of the major stress-inducible member. *J. Bacteriol.* **177**, 212–221.

CHAPTER 30

Transformation of *Candida maltosa* by Electroporation

Dietmar Becher and Stephen G. Oliver

1. Introduction

The genus *Candida* comprises a group of yeasts united by nothing more than the fact that none has a natural sexual cycle. If these yeasts do not form a natural grouping, they nonetheless contain a number of species of considerable scientific and practical interest. *Candida albicans* is an important opportunistic pathogen of humans, it is a dimorphic organism, and the transition from the yeast to the mycelial form is an essential prerequisite for the establishment of an infection. Commercially important species include *Candida utilis* and *Candida maltosa,* which have applications as food and feed organisms and can be used in the processing of a wide range of substrates, including, in the latter case, crude oil. *Candida* species are diploid or aneuploid in their genetic constitution and have proven to be difficult organisms to study in terms of their genetics and molecular biology. Traditional approaches to mutation and genetic mapping have provided very limited information, and only a few of the 166 species of *Candida (1)* have yielded to molecular genetic methods. Nevertheless, recombinant DNA technology represents the only efficient route to analyzing their genomes, and this requires an efficient DNA transformation method. Electroporation has become the preferred method for gene transfer owing to its ease and efficiency in comparison to alternative techniques. Moreover, electroporation permits the introduction of DNA into organisms that are refractory to other transformation techniques

From: *Methods in Molecular Biology, Vol. 47: Electroporation Protocols for Microorganisms*
Edited by: J. A. Nickoloff Humana Press Inc., Totowa, NJ

(2). However, experience with *Saccharomyces cerevisiae* has indicated that the electroporation conditions may need to be adapted for each strain used *(3–6)*, and our findings with *Candida maltosa* support this view *(7)*. We have found that optimal electroporation conditions show significant variation even between different mutants derived from a single strain, and thus, it is impossible to give a general transformation protocol for the extremely heterogeneous genus *Candida*. Therefore, in this chapter, we will focus on the hydrocarbon-utilizing yeast *C. maltosa,* a member of the *Candida* genus whose molecular genetics has advanced significantly of late *(8–18)*.

Despite the foregoing, there are a number of general principles that may be used in the selection of appropriate electroporation parameters for the transformation of a novel strain or species of yeast: (1) It is necessary to strike a balance between a sufficiently high rate of survival of recipient cells and an electrical field strength that creates the highest possible number of pores in the cell membrane. The electric field strength required increases in inverse proportion to cell size. (2) The optimal pulse duration, which allows DNA entry while avoiding irreversible damage to the cells, must be found. The higher the field strength employed, the shorter the acceptable pulse duration. Table 1 lists the parameters employed for the electroporation of DNA into a number of imperfect and industrially important yeasts.

2. Materials

2.1. Recipient Strains

C. maltosa, like many other *Candida* species, has a diploid or highly aneuploid genome *(10,11,18,19)*. A number of different host strains have been described *(8–11,14)*, but since *C. maltosa* is extremely resistant to most common antibiotics and uses a wide range of carbon sources, transformation of prototrophic wild-type strains using vectors carrying dominant, selectable marker genes is difficult. Therefore, most host-vector systems are based on strains with recessive auxotrophic mutations, and a large collection of such strains has been created at Greifswald. These mutants possess blocks in the pathways for the synthesis of amino acids, vitamins, or nucleotides, or in those responsible for alkane assimilation. Strains may be obtained from I. A. Samsonsova or D. Becher (Institute of Genetics and Biochemistry, F. R. Biology, E. M. Arndt University, D17487 Greifswald, Germany).

Table 1

Experimental Parameters for Transformation of Imperfect and Industrial Yeasts by Electroporation

Species	Cell number of sample	Sample volume, μL	Plasmid DNA, μg	Carrier[a] DNA, μg	Field strength, kV/cm	Computed pulse in ms $R \times C$	Electroporation medium[b] i/n/P	Pretreatment of cells	Reference
C. albicans	3×10^7	100	10/20	0.0	7.50	2	+/-/-	Yes	R. Swoboda pers. comm.
C. maltosa	5×10^8	50	0.1	0.0	7.50	32	-/+/-	No	7
C. tropicalis	5×10^7	100	1.0	5.0	2.25	50	-/+/+	No	29
Y. lipolytica	1×10^{8c}	20	1.0	0.0	10.56	Not indic.	-/+/-	Yes	30
K. lactis	5×10^8	50/100	0.1/0.2	0.5	5.00	10	+/-/-	Yes	31,32
K. marxianus	1×10^{8c}	50	0.1	0.0	2.50	Not indic.	+/-/-	Yes	33
S. bayanus	4×10^8	40	0.1	0.0	3.75	5	-/+/-	No	2

[a]Type of carrier DNA used: C. tropicalis, sonicated calf thymus DNA; K. lactis, salmon sperm DNA.

[b]i: ionic medium, n: nonionic medium, P: polyethylene glycol.

[c]Cell titer estimated from data on optical density.

Table 2
Examples for Plasmid Vectors Utilizing Homologous Genes
Encoding Biosynthetic Enzymes as Selectable Markers

Vector	Type	Marker gene[a]	Bacterial resistances[b]	References
CipA15	Cip	Cm-ADE1	Km[r], Amp[r]	7
CrAp2	Crp	Cm-ADE1	Amp[r]	7
CipL1	Cip	Cm-LEU2	Amp[r]	11
CrLp1	Crp	Cm-LEU2	Amp[r]	11
PCCD5	Cdp	Cm-ADE1	Km[r]	7
pA3	Cip	Cm-ADE1	Amp[r]	14
pRJ1	Crp	Cm-ADE1	Amp[r]	14
pBTH10B	Crp	Cm-HIS5	Amp[r], Tc[r]	9
pUD45H	Cdp	Cm-HIS5	Amp[r]	13
pUD45A	Cdp	Cm-ADE1	Amp[r]	13

[a]Cm-ADE1 encodes phosphoribosylaminoimidazolesuccinocarboxamide synthetase; Cm-LEU2 specifies β-isopropylmalate dehydrogenase; Cm-HIS5 codes for L-histidinol phosphate aminotransferase.

[b]Amp[r], ampicillin resistance; Km[r], kanamycin resistance; Tc[r], tetracycline resistance.

2.2. Plasmid Vectors

No natural plasmids have been discovered in *C. maltosa,* and the species is unable to replicate the 2-μm plasmid of *S. cerevisiae.* However, *C. maltosa* autonomously replicating sequence elements *(Cm-ARS)* have been cloned *(11,14,20),* and these elements permit the construction of self-replicating plasmid vectors for the organism. Such vectors exhibit a transformation efficiency that is some 100- to 1000-fold greater than that obtained with integrative plasmids, but the transformants are mitotically unstable. Three types of plasmid vectors are now available. All are "shuttle" vectors, and possess a replication origin and an antibiotic resistance gene that permit their propagation and maintenance in *Escherichia coli.* Table 2 lists examples of the different vector types.

1. Integrative plasmids (Cip): These plasmids possess at least one intact homologous marker gene and become integrated into the genome, following transformation, via a single crossover with a nonfunctional homolog on a *C. maltosa* chromosome. The efficiency of integration can be increased by linearization of the plasmid by cleavage at a unique restriction site in the marker gene prior to transformation.
2. Integrative plasmids with gene-disrupting or gene-displacing function (Cdp): Cdp plasmids possess an intact marker gene and a manipulated copy

of the gene whose function is to be destroyed. In the case of disruption, the manipulated gene is usually truncated at its 5' and 3' ends, leading to the generation of two nonfunctional gene copies after plasmid integration. For gene displacement, linearized plasmids are generally employed. They are constructed in such a way that, after linearization, the selectable marker and any contiguous bacterial sequences are flanked by two noncontiguous segments of the gene that is to be replaced. The ends so generated are highly recombinogenic and facilitate the integration of the construct via a double crossover event, thus leading to the displacement of the chromosomal copy of the gene of interest.

3. Autonomously replicating plasmids (Crp): The average plasmid copy number per cell, and its maintenance during successive mitotic divisions, depends on the efficiency of the *Cm-ARS* and on the presence of other stabilizing elements *(21)*. Plasmids that carry large fragments of the *C. maltosa* genome tend to become integrated into a chromosome, where they can create extensive rearrangements *(7)*.

2.3. Solutions and Reagents for Electroporation and Cell Growth

(*See* Note 1 before preparing these reagents.)

1. Double-distilled water (ddH$_2$O): Sterilize by autoclaving (15 min, 121°C, 1 bar). The pH should be ~4.2.
2. Sorbitol solution: Add 182.2 g of D-sorbitol (analytical-grade) to ddH$_2$O to give a final volume of 1 L.
3. YPD medium: Dissolve 10 g yeast extract, 20 g peptone, and 20 g dextrose in ddH$_2$O to a final volume of 1 L. Divide the solution into aliquots, and sterilize by autoclaving. For solid media, add bacto-agar to 2% w/v.
4. YPDS medium: Add 182.2 g of D-sorbitol to YPD medium to a final volume of 1 L. Filter-sterilize by passage through a 0.4-μm cellulose acetate membrane or glass-fiber filter. Divide into aliquots and store at 4°C.
5. Selective medium: 0.67% w/v yeast nitrogen base (Difco, Detroit, MI), 2% w/v dextrose, 18.2% w/v D-sorbitol, and 2% w/v bacto-agar. Add supplements (amino acids, purines, or pyrimidines) to a final concentration of 50 mg/L, according to the requirements of the recipient strains and the marker carried on the transforming plasmid.
6. Plasmid DNA: Purified plasmid DNA is ethanol-precipitated, resuspended in ddH$_2$O at a concentration of 1 μg/μL, and stored at –20°C as a stock solution. To obtain a working solution, dilute 1 μL of this stock with 9 μL of sorbitol solution to give a final concentration of 100 ng/μL (*see* Note 2).

3. Methods

3.1. Growth Conditions
and Preparation of Electrocompetent Cells

1. For inoculum preparation, obtain individual colonies of the recipient strain by streaking a colony onto YPD agar and then replica-plating *(22)* onto selective agar with and without the required supplements.
2. Pick those colonies that display the phenotype expected of the recipient strain. (These are assumed to be free of revertants or contaminants, and may be stored at 4°C for up to 1 mo.)
3. Inoculate liquid YPD cultures (100 mL of medium in 500-mL Erlenmeyer flask) with a single colony.
4. Grow at 30°C, with shaking, to a cell density of 1×10^8 cells/mL, which corresponds to the last third of the exponential growth phase of many auxotrophic mutants derived from *C. maltosa* strain L4P (*see* Note 3).
5. Harvest cells by centrifugation at 2000*g* for 5 min at 4°C. Resuspend the cell pellet in an equal volume of ice-cold ddH$_2$O by pipeting up and down with a wide-mouth pipet during the stepwise addition of water (*see* Note 4).
6. Wash the cells twice by centrifugation at 2000*g* for 5 min at 4°C, and resuspend in ice-cold 1*M* D-sorbitol, as described above.
7. After the final centrifugation step, decant the supernatant and resuspend the cell pellet in the remaining traces of sorbitol (supernatant). This gives a final cell density of ~2×10^{10} cells/mL (*see* Note 5).

3.2. Electroporation Protocol

1. Place the electroporation cuvets (electrode distance, 2 mm) on ice for 5 min to precool (*see* Note 6).
2. Place 1.5-mL tubes in an ice bath, and transfer 40 µL of cell suspension (~5×10^8 cells) to each tube using sterile glass Pasteur pipets. Leave the pipets in the opened tubes.
3. For transformation, add 1 µL of the DNA working solution (100 ng) to each tube. For control tubes, add 1 µL of 1*M* D-sorbitol. Mix the cells with the DNA by swirling with the inserted Pasteur pipet.
4. Beginning with the control tube (i.e., the one to which no transforming DNA was added), transfer the cell mixtures to the bottom of the precooled electroporation cuvets, using the Pasteur pipets. Sometimes, the cell suspension has to be tapped to spread it onto the bottom of the cuvet.
5. Transfer the cuvet into the electroporation device. Because of the temperature difference between the cuvet and its surroundings, the cuvet becomes wet with condensation and should be dried with a tissue before it is inserted into the electroporation chamber.

6. For a routine experiment, pulse once at 1.5 kV, internal resistance of 100 Ω, and a capacitance of 40 μF (*see* Note 7).
7. Immediately after transmitting the pulse, add 1 mL of 1*M* D-sorbitol to the electroporation cuvet, and mix by pipeting up and down with a Pasteur pipet.
8. Place the cuvet with the electroporated cells at room temperature for at least 30 min in order to allow the cell membranes to recover (*see* Note 8).
9. Spread the electroporated cells of the transformation samples onto selective agar plates. Use at least 5 plates/electroporation; the actual number of cells spread per plate will depend on the expected efficiency of transformation.
10. Incubate plates at 30°C (lids down). The first transformant colonies should be visible after 36 h (*see* Notes 9–11).

3.3. Control Experiments

It is important both to carry out the following control experiments and to investigate the growth characteristics of the recipient strain thoroughly. Although these may appear tedious, they will save time in the long run—particularly when dealing with an unfamiliar organism or strain.

1. Within 10 s after step 7 of the electroporation protocol (Section 3.2.), transfer 100 μL of the control sample to 9.9 mL of ddH$_2$O, and incubate at room temperature for 1 h. Thereafter, make serial dilutions in ddH$_2$O, and spread the corresponding samples onto YPD agar plates. Following incubation, count these plates, and calculate the concentration of colony-forming units in the control sample.
2. Take a second 100-μL aliquot of the electroporated control sample through the same procedure as in step 1 (above), but use 1*M* sorbitol for the dilutions, and plate on YPDS agar. Calculate the survival rate of the electroporated cells and the proportion of osmotically sensitive cells in the electroporation sample using the data gained from steps 1 and 2. Under the electroporation conditions given above (Section 3.2.), ~80% of the surviving *C. maltosa* cells will be osmotically sensitive.
3. Spread the remainder of the electroporated control sample onto selective agar plates to determine the frequency of revertants, contaminants, or recombinants.
4. Dilute a 40-μL sample of electrocompetent, but not electroporated, cells with 1*M* sorbitol, plate on YPD agar, and determine the concentration of colony-forming units. The difference between values determined in steps 2 and 4 gives the killing rate of electroporation. Under the standard conditions (Section 3.2.), a killing rate of 50% is expected (*see* Note 12).

4. Notes

1. Purity of reagents: Transformation efficiency is reduced by poor-quality water, yellow-colored sorbitol solutions, remnants of the growth medium, and remnants of phenol or bacterial proteins in the plasmid DNA solutions. Obviously, many compounds surrounding the recipient cells may be taken up during electroporation, but losses from the cells also occur *(23)*. Both of these effects probably play a significant role in the increased cell mortality that is observed as the duration of the electric pulse is extended.

2. The fact that DNA is a charged polymer probably facilitates its transfer into the cell by a process analogous to electrophoresis. For uncharged contaminating molecules, diffusion appears to play an important role in their uptake *(23)*. However, the transfer of such contaminants rarely proceeds to equilibrium—perhaps because of their failure to penetrate intracellular compartments.

3. It is important to harvest cells in late-exponential growth phase; the investigation of the growth characteristics of the recipient strain is the final prerequisite for successful electroporation experiments. The cell titer is calculated using a hemocytometer (Thoma). We do not use optical density measurements because different auxotrophic strains can vary considerably in both cell size and shape.

4. Harvesting conditions will vary between species and strains, as explained above *(1)*. The cells **must not** be mixed or resuspended by vortexing, and centrifuge tube should be kept in an ice bath during resuspension of the cells.

5. In our hands, ~2×10^{10} cells/mL was the highest cell concentration that could be handled. The electrocompetence of such a cell preparation decreases with storage time; after 24 h of storage at 4°C, the transformation efficiency was found to decrease to 20% of the original value. Prolonged storage of highly concentrated cell suspensions results in autolysis. This has two important consequences: The release of intracellular nucleases may damage the transforming DNA, and the risk of arcing is increased. Cells stored for 48 h need a longer recovery phase after electroporation, and control and transformant colonies take at least 12 h longer to appear.

6. Electric arcing: Transformation of *C. maltosa* by electroporation is carried out at high electric field strength, and there is a risk of electric arcing within the electroporation cuvet. If the electric field exceeds a critical value, the pulse current flows with very high density through a small channel. As a result, localized melting of the electrode material may occur, generating vapor in an explosive manner. Apart from safety considerations, this arcing may destroy the electroporation device. The risk of arcing can be reduced by a correspondingly high value of impedance. If a high-voltage

pulse is required for pore formation, the resistance of the electroporation medium should be increased by reducing the concentration of ions in the medium, increasing the interelectrode distance (e.g., by using 4-mm electroporation cuvets), and cooling to temperatures below 4°C. Using cuvets several times can introduce uncertainties, and we find that cuvets in which arcing has occurred show an increased risk of arcing in subsequent experiments; they are best discarded.

7. The required electrical conditions will vary between species and strains, as described in Section 1. and Table 1.

8. Osmotic conditions: Electroporated cells remain osmotically sensitive for a considerable time, which, it is assumed, is necessary for membrane recovery. Certainly, cells remain permeable for significant periods after the pulse. We were able to obtain up to 20 transformants with Crp DNA when high DNA concentrations (10 μg in 1 mL sorbitol) were added 10 s after the electric pulse was delivered. Therefore, the osmolality and osmolarity of the electroporation medium should be controlled carefully.

9. Selection and identification of transformants: Transformants are identified by their ability to grow on an appropriate selective agar owing to complementation of an auxotrophic mutation in the recipient strain by the corresponding wild-type gene on the transforming plasmid. However, to exclude other explanations for the change in phenotype, the presence of the plasmid in the recipient cells must be proven. In the case of Cip vectors, this is possible only by Southern hybridization (24) with corresponding DNA probes. This is laborious, but gives considerable of additional information on the site and mode of integration of the plasmid. With the standard protocol, we obtained up to 20 integrative transformants/transformation sample. The presence of plasmids of the Cdp type can be detected unambiguously by the occurrence of the required unique phenotype of the transformants. For this purpose, a two-step selection is necessary. Transformants are first selected on the basis of complementation of an auxotrophic marker in the recipient strain under conditions in which the function of the target gene is not required. In the second selection step, putative transformants are replicated onto a medium where the function of the target gene is essential for growth. Colonies that fail to grow on the second medium are then identified as having the function of the target gene destroyed by integration of the Cdp plasmid bearing the selectable marker.

Plasmids of the Crp type produce transformants at high frequency (~2000 will be produced by the standard procedure). The acquisition of a Crp plasmid confers a mitotically unstable phenotype on the host. Loss of the plasmid, and its marker gene, may be monitored in the following way:

    a. A transformant colony, which was identified by complementation of the corresponding marker gene, is streaked onto selective agar in order to isolate separate single colonies that should be free of contaminating untransformed cells.

    b. Such a colony is streaked onto a YPD agar plate (master plate) and incubated to produce well-isolated, single colonies again.

    c. The master plate is replica plated onto nonselective and selective media. Those colonies that fail to grow on selective agar are judged to have lost the plasmid.

10. Twin-pulse technique: In 1990, the "Twin-Pulse™" technique was introduced by EquiBio (Eurogentec, Seraing, Belgium) *(25)*. This technique is based on the delivery of two different electric pulses in rapid succession. The first pulse is characterized by high electric field strength and short pulse duration, and is intended to create a large number of pores in the cell membrane. The second pulse is of low electric field strength and long pulse duration, a combination that is assumed to favor the entry of exogenous molecules into the recipient cells *(25)*. In our hands, this technique was not superior to the standard single-pulse technique. Nevertheless, the twin-pulse approach may be worth trying if the single-pulse method proves unsatisfactory. However, extensive trials are likely to be required in order to determine the correct parameters for this technique.

11. Cotransformation: It has been found that electroporated cells can take up a considerable number of DNA molecules *(26)*. Frequencies of cotransformation vary with different plasmid types and also increase as the efficiency of integrative transformation increases. This is especially important for gene-disruption experiments in diploid and aneuploid strains *(7)*.

12. Genetic consequences of the electric pulse: A large proportion of the cell population is killed by electroporation, even under optimal physiological conditions (iso-osmotic media, short pulse duration), and it is likely that the survivors have suffered some consequent genetic damage *(27,28)*. This damage results in a stimulation of intrachromosomal and interchromosomal mitotic recombination, which may have profound effects on the outcome of the transformation, particularly where plasmid integration or cotransformation is involved.

References

1. Barnett, J. A., Payne, R. W., and Yarrow, D. (1990) *Yeasts: Characteristics and Identification,* 2nd ed. Cambridge University Press, Cambridge.
2. Ogata, T., Okumura, Y., Tadenuma, M., and Tamura, G. (1993) Improving transformation method for industrial yeasts: construction of *ADH1-APT2* gene and using electroporation. *J. Gen. Appl. Microbiol.* **39,** 285–294.

3. Meilhoc, E., Masson, J. M., and Teissie, J. (1990) High efficiency transformation of intact yeast cells by electric field pulses. *Bio/Technology* **8**, 223–227.
4. Grey, M. and Brendel, M. (1992) A ten-minute protocol for transforming *Saccharomyces cerevisiae* by electroporation. *Curr. Genet.* **22**, 335,336.
5. Delorme, F. (1989) Transformation of *Saccharomyces cerevisiae* by electroporation. *Appl. Environ. Microbiol.* **55**, 2242–2246.
6. Becker, D. M. and Guarente, L. (1991) High-efficiency transformation of yeast by electroporation. *Methods Enzymol.* **194**, 182–187.
7. Kasüske, A., Wedler, H., Schulze, S., and Becher, D. (1992) Efficient electropulse transformation of intact *Candida maltosa* cells by different homologous vector plasmids. *Yeast* **8**, 691–697.
8. Takagi, M., Kawai, S., Chang, M. C., Shibuya, I., and Yano, K. (1986) Construction of a host-vector system in *Candida maltosa* by using an ARS site isolated from its genome. *J. Bacteriol.* **167**, 551–555.
9. Hikiji, T., Ohkuma, M., Takagi, M., and Yano, K. (1989) An improved host-vector system for *Candida maltosa* using a gene isolated from its genome that complements the *his5* mutation of *Saccharomyces cerevisiae*. *Curr. Genet.* **16**, 261–266.
10. Kawai, S., Hikiji, T., Murao, S., Takagi, M., and Yano, K. (1991) Isolation and sequencing of a gene, *C-ADE1*, and its use for a host-vector system in *Candida maltosa* with two genetic markers. *Agric. Biol. Chem.* **55**, 59–65.
11. Becher, D., Wedler, H., Schulze, H., Bode, R., Kasüske, A., and Samsonova, I. (1991) Correlation of biochemical blocks and genetic lesions in leucine auxotrophic strains of the imperfect yeast *Candida maltosa*. *Mol. Gen. Genet.* **227**, 361–368.
12. Ohkuma, M., Tanimoto, T., Yano, K., and Takagi, M. (1991) *CYP52* (Cytochrome P450alk) multigene family in *Candida maltosa*. Molecular cloning and nucleotide sequence of two tandemly arranged genes. *DNA and Cell Biol.* **10**, 271–282.
13. Ohkuma, M., Hikiji, T., Tanimoto, T., Schunck, W. H., Müller, H. G., Yano, K., and Takagi, M. (1991) Evidence that more than one gene encodes n.alkane-inducible cytochrome P450s in *Candida maltosa*, found by two step gene disruption. *Agric. Biol. Chem.* **55**, 1757–1764.
14. Sasnauskas, K., Jomantiene, R., Lebediene, E., Lebedys, J., Januska, A., and Janulaitis, A. (1992) Molecular cloning and analysis of autonomous replicating sequence of Candida maltosa. *Yeast* **8**, 253–259.
15. Sasnauskas, K., Jomantiene, R., Januska, A., Lebediene, E., Lebedys, J., and Janulaitis, A. (1992) Cloning and analysis of a *Candida maltosa* gene which confers resistance to formaldehyde in *Saccharomyces cerevisiae*. *Gene* **122**, 207–211.
16. Kamiryo, T., Sakasegawa, Y., and Tan, H. (1989) Expression and transport of *Candida tropicalis* peroxisomal acyl-coenzyme A oxidase in the yeast *Candida maltosa*. *Agric. Biol. Chem.* **53**, 179–186.
17. Tanaka, H., Takagi, M., and Yano, K. (1987) Separation of chromosomal DNA molecules of *Candida maltosa* on agarose gels using the OFAGE technique. *Agric. Biol. Chem.* **51**, 3161–3163.
18. Becher, D., Schulze, S., Kasüske, A., Schulze, H., Oliver, S. G., and Samsonova, I. A. (1994) Molecular analysis of a *leu2* mutant of *Candida maltosa* demonstrates presence of multiple alleles. *Curr. Genet.* **26**, 208–216.

19. Chang, M. C., Jung, H. K., Suzuki, T., and Takagi, M. (1984) Ploidy in the asporogenous yeast *Candida maltosa.* Isolation of its auxotrophic mutants and their cell fusion. *J. Gen. Appl. Microbiol.* **30,** 489–497.

20. Kawai, S., Hwang, C. W., Sugimoto, M., Takagi, M., and Yano, K. (1987) Subcloning and nucleotide sequence of an ARS site of *Candida maltosa* which also functions in *Saccharomyces cerevisiae. Agric. Biol. Chem.* **51,** 1587–1591.

21. Umek, R. M., Linskens, M. H. K., and Kowalski, D. (1989) New beginnings in studies of eukaryotic DNA replication origins. *Biochim. Biophys. Acta* **1007,** 1–14.

22. Lederberg, J. and Lederberg, E. M. (1952) Replica plating and indirect selection of bacterial mutants. *J. Bacteriol.* **63,** 399–406.

23. Marcil, R. and Higgins, D. R. (1992) Direct transfer of plasmid DNA from yeast to *Escherichia coli* by electroporation. *Nucleic Acids Res.* **20,** 917.

24. Southern, E. M. (1975) Detection of specific sequences among DNA fragments separated by gel electrophoresis. *J. Mol. Biol.* **98,** 503–517.

25. EASYJECT (1992) *User's Manual.* **2,** 1–46.

26. Bartoletti, D. C., Harrison, G. I., and Weaver, J. C. (1989) The number of molecules taken up by electroporated cells: quantitative determination. *FEBS Lett.* **256,** 4–10.

27. Danhash, N., Gardner, D. C. J., and Oliver, S. G. (1991) Heritable damage to yeast caused by transformation. *Bio/technology* **9,** 179–182.

28. Higgins, D. R. and Strathern, J. N. (1991) Electroporation stimulated recombination in yeast. *Yeast* **7,** 823–832.

29. Rohrer, T. L. and Picataggio, S. K. (1992) Targeted integrative transformation of *Candida tropicalis* by electroporation. *Appl. Microbiol. Biotechnol.* **36,** 650–654.

30. Nutley, W. M., Brade, A. M., Gaillardin, C., Eitzen, G. A., Glover, J. R., Aitchinson, J. D., and Rachubinski, R. A. (1993) Rapid identification and characterization of peroxisomal assembly mutants in *Yarrowia lipolytica. Yeast* **9,** 507–517.

31. Russel, C., Jarvis, A., Yu, P., and Mawson, J. (1993) Optimization of an electroporation procedure for *Kluyveromyces lactis* transformation. *Biotechnol. Tech.* **7,** 417–422.

32. Sanchez, M., Iglesias, F. J., Santamaria, C., and Dominguez, A. (1993) Transformation of *Kluyveromyces lactis* by electroporation. *Appl. Environ. Microbiol.* **59,** 2087–2092.

33. Iborra, F. (1993) High efficiency transformation of *Kluyveromyces marxianus* by a replicative plasmid. *Curr. Genet.* **24,** 181–183.

CHAPTER 31

Electroporation
of *Physarum polycephalum*

Timothy G. Burland and Juliet Bailey

1. Introduction

The protist *Physarum polycephalum* is a convenient system for studies of molecular and cellular biology of fundamental eukaryotic processes, including DNA replication, mitotic regulation, single-cell development, the cytoskeleton, and motility. The life cycle of this acellular slime mold exhibits a variety of developmental transitions, and the two vegetative cell types, amoeba and plasmodium, alternate via a sexual cycle *(1)*. The plasmodium is a multinucleate syncytium, in which the nuclei transit the mitotic cycle in perfect natural synchrony. This is convenient for analysis of the relative timing of biochemical events in the unperturbed mitotic cycle, since huge quantities of material can be isolated from multiple stages of consecutive mitotic cycles. The haploid amoebae are by contrast uninucleate and, in addition to vegetative growth, are capable of acting as gametes in sexual crosses, giving rise to diploid plasmodia. Meiosis occurs when the plasmodium sporulates, and the spores hatch to yield haploid amoebae. Mutations in *gadA*, tightly linked to the major mating-type locus *matA*, permit asexual development of haploid plasmodia ("selfing") from clones of haploid amoebae *(2)*. Thus, methods of classical genetic analysis and mutant isolation in *Physarum* are straightforward.

To exploit the rich cell biological characteristics of *Physarum* fully, we needed to develop reliable and efficient methods for DNA transformation. Since many of the experiments would require production of

From: *Methods in Molecular Biology, Vol. 47: Electroporation Protocols for Microorganisms*
Edited by: J. A. Nickoloff Humana Press Inc., Totowa, NJ

clones of stable transformants, a multinucleate syncytium, such as the plasmodium, is clearly inappropriate as a target cell type. We therefore developed methods for transforming the haploid uninucleate amoebae *(3–6)*. The method we have found to be routinely successful for introducing functional DNA into amoebae is electroporation.

We have successfully expressed in *Physarum* amoebae all three heterologous genes tested: *cat* (encoding chloramphenicol acetyltransferase; ref. *6*), *luc* (encoding firefly luciferase; ref. *3*) and *hph* (encoding hygromycin phosphotransferase; ref. *5*). The *cat* and *luc* genes are used for transient expression, where gene activity is determined via assays of activity of the gene product. The *hph* gene is used for selection of stable transformants, where expression of *hph* confers resistance to hygromycin B *(7)*. The latter system is analogous to use of the *neo* gene and selection for G418 resistance *(8)*. Most of our method development has utilized transient expression of *luc,* since the gene product is quickly a:.d simply measured by luminometry of cell lysates *(9)*.

2. Materials
2.1. Media, Solutions, and Strains

1. LB: To 800 mL of distilled water (dH$_2$O), add 10 g Difco (Detroit, MI) bacto-tryptone, 5 g Difco yeast extract, and 5 g NaCl. Dissolve solids, adjust pH to 7.4 with NaOH, make to 1000 mL with dH$_2$O, and autoclave to sterilize. After autoclaving, add 5 mL of 20% glucose. For plates, add 15 g of agar/L before autoclaving.

2. Superbroth: To 800 mL dH$_2$O, add 32 g Difco bacto-tryptone, 20 g Difco yeast extract, and 5 g NaCl. Dissolve solids, pH to 7.4 with NaOH, make to 1000 mL with dH$_2$O, and autoclave to sterilize. After autoclaving, add 5 mL of 20% glucose.

3. SM plates: To 1000 mL dH$_2$O, add 0.7 g Difco bacto-tryptone, 0.2 g Difco yeast extract, 0.6 g D-glucose, 0.8 g sodium phosphate monobasic dihydrate, 0.7 g sodium phosphate dibasic, and 15 g Difco bacto-agar. Autoclave to sterilize and then pour plates.

4. Hemin solution: Dissolve 10 g NaOH in 900 mL of dH$_2$O. Add 500 mg hemin and allow to dissolve. Make to 1000 mL. Dispense into aliquots. Autoclave for 20 min. Store at 4°C.

5. SDM: (per liter) 10 g glucose, 10 g Difco bacto-soytone, 3.54 g citric acid monohydrate, 2 g KH$_2$PO$_4$, 1.026 g CaCl$_2$ · 2H$_2$O, 0.6 g MgSO$_4$ · 7H$_2$O, 34 mg ZnSO$_4$ · 7H$_2$O, 42.4 mg thiamine-HCl, and 15.8 mg biotin. Adjust pH to 4.6 with KOH, autoclave for 15 min, and store at room temperature. Before use, add 10 mL hemin solution/L.

6. DSDM plates: To 1000 mL dH$_2$O, add 15 g Difco bacto-agar. Autoclave to sterilize and then add 65 mL SDM, mix, and dispense into plates.

7. HBS: 40 mM ultra-pure sucrose (Gibco-BRL Gaithersburg, MD) and 10 mM ultra-pure HEPES (Gibco-BRL). Adjust pH to 8.2 with KOH. Sterilize by filtration or by autoclaving and store at room temperature.

8. BSS: (per liter) 3 g citric acid monohydrate, 4.20 g K$_2$HPO$_4$ · 3H$_2$O, 0.25 g NaCl, 0.21 g MgSO$_4$ · 7H$_2$O, and 0.05 g CaCl$_2$ · 2H$_2$O. Adjust pH to 5.0 with KOH, and autoclave to sterilize.

9. Formalin-killed bacteria (FKB): Inoculate 6 L superbroth (1000 mL/2-L flask) with 1 mL/L of overnight culture of *Escherichia coli* HB101. Shake vigorously overnight at 37°C. Add 15 mL of 37–40% formaldehyde solution/L, and shake for 60 min. Centrifuge the cells at 7250g for 25 min at room temperature. Pour off supernatants, and resuspend cells in 300 mL of 0.5M glycine. Shake for 60 min at room temperature. Then centrifuge the cells at 7250g for 10 min at room temperature. Decant supernatant. Wash cells once in 0.5M glycine, once in BSS, and then resuspend cells in 120 mL of BSS. Store in 10-mL aliquots at 4°C. Check sterility by streaking out a loopful of suspension from each aliquot on an LB plate, and incubate 3 d at 30°C. Nothing should grow.

10. Live bacteria: Streak *E. coli* HB101 over an LB plate, and incubate at 37°C overnight or at 26°C for 2–4 d. Flood the plate with 10 mL of dH$_2$O, and then scrape off bacteria with a glass spreader. Transfer suspension to a test tube, and vortex to disperse clumps. Use 0.1 mL of suspension/plate the same or next day.

11. Luciferase assay reagents: We use the three stock reagents supplied with the luciferase assay kit (Promega Inc, Madison, WI). It is vital to supplement the cell lysis reagent with protease inhibitors as indicated below.

    a. 7 mg Lyophilized luciferase substrate: When rehydrated with assay buffer, the concentrations are 270 μM coenzyme A (lithium salt), 470 μM luciferin, and 530 μM ATP.

    b. 10 mL luciferase assay buffer: This contains 20 mM tricine, 1.07 mM (MgCO$_3$)$_4$Mg(OH)$_2$ · 5H$_2$O, 2.67 mM MgSO$_4$, 0.1 mM EDTA, and 33.3 mM DTT. The final pH is 7.8 when combined with substrate.

    c. 5X cell lysis reagent: This contains 125 mM Tris, pH 7.8, with H$_3$PO$_4$, 10 mM EDTA, 100 mM DTT, 50% glycerol, and 5% Triton X-100.

    To prepare for use, add 10 mL assay buffer to the bottle of substrate. Dissolve to make the assay reagent, and store at –70°C in small volumes. Promega recommends using 100 μL reagent/20 μL of cell lysate, although we have also successfully used 50 μL reagent/10 μL of cell lysate. Unless otherwise stated, all experiments reported here used 20 μL cell lysate assayed in 100 μL of reagent.

The cell lysis reagent is diluted to 1X with water and protease inhibitors are then added: leupeptin to 25 µg/mL (10 µL/mL of 2.5 mg/mL stock in water), pepstatin to 5 µg/mL (2 µL/mL of 2.5 mg/mL stock in water), and chymostatin to 5 µg/mL (2 µL/mL of 2.5 mg/mL stock in DMSO). Final concentration of inhibitors will be about half these values when mixed with cells. This protease-containing lysis reagent is then stored in 1-mL aliquots at –70°C.

12. Hygromycin B: 400,000 U/mL (CalBiochem).
13. *P. polycephalum* strains:

LU352 *(10)*,	*matA2 gadAh npfC5 matB3 fusA1 axe;*
LU353 *(10)*,	*matA3 matB1 fusA2 axe;*
MA460 (unpublished),	*matA2 gadAh npfC5 matB2 fusA2 whiA1 benD210 axe;*
DIP21 (unpublished),	*matA2/A2 gadAh/Ah npfC5/C5 matB1/ B3 fusA1/A2 whiA⁺/A1 benD⁺/D210 axe/axe.*

2.2. Promoter-Gene Fusions

We have observed expression of transformed genes in *Physarum* amoebae only when the genes are under control of *Physarum* promoters. Absence of a promoter or use of bacteriophage or Rous sarcoma virus promoters has not yielded detectable gene expression *(6)*.

For expression of the *luc, cat,* and *hph* genes in *Physarum,* we have used plasmids that carry a *Physarum* actin gene promoter upstream of the coding region. The strongest promoter tested so far is *PardC,* the promoter for the *ardC* actin gene. This gene is the most highly expressed of the *Physarum* actin multigene family *(11)*, and is expressed in both amoebal and plasmodial stages of the life cycle. Thus, *PardC*-driven genes transformed stably into amoebae should also be expressed in plasmodia that develop apogamically from them. Among the stable transformants so far analyzed, this is clearly the case *(5)*. In all vectors tested so far, we have included the first few codons of the *ardC* gene in fusions between *PardC* and *cat, luc,* or *hph* *(3,5,6,12)*. Except where noted, the experiments described here used vector p*PardC-luc* *(3)*, which consists of vector pGEM-*luc* (Promega Inc.) with the *PardC* promoter upstream of the *luc* gene.

2.3. Plasmid DNA

In our initial experiments, we found that plasmid DNA purified using the "Magic-Prep" method and resin supplied by Promega gave the best results in experiments involving transient expression of the *cat* gene, when compared to cesium chloride-purified DNA or more crude preparations *(6)*. However, we find that fairly crude preparations of plasmid DNA are comparable to the Magic-Prep DNA in performance if care is taken to remove RNA, and a variety of other commercial and standard methods we have tested all yield DNA that transforms amoebae effectively.

For transient expression experiments, either linear or circular plasmid DNAs can be used, with circular plasmid giving slightly higher expression levels *(6)*. However, for selection of stable transformants that carry the transforming DNA integrated into the nuclear genome, it appears that linear DNA is more efficient *(5)*.

3. Methods
3.1. Storage of Amoebae

We maintain stocks of amoebae as suspensions in 10% glycerol at –70°C. However, viability of axenic amoebae grown in broth is not preserved well when they are frozen directly. Rather, amoebae are coinoculated with live *E. coli* onto SM plates, incubated for 7–10 d at 26°C until confluent, and then harvested by flooding the plates with 5–10 mL of 10% glycerol. This suspension is then frozen in 2-mL aliquots. Each aliquot can be thawed and refrozen many times.

3.2. Growth of Amoebae

1. To begin growing axenic amoebae in broth, first thaw freezer stocks at room temperature, and then coinoculate 20–50 µL of stock with 0.2 mL of undiluted FKB onto SM plates containing 250 µg/mL each of streptomycin sulfate and penicillin G. Incubate plates at 26–30°C until amoebae first become confluent.
2. Harvest cells by flooding the plates with 10 mL of SDM, and make to 50 mL with SDM containing 250 µg/mL each streptomycin and penicillin. There is no need to remove the residual bacteria.
3. Shake cell suspensions at 26–30°C in 500-mL flasks on a gyrotory or reciprocating shaker at around 2 Hz. Growth rates are initially slow, and cells do not transform well at this early stage of culture. Therefore, before using amoebae for experiments, wait until growth rates increase to a doubling time of around 18–24 h, which usually occurs 1–2 wk after first inoculating the culture (*see* Note 1).
4. To maintain amoebal broth cultures, dilute as needed in SDM before cell density exceeds 10^7/mL. Penicillin and streptomycin can be omitted after the initial inoculation into SDM.

3.3. Preparation of Electrocompetent Amoebae

1. Estimate how many amoebae are needed. Approximately 5×10^7 amoebae are electroporated in one electrical discharge, with 10–80% of cells surviving, depending on the electrical conditions. For subsequent assay of *luc* expression, use 10^7 surviving amoebae/sample *(3)*; for assay of *cat* expression, use 5×10^7 surviving amoebae/sample *(6)*; for selection of hygromycin resistant amoebae following transformation with the *hph* gene, use $0.1–1 \times 10^8$ survivors/sample *(4,5)*. Grow amoebae exponentially in SDM to $2–8 \times 10^6$/mL. Check that the doubling time is 24 h or less.
2. Harvest amoebae from SDM broth by centrifugation for 2 min at 200*g*.
3. Wash the amoebae once in HBS at room temperature, using 1 mL of HBS for every 5×10^6 cells, centrifuging again at 200*g* for 2 min (*see* Notes 2–4).
4. Resuspend the washed amoebal pellet in HBS at 5×10^7 cells/500–800 µL. Immediately place cells on ice for 2 h (*see* Notes 5 and 6).

3.4. Electroporation of Amoebae

1. Immediately before electroporation, add DNA to the cell suspension (100–500 ng DNA for luciferase vectors; 1–20 µg for stable transformation with *hph* vectors).
2. Transfer about 600 µL of ice-cold cells (*see* Note 7) to an ice-cold 4-mm electroporation cuvet (Bio-Rad [Richmond, CA]; *see* Note 8). Use a Pasteur pipet to mix the cells and DNA by gentle aspiration.
3. Using a Gene Pulser (Bio-Rad), electroporate cells at about 0.8 kV with approx 800-Ω resistance in parallel on the controller, at 25-µF capacitance (*see* Note 9). Optimal electrical parameters may vary between different strains (*see* Notes 10 and 11).
4. Immediately after electroporation, transfer cells to 15-mL screw-cap centrifuge tubes, and incubate for 20 min at 30°C (*see* Note 12).
5. Add 5 mL of SDM at 26°C to the cells, mix, and then shake them at 26–30°C.

3.5. Assay for Luciferase

1. Grow transformed cells for approx 2.5–5 h following electroporation, and then count the cells using a hemacytometer. Some cells lyse following electroporation, leaving debris in the culture, whereas others die but remain intact. To distinguish live and dead whole cells, use 0.1% eosin yellow (dissolved in BSS) as a viable dye. However, after some experience, it is fairly simple to distinguish the refractile, live cells from the nonrefractile dead cells under phase-contrast optics without staining.
2. Pellet 10^7 live cells for each sample by centrifugation for 2 min at 200*g*, resuspend in 1 mL BSS, and transfer to a 1.5-mL microcentrifuge tube.
3. Pellet cells by spinning in a microcentrifuge for 20 s, remove the supernatant, and place tubes on ice for 1 min before lysis to slow protease activity.

4. Resuspend cooled cells in 100 μL of ice-cold cell lysis buffer by pipeting the cells up and down several times, and incubate on ice for 2 min to complete lysis. Lysates can either be frozen and stored at –70°C, or assayed immediately.

5. To assay, place 20 μL of cell lysate into a luminometer tube, add 100 μL of assay reagent, and immediately place the mixture into a luminometer to measure relative light units (RLU). Previous methods required that the assay reagent be added by injection directly into the sample tube in the instrument for immediate measurement, since the light emission decayed rapidly. However, the light emission with the Promega reagents is sufficiently sustained that no loss of signal results from the beginning of the 30-s duration measurements several seconds after mixing reagent and lysate, so we do not use the injection capability of the instrument. This saves losing valuable reagent in the injector. For 5×10^7 LU352 amoebae electroporated with 0.5 μg p*PardC-luc* DNA, harvested 2.5–5 h later, and lysed in 100 μL of lysis buffer, 20 μL of cell lysate assayed in 100 μL of reagent should give at least 100,000 RLU (*see* Note 13). However, if the luciferase expression vector carries transcriptional enhancers, the RLU obtained in these assays can be dramatically increased (*see* Note 14). With longer outgrowth times following electroporation, the level of luciferase expression falls (*see* Note 15).

3.6. Selection of Stable
Hygromycin Resistant Transformants

1. Grow cells for 1–2 d after electroporation before selective plating. Count cells after 1 d of incubation, and dilute in SDM if necessary to prevent amoebae from exceeding 10^7/mL and reaching stationary phase before plating.

2. Filter-sterilize the solution of hygromycin B (400,000 U/mL; *see* Note 16), and store at 4°C. Make selective plates by adding stock hygromycin B to 100 U/mL of molten DSDM agar after autoclaving. Dispense into plates, let solidify overnight, and then store plates in plastic bags in a 20°C incubator.

3. Count transformed amoebae again using a hemacytometer. Harvest amoebae by centrifugation at 200*g* for 2 min, and resuspend in FKB at 1–50 × 10^6 amoebae/200 μL.

4. For viable cell counts, dilute amoebae ×10^4 using SDM, and then spread about 100 cells (judged by hemacytometer count) with 150 μL of FKB in duplicate on DSDM plates.

5. For selection of hygromycin-resistant transformants, spread 200 μL of undiluted transformed cell suspension in FKB (1–50 × 10^6 amoebae) onto each 100 U/mL hygromycin B-DSDM plate. Incubate all plates 26°C.

6. Count the number of colonies on the viable count plates after 5 and 7 d of incubation. Examine the selective plates for resistant colonies after 10–21 d of incubation.

4. Notes

1. Wild-type *Physarum* amoebae grow readily in the laboratory on lawns of bacteria on dilute nutrient plates, but do not grow in axenic liquid media. However, mutant strains have been isolated that are capable of axenic growth *(10,13)*, and the convenience of this characteristic led us to focus our electroporation efforts on axenic strains.

   We used strain LU352 to develop electroporation methods. This strain carries a *gadAh* mutation that permits asexual plasmodium development, but this development rarely occurs owing to another mutation, *npfC5*, that blocks selfing. However, plasmodia can be obtained under appropriate conditions by asexual selfing of rare revertants of *npfC5* to *npfC⁺*. Thus, transformants of LU352 can in principal be analyzed in haploid form at any stage of the life cycle, though it is only the amoebae that we have transformed by electroporation.

   Growth at 30°C is preferred, since it prevents selfing of *gadAh* amoebae, such as LU352; any (rare) plasmodia that might develop at 26°C would soon outgrow the amoebae in liquid culture. After several months in culture, growth rates may increase further, and transformation efficiency tends to improve as the culture ages. At this stage, cell survival also improves after electroporation, allowing higher voltages to be used with concomitant increases in transformation efficiency. However, the increased growth rate is a sign that the culture is drifting genetically, and to maintain genetic consistency, such a culture should be replaced with a newly started stock.

2. Using expression of the *cat* gene to estimate transformation efficiency, we found that HEPES buffer was superior to PIPES, CHES, and citrate-phosphate, and that the pH optimum is approx 8.2 (ref. *6* and unpublished data), which is significantly higher than used for amoebae of the cellular slime-mold *Dictyostelium (14)*.

3. We find the optimum sucrose concentration for electroporation of LU352 amoebae to be 40 m*M*, although hypotonic concentrations from 30 to 50 m*M* work well (Fig. 1A). This compares with 50 m*M* sucrose used for electroporation of amoebae of *Dictyostelium*, but for *Dictyostelium*, a substantially lower capacitance is used for the electrical discharge *(14)*. Optimal electrical parameters for *Physarum* vary slightly at 30–50 m*M* sucrose, but 40 m*M* gives superior luciferase expression for p*PardC-luc*-transformed cells at the optimal electrical conditions for all three concentrations (Fig. 1A).

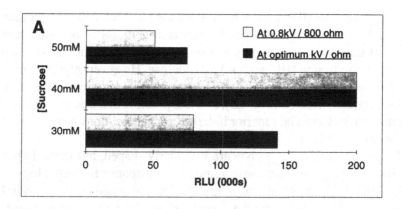

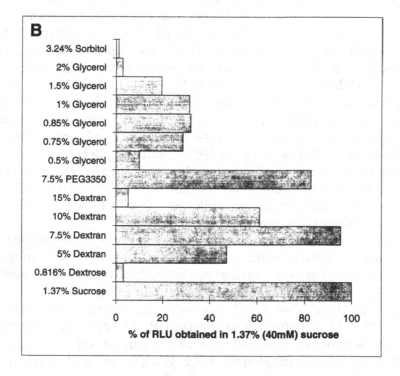

Fig. 1. Effects of osmotic stabilizers. Each cell suspension was electroporated with 1 μg p*PardC-luc* DNA. (**A**) Effect of varying sucrose concentration. The optimum electrical conditions for cells electroporated in 30 and 50 m*M* sucrose in this experiment were 1.0 kV/4 mm and 400 Ω in parallel, and for 40 m*M* sucrose, 0.85 kV/4 mm, 1000 Ω. (**B**) Effects of various alternative osmotic stabilizers. All cells electroporated at 0.85 kV/4 mm and 1000 Ω. Buffers were made by replacing the sucrose in HBS with the indicated solute. RLU = relative light units.

4. Using the optimal electrical conditions for HBS with 40 m*M* sucrose, we tested several alternative osmotic stabilizers (Fig. 1B). None gave superior transformation, as judged by transient expression of luciferase, than did electroporation in HBS with 40 m*M* sucrose. However, glycerol, polyethylene glycol (PEG) 3350, and Dextran all yielded substantial transient luciferase expression (Fig. 1B), and it is possible that one or more of these alternatives would be superior if the electrical conditions were reoptimized for each stabilizer.

5. In growth medium, amoebae are irregularly shaped, but in the hypotonic HBS, they assume a uniform, approximately spherical morphology.

6. We previously reported that cells should be harvested and washed in HBS at room temperature, and then cooled slowly by transfer to a refrigerator rather than directly to ice *(5)*. However, using luciferase assays as an indicator for transformation efficiency, we now find that direct transfer of cells to ice after washing is preferable (Fig. 2). This appears to be true for both apogamic strain LU352 and heterothallic strain LU353 (Fig. 2A,B). High levels of luciferase expression can be detected when cells are electroporated from 1 to 3 h after transfer to ice (Fig. 2). For convenience, we have standardized our protocol using 2 h on ice before electroporation, although longer times may increase transformation efficiency somewhat. However, incubation of cells in HBS overnight on ice before electroporation is deleterious (unpublished results). Incubation of cells at room temperature prior to electroporation is also deleterious, whereas using cold rather than room-temperature HBS for washing cells appears to have no effect on transformation efficiency.

7. Using transient expression of *luc* as an indicator of transformation efficiency, we tested the effects of the volume of cell suspension during electroporation. For each electrical discharge, 5×10^7 cells were used and resuspended in HBS to the indicated volumes (Fig. 2C); 0.8 mL is the capacity of the 4-mm cuvets. It is evident that volumes from 0.5–0.8 mL $(6–10 \times 10^7$ cells/mL) give substantially higher levels of transient expression of luciferase, and there are no significant differences between different volumes tested in this range. Using 0.6 mL gives ample margin for error in measuring the volume of cell suspension.

8. Although the individually packaged cuvets bear the instruction "for single use only," we use each cuvet dozens of times, rinsing copiously with high-quality dH_2O after use and sterilizing by rinsing with 95% ethanol. However, the electrical discharge does change over time when cuvets are used repeatedly. This is manifested in a higher decay time for a given electrical pulse and higher survival rate for the cells. One can compensate by increasing the resistance in parallel with the discharge, effectively increas-

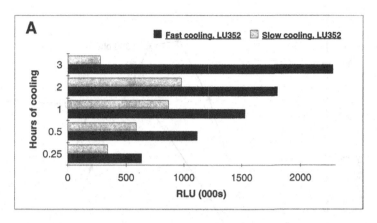

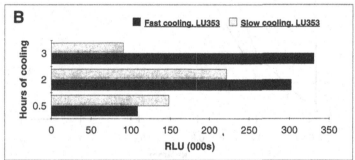

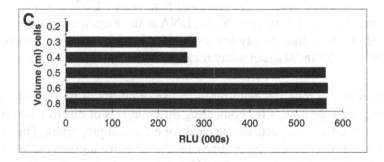

Fig. 2. Transient luciferase expression following different cell pretreatments. Cells were electroporated with 1 µg p*PardC-luc* DNA; LU352 at 0.95 kV/4 mm, 800 Ω; LU353, 0.75 kV/4 mm, 800 Ω. (**A,B**) Effect of cooling amoebae quickly on ice versus cooling them slowly by transfer to a refrigerator; (A) LU352; (B) LU353. (**C**) Effect of cell concentration during electroporation on subsequent transient gene expression.

ing the discharge passing through the cuvet, aiming to produce a constant cell survival rate in the region of 50%. Alternatively, a new cuvet can be used as soon as changes in the decay times are noted. Because of the change

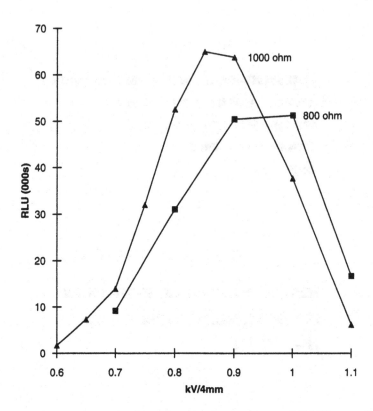

Fig. 3. Effects of varying the electrical parameters. LU352 amoebae were electroporated with 0.7 µg p*PardC-luc* DNA at the electrical conditions indicated. After harvesting and lysing cells, 10 µL of lysate were assayed for luciferase activity in 50 µL of assay reagent.

in electrical properties with repeated use of cuvets, it is important when using separate cuvets for comparing different sets of electroporated cells that the different cuvets be matched for electrical properties. This can be achieved by ensuring decay times are the same for a given electrical discharge across cuvets containing HBS.

9. Use of the appropriate electrical discharge is essential for successful electroporation of *Physarum* amoebae. Using the 4-mm cuvets for LU352 amoebae, we obtain optimal electroporation using a capacitance of 25 µF, 400–1000 Ω resistance in parallel with the cuvet, and 0.8–1.0 kV potential across the 4-mm cuvet. The voltage gradient is critical, although the optimum varies according to the resistance in parallel (e.g., Fig. 3). In the example shown, the highest transient expression of luciferase was obtained after LU352 amoebae were electroporated in HBS at 1000 Ω and 0.85 kV

(Fig. 3). The optimum electrical potential at 800 Ω in this experiment was between 0.9 and 1.0 kV/4 mm, although this gave a lower level of luciferase expression than did the peak at 1000 Ω. In general, for lower resistance in parallel (which permits more of the discharge to bypass the cells, equivalent to lowering the capacitance of the discharge), higher voltage is required to reach peak electroporation conditions. Since the peak transformation (i.e., luciferase expression in our tests) obtained varies for different resistances, it is important to optimize for both resistance and voltage.

10. Some elements of the electroporation protocol are consistent between different strains. For example, rapid cooling of cells rather than slow cooling prior to electroporation is clearly the superior method for both LU352 and LU353 (Fig. 2A,B). Furthermore, the strikingly rapid pattern of decay in luciferase activity during outgrowth after electroporation is indistinguishable between LU353 (Fig. 4) and LU352 *(3)*. Optimal electrical parameters vary only slightly for different haploid strains of *Physarum* amoebae. For example, a slightly lower voltage gradient is optimal for the apogamic strain LU352 compared with another apogamic strain, MA460, and for a given voltage gradient, different haploid strains show similar profiles of transformation efficiency with varying resistance in parallel (not shown). The differences in electrical parameters required to reach peak transient expression values for different strains, although sometimes small, nevertheless suggest that it may be worth the rather small effort needed to optimize for a particular strain using the luciferase transient expression system before proceeding with experiments, such as gene disruptions, that are large-scale or that require the highest possible transformation efficiency.

11. In the future, it is likely that gene disruption experiments in *Physarum* will become common, and for many of these experiments, diploid amoebae will be needed so that recessive lethal disruptions can be established in diploid heterozygous form. This can be achieved in *Physarum* using diploid amoebae of apogamic strains that, like LU352, carry the *gadAh* and *npfC5* mutations, which allow amoebae to be propagated stably, but also permit selfed plasmodia to develop when needed. When such amoebae self to form plasmodia, the plasmodia can sporulate and undergo meiosis, reducing the genome to haploid form in the resulting spores. We constructed a diploid apogamic strain of amoebae, DIP21, similar to LU352, but carrying additional genetic markers. This was tested in electroporation experiments using p*PardC-luc* DNA. It is evident that the optimum voltage gradient for the diploid strain is significantly lower than for the haploid strain (Fig. 5), as one would expect if a critical parameter is the electrical potential difference across the cell; a constant voltage across the cuvet creates a higher potential difference across a larger diploid cell than

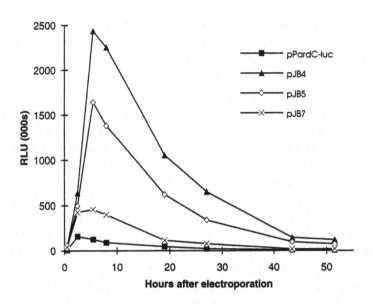

Fig. 4. Effect of enhancer-like genetic elements cloned downstream of *luc,*
and decay of transient *luc* expression during outgrowth. LU353 amoebae were
electroporated with 1 µg plasmid DNA at 0.75 kV/4 mm, 800 Ω. p*PardC-luc* is
the standard luciferase vector, which does not carry enhancers downstream of
luc. pJB4 has the entire 1180-bp sequence of *Physarum* genomic DNA fragment AC1A (ref. *3*; Genbank accession #X74751) cloned immediately downstream of *luc.* This fragment carries several near-consensus sequences for a
budding yeast *ARS* element, although there is no evidence that these elements
affect transformation or replication of the plasmid in *Physarum.* Rather, AC1A
appears to function as an enhancer element. AC1A can be divided into a 420-bp
left-hand *Eco*RV fragment, which is cloned downstream of *luc* in pJB7, and a
right-hand 760-bp *Eco*RV fragment, which is cloned downstream of *luc* in pJB5
(3). When a bacteriophage DNA fragment is cloned downstream of *luc* in
p*PardC-luc,* no such enhancement of expression is observed *(3).*

across a smaller haploid cell. The transformation efficiency appears to be
much lower for DIP21, since the peak luciferase expression was 40-fold
lower than for LU352. It is not yet clear if this is a general feature of diploid *Physarum* amoebae or a characteristic specific to DIP21. However,
DIP21 amoebae grow more slowly in SDM broth than do LU352, and we
generally observe superior transient expression in faster-growing cells.
12. Immediately after electroporation, cells are transferred to 15-mL centrifuge tubes at 30°C for 20 min before adding 5 mL SDM for outgrowth.
The temperature of this postelectroporation incubation is critical; at 0°C,

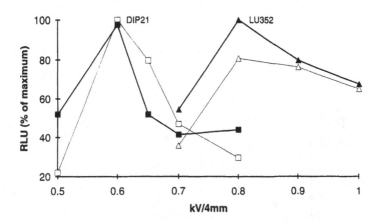

Fig. 5. Comparison of haploid and diploid amoebae. Diploid DIP21 amoebae were electroporated with 1 μg and haploid LU352 with 0.5 μg p*PardC-luc* DNA at the voltages indicated. Thick lines and solid shapes at 800 Ω, thin lines and open shapes at 600 Ω.

more than 50% of the subsequent gene expression is lost, and even at 26°C, subsequent transient gene expression is significantly reduced compared with that obtained after incubation at 30°C *(6)*. However, a 37°C heat shock appears to be deleterious, which is not surprising considering that amoebae do not grow above 31–32°C. The time of postelectroporation incubation before adding medium is less critical. Although the optimum is 20 min, incubation times of 10 and 40 min give almost as much subsequent gene expression *(6)*.

13. The luciferase assay yields higher RLU for increasing volumes of transformed cell lysate added to the assay reaction, at least over the range 1–20 μL of lysate (Fig. 6A). As expected, if the luciferase assays reflect electrotransformation efficiency, increasing the mass of DNA added to the electroporation mixture increases the RLU measured subsequently (Fig. 6B). We find no improvement when 20 μg carrier plasmid DNA are added to electroporations when 1 μg or less of *luc*-bearing plasmid DNA is used for transformation. Using optimum electroporation and outgrowth conditions, 100 ng p*PardC-luc* DNA gives amply measurable RLU among the equivalent of 2×10^6 lysed cells (i.e., 20 μL of lysate from 10^7 cells), a convenience when many sets of transformed cells need to be compared.

14. Transient expression of *luc* in LU352 amoebae can be dramatically enhanced by electroporating amoebae with plasmids that carry all or part of the 1.2-kbp *Physarum* sequence AC1A downstream of the *luc* coding region *(3)*. These sequences act similarly in LU353 amoebae (Fig. 4),

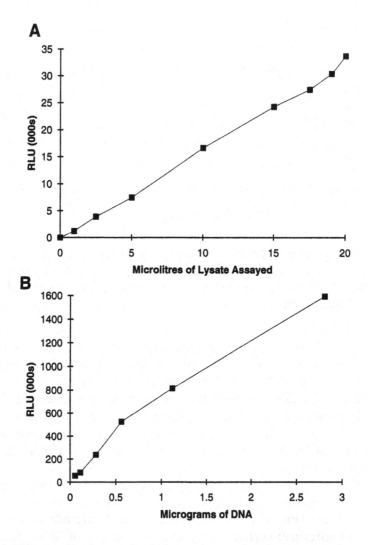

Fig. 6. Effects of volume of cell lysate assayed and concentration of DNA added. The background RLU obtained from mock-transformed cell lysates varied from 350–5000, depending on the luminometer used to measure photon emission, but any one instrument gave a consistent background. (A) Relationship between volume of electrotransformed cell lysate assayed and RLU measured in the luciferase assay. LU352 amoebae were electroporated with 1 µg p*PardC-luc* DNA. Volumes of transformed cell lysate assayed were made to 20 µL with a lysate of cells electroporated without DNA. (B) Increasing the concentration of p*PardC-luc* DNA added to electroporations increases subsequent transient expression of *luc*. Even the smallest mass of DNA added (56 ng) yielded over 50,000 RLU/20 µL cell lysate after subtracting the background of 5000 RLU.

indicating that for enhancement of gene expression following electro-transformation, different strains of *Physarum* act similarly.

15. So far we have not found a way to prevent the dramatic fall in transient gene expression that we observe over time after postelectroporation outgrowth. The decline in expression level is faster than the growth rate, indicating an active process of degradation. This phenomenon is clearly not specific to LU352, where we originally detected it, since the same temporal decline in transient *luc* expression observed in LU352 *(3)* is found in LU353 (Fig. 4).

   The next modification to vectors needed is addition of elements that allow plasmids to replicate autonomously. Although the AC1A element carries a consensus sequence for budding yeast *ARS* elements *(3)*, there is no evidence that AC1A allows plasmids to replicate autonomously in *Physarum.* However, chromosomal replicons can readily be localized in the *Physarum* plasmodium *(1),* and it is expected that some of these will soon be cloned and tested for activity in existing plasmids. Plasmids with the capacity to replicate autonomously will further enhance the experimental opportunities for studying the rich cell biology of this protist.

16. These methods were established using hygromycin B powder (CalBiochem cat. #400050) at about 1000 U/mg. This product is no longer available. Instead, hygromycin B is supplied as a solution (CalBiochem cat. #400051) at around 400,000 U/mL. Preliminary experiments indicate that the concentration we previously used successfully for selection, 100 µg/mL, can be substituted by 100 U/mL of the new product, but further experiments are needed to prove this.

Acknowledgments

This work has been supported by Program Project Grant CA23076 and Core Grant CA07175 from the National Cancer Institute. T. G. B. was subsequently supported by The University of Wisconsin Graduate School. T. G. B. is currently supported by grant MCB-9405605 from the National Science Foundation, and J. B. by grant 034879 from the Wellcome Trust. We thank W. F. Dove and J. Dee for support, critique, and encouragement during the course of this work, J. L. Foxon and R. Barber for technical assistance, and K. Lindsey, D. Twell, and B. Sugden for use of their luminometers. This is paper 3428 from the Laboratory of Genetics.

References

1. Burland, T. G., Solnica-Krezel, L., Bailey, J., Cunningham, D., and Dove, W. F. (1993) Patterns of inheritance, development and the mitotic cycle in the protist *Physarum. Adv. Microbial Physiol.* **35**, 1–69.

2. Anderson, R. W., Hutchins, G., Gray, A., Price, J., and Anderson, S. E. (1989) Regulation of development by the *matA* complex locus in *Physarum polycephalum. J. Gen. Microbiol.* **135,** 1347–1359.

3. Bailey, J., Benard, M., and Burland, T. G. (1994) A luciferase expression system for *Physarum* that facilitates analysis of regulatory elements. *Curr. Genet.* **26,** 126–131.

4. Burland, T. G. and Pallotta, D. (1995) Homologous gene replacement in *Physarum. Genetics* **139,** 147–158.

5. Burland, T. G., Bailey, J., Pallotta, D., and Dove, W. F. (1993). Stable, selectable, integrative DNA transformation in *Physarum. Gene* **132,** 207–212.

6. Burland, T. G., Bailey, J., Adam, L., Dove, W. F., and Pallotta, D. (1992) Expression of a chloramphenicol acetyltransferase gene under the control of actin gene promoters in *Physarum. Curr. Genet.* **21,** 393–398.

7. Gritz, L. and Davies, J. (1983) Plasmid-encoded hygromycin resistance: the sequence of the hygromycin B phosphotransferase gene and its expression in *Escherichia coli* and *Saccharomyces cerevisiae. Gene* **25,** 179–188.

8. Jimenez, A. and Davies, J. (1980) Expression of a transposable antibiotic resistance element in *Saccharomyces. Nature* **287,** 869–871.

9. DeWet, J. R., Wood, K. V., DeLuca, M., Helinski, D. R., and Subramani, S. (1987) Firefly luciferase gene: structure and expression in mammalian cells. *Mol. Cell Biol.* **7,** 725–737.

10. Dee, J., Foxon, J. L., and Anderson, R. W. (1989) Growth, development and genetic characteristics of *Physarum polycephalum* amoebae able to grow in liquid, axenic medium. *J. Gen. Microbiol.* **135,** 1567–1588.

11. Hamelin, M., Adam, L., Lemieux, G., and Pallotta, D. (1988) Expression of three unlinked isocoding actin genes of *Physarum polycephalum. DNA* **7,** 317–328.

12. Burland, T. G., Pallotta, D., Tardif, M. C., Lemieux, G., and Dove, W. F. (1991) Fission yeast promoter-probe vectors based on hygromycin resistance. *Gene* **100,** 241–245.

13. McCullough, C. H. R., Dee, J., and Foxon, J. L. (1978) Genetic factors determining the growth of *Physarum polycephalum* amoebae in liquid axenic medium. *J. Gen. Microbiol.* **106,** 297–306.

14. Howard, P. K., Ahern, K. G., and Firtel, R. A. (1988) Establishment of a transient expression system for *Dictyostelium discoideum. Nucleic Acids Res.* **16,** 2613–2623.

CHAPTER 32

Electroporation
of *Dictyostelium discoideum*

David Knecht and Ka Ming Pang

1. Introduction

Dictyostelium discoideum has long been an intriguing model system for the study of cell type divergence during development, cell signaling, gene expression, and other cell biological problems (*1*). With the development of DNA-mediated transformation, research in *Dictyostelium* has entered a new era. The ability to introduce molecules into cells has allowed the analysis of promoters by transient and stable expression of reporter genes, such as luciferase and β-galactosidase (*2,3*). Stable transformants have also been created that overexpress normal and altered copies of important structural and regulatory proteins (*4–9*). Most importantly, the development of homologous gene targeting and antisense RNA inhibition of gene expression has allowed the creation of mutant cell lines lacking specific gene products (*10,11*). These techniques, coupled with the recently created ordered yeast artificial chromosome library, will allow large-scale mapping, fine structure mapping, and eventually sequencing of the entire genome (*12*).

All of these recent developments in *Dictyostelium* research have come about because of our ability to introduce DNA molecules into amoebae efficiently. The first successes with this technology were accomplished using $CaPO_4$ coprecipitated DNA and a protocol nearly identical to that used for mammalian cells (*13*). In fact, one of the appeals of this system is the remarkable similarity of the amoebae to mammalian cells, especially to the leukocyte (*14*). Since the first successful transformation

From: *Methods in Molecular Biology, Vol. 47: Electroporation Protocols for Microorganisms*
Edited by: J. A. Nickoloff Humana Press Inc., Totowa, NJ

experiments, many new vectors and selections have been developed that expand the repertoire of possible experimental manipulations of the genome. Among these techniques is electroporation of cells, which has been used not only to introduce DNA molecules into cells, but also to introduce labeled tracers and fluorescent probes *(15–19)*. Currently, electroporation has replaced $CaPO_4$ coprecipitation as the favored method of introducing DNA into *Dictyostelium* cells. Anecdotal evidence indicates that when integrating vectors are DNA introduced into cells by electroporation, the copy number of vectors in stable transformants is very low, and the efficiency of homologous gene targeting is very high. $CaPO_4$ coprecipitated DNA tends to result in a high copy number tandem repeat of integrated DNA. This chapter describes the current state of the art in electroporating *Dictyostelium* amoebae, focusing particularly on different DNA vectors and selection systems.

2. Materials

1. Electroporation buffer: 10 mM $NaPO_4$, pH 6.1, 50 mM sucrose; sterilize by autoclaving.
2. HL5 medium: 5 g Proteose peptone #2 (Difco, Detroit, MI), 5 g thiotone E (BBL), 5 g yeast extract (Gibco [Gaithersburg, MD] or Difco), 10 g glucose, 0.35 g Na_2HPO_4, and 0.35 g KH_2PO_4; bring to 1 L with water, adjust pH to 6.7, and autoclave.
3. HL5 plus thymidine: Add 40 mg of thymidine to HL5 medium before autoclaving.
4. HL5 plus uracil: Add 100 mg of uracil to HL5 medium before autoclaving.
5. 100X Ampilcillin (–20°C) + dihydrostreptomycin: 10 mg/mL each in water; store at –70°C. Dilute to 1X in all HL5 and agar plates.
6. 1000X G418 stock: Dissolve 100 mg G418 (Geneticin; Sigma [St. Louis, MO]) in 10 mL of 10 mM HEPES, pH 7.2; filter-sterilize, and store at 4°C (stable for months).
7. 1000X Hygromycin stock: Dissolve 375 mg of hygromycin (Calbiochem Inc., La Jolla, CA) in 10 mL of 10 mM HEPES, pH 7.2, filter-sterilize, and store at 4°C (stable for months).
8. 1000X FOA stock: Dissolve 100 mg of 5-fluoro-orotic acid in 1 mL of DMSO, and store at –20°C in small aliquots.
9. FM medium: This is a completely defined medium *(20)* available from Gibco. A detailed protocol for making the medium can be found in the *Dictyostelium* Internet archive at worms.cmsbio.nwu.edu (\pub\dicty\ fm_medium).
10. HL5 Base agar: 500 mL of HL5 medium with 5 g of Difco bacto-agar; autoclave and store at 4°C.

Table 1
Dictyostelium Transformation Vectors

Vector	Chromosomal location	Selection	Reference
pA15TX	Integrating	G418	24
pA6NPTII	Integrating	G418	23
pB10SX	Integrating	G418	22
pnDeI	Extrachromosomal	G418	33
pDE102	Integrating	Hygromycin	25
pDE104	Integrating	Hygromycin	25
pDE109	Extrachromosomal	Hygromycin	25
pDU3B1	Integrating	Pyr5-6	31
pRHI13	Integrating	Pyr5-6	—[a]
pMYC10	*pyr5-6* knockout	FOA	—[a]
pRG24	*pyr5-6* knockout	FOA	—[a]
pGEM25	Integrating	Thy	32
pDNeoII	Integrating/expression	G418	34

[a]These vectors were constructed in the laboratory of Peter Devreotes and are as yet unpublished. They can be obtained from the author or from Devreotes.

11. HL5 Soft agarose: 100 mL of HL5 medium with 1.2 g of Ultralow gelling temperature agarose (Fisher or FMC, gelling point below 25°C). Autoclave with a stir bar in bottle; after autoclaving, stir for 10–15 min to dissolve agar completely, pipet 5-mL aliquots into sterile tubes and store at 4°C.
12. Plasmid DNA: Suitable vectors are listed in Table 1.

3. Methods

3.1. Electroporation of DNA into Cells

A number of protocols have been used to electroporate DNA into cells (e.g., *3,16,21*). We have recently tested several protocols and compared the efficiency of transformation with each. We found a modified version of the protocol described by Howard et al. *(3)* to be the most efficient (R. Insall and P. Devreutes, unpublished). This protocol works with all *Dictyostelium* strains and selections that we have tested.

1. Grow cells in liquid medium (HL5) as a monolayer in Petri dishes or in shaking suspension (*see* Note 1).
2. Pellet cells by centrifugation at 500*g* for 5 min at 4°C in a 15-mL conical sterile tube.
3. Aspirate the supernatant, carefully removing as much media as possible.
4. Resuspend the cells in ice-cold electroporation buffer to a density of 4×10^7 cells/mL (*see* Note 2).

5. Add 10–30 µg of DNA to an ice-cold cuvet with a 0.2-cm electrode gap, and then add 0.4 mL of the cell suspension (*see* Note 3).
6. Immediately pulse twice, 5 s apart, at 1.1 kV and 3 µF. A 5-Ω resistor should be placed in series. The time constant will be 0.6–0.7 ms.
7. Allow cells to recover for 5 min on ice, and then transfer the contents of the cuvet to 10 mL of HL5 medium in a 100-mm Petri dish (*see* Note 4).

3.2. Antibiotic Selections

Transformation vectors have been constructed that take advantage of antibiotic resistance genes from a variety of organisms. These resistance genes have been fused to a *Dictyostelium* actin 6 or actin 15 promoter in order to drive expression in amoebae. These vectors generally insert into the genome as either single copies or multicopy tandem repeats. However, when fused to an endogenous *Dictyostelium* plasmid sequence, they can be maintained as stable extrachromosomal vectors.

Selections engineered in this way include G418 *(22–24)*, hygromycin *(25)*, bleomycin *(26,27)*, and blasticidin-S *(28)* (*see* Table 1).

3.3. Selection of Transformants in Liquid Medium

1. Add the cells from the electroporation cuvet to 10 mL of HL5 medium, and plate no more than 10^7 cells/100-mm dish.
2. Allow cells to recover overnight in HL5 medium.
3. Aspirate the medium, and add 10 mL of fresh HL5 to each plate.
4. Add selective agent to each plate from concentrated stock solution (final concentration 10 µg/mL for G418 and 37.5 µg/mL for hygromycin) (*see* Note 5).
5. After 2–3 d, most cells will round up as they die and detach from the dish. When most cells are detached (usually on d 3), gently swirl the dish to suspend the detached cells, and aspirate the medium and cells. The media should be changed thereafter at approx 3-d intervals (use 10 mL of HL5 with selective agent). By this time, small colonies of transformed cells should be visible with an inverted microscope. Because some of these colonies die on further incubation, wait about 1 wk before picking transformed cells.
6. Cells can be removed directly from colonies with a Pipetteman and cloned in 96-well microtiter plates in HL5, or on SM plates in association with bacteria *(29)*. Alternatively, cells from the entire population can be recovered by washing the amoebae off the plate with a gentle stream of media using a 10-mL pipet.

3.4. Selection of Transformants in Soft Agarose

Cloning cells directly in soft agarose has the advantage of generating single-cell-derived clonal transformants directly after transformation

(21). Also, in situations where some transformants have an impaired growth rate (for instance, in gene knockout experiments), a mutant cell line is not overwhelmed by other transformants in the dish. The efficiency of transformation using this selection procedure is at least as high as that obtained in liquid and gives a more accurate reflection of the number of transformants.

1. The day after transformation, aspirate the medium, and add 10 mL of fresh HL5 medium, including selective agent.
2. Melt HL5 base agar in a microwave or steam oven and pipet 15 mL of agar into 100-mm Petri dishes.
3. Allow to cool on a level table.
4. If desired, add G418 (final: 20 µg/mL), hygromycin (final: 100 µg/mL), or ampicillin + dihydrostreptomycin sulfate (final: 100 µg/mL each).
5. Melt HL5 soft agarose by brief microwaving (about 15 s), cool in a water bath to 25°C, and add antibiotics to the same concentration as in the HL5 base agar.
6. Wash the cells off the bottom of the plate with a vigorous stream in 5–10 mL of HL5 medium using a 10-mL pipet.
7. Add 1 mL of cells (1–2 × 10^6 total) to a tube containing the HL5 soft agarose, and gently mix.
8. Pipet or pour onto base agar plate, and tilt to spread the agarose evenly.
9. Place the plates in a cold room on a level table to harden for 15–20 min.
10. Incubate in a humid environment at about 22°C. Colonies will be visible under the microscope within 1 wk. After about 2 wk, the colonies can be picked with a sterile toothpick by stabbing the colony, and then transferring the cells to a 24-well tissue-culture plate containing 0.5 mL of HL5 medium.

3.5. Positive and Negative pyr5-6 Selection

The *pyr5-6* gene encoding orotidine-5'-phosphate decarboxylase and orotate phosphoribosyl transferase were cloned from *Dictyostelium* by complementation of Ura⁻ yeast *(30)*. The *Dictyostelium pyr5-6* gene has been used to create transformation vectors that will alter the endogenous locus by homologous recombination *(31)*. Transformants are selected by growing cells in 5-fluoro-orotic acid (FOA) plus uracil. Only mutant cells lacking Pyr5-6 can grow in FOA and these cells are auxotrophs for uracil. The *pyr5-6* gene can be reintroduced into these cells by transformation selecting for the ability of the cells to grow without uracil supplementation. It is possible with the alternating selections to mutate a series of genes sequentially in the same cell line.

3.5.1. FOA Selection

1. Electroporate cells with 10–20 µg of pMYC10 or pRG24 DNA (Table 1).
2. Remove cells from the cuvet, and plate in HL5 containing 100 µg/mL uracil.
3. After 3 d, cells are split into multiple dishes at a density of about 2×10^6 cells/dish in HL5 plus uracil (*see* Notes 6 and 7).
4. Add FOA to a concentration of 100 µg/mL.
5. Replace the media every 4–5 d until colonies begin to appear in the dish.

3.5.2. Uracil Selection

1. Grow *pyr5-6⁻* cells in HL5 containing 100 µg/mL uracil.
2. Harvest cells, and electroporate with 10–20 µg of pRHI13 DNA (Table 1).
3. Add the cells to 10 mL FM medium.
4. Split cells into multiple dishes containing FM medium at a maximum density of 2×10^6 cells/dish (*see* Notes 8–11).
5. Change the media every 4–5 d until colonies become obvious (1–2 wk).

3.6. Thy Selection

A cell line (JH10) that is auxotrophic for thymidine has been created by deletion of the *Thy1* locus (*32*, and R. A. Firtel, personal communication). These cells can be transformed by electroporating the pGEM25 vector containing the *Thy1* gene (Table 1) and selecting for growth in normal HL5 (*see* Note 12).

1. Grow JH10 cells in HL5 containing 40 µg/mL thymidine.
2. Harvest cells, and electroporate with 10–20 µg pGEM25 DNA.
3. Add the cells from the cuvet to 10 mL of HL5.
4. Split cells into multiple dishes containing HL5 medium at a maximum density of 2×10^6 cells/dish.
5. Media in the dishes is changed every 4–5 d until colonies become obvious (1–2 wk).

4. Notes

4.1. Electroporation

1. We have seen a minimal effect using cells from log through stationary phase on the efficiency of transformation.
2. The cell density in the cuvet is not critical. We have used from 10^7–10^8 cells/mL. In general, the more cells electroporated, the more transformants are obtained.
3. The amount of DNA is also not critical. Generally, the more DNA, the more transformants are obtained, up to about 50 µg of DNA. For extrachromosomal vectors, the efficiency of transformation is 100–1000 times

higher than with integrating vectors, so 0.1–1 µg of DNA can be used or fewer cells treated when extrachromosomal vectors are used.

4. Some protocols suggest adding healing buffer (4 µL of 100 mM $CaCl_2$ + 100 mM $MgCl_2$) directly to cells in the cuvet and incubating for 15 min before plating. We have found the transformation efficiency to be no better and often worse when healing buffer is used.

4.2. Antibiotic Selections

5. Selection using bleomycin, phleomycin, and blasticidin-S has been reported, but has not been tested in our laboratory.

4.3. pyr5-6 Selections

6. The 3-d recovery is intended to allow the activity of the Pyr5-6 enzyme to decay following gene knockout. Otherwise, FOA kills cells that have the gene knocked out, but still retain enzyme. However, some labs do not use this procedure and add the cells immediately to HL5 containing uracil and FOA. Presumably, the rate of killing by FOA is slow enough to allow the enzyme to decay. The efficiency of the two protocols has not been compared.

7. It is important to do both the FOA and the uracil selections at a sufficiently low cell density. If the density is too high, the cells will initiate their developmental program and aggregate together to form mounds. The aggregates can be confused with transformed colonies. However, aggregates appear a few days after transformation, whereas transformed colonies require several weeks to become apparent. It is important to perform a control without DNA so that the difference between aggregates and colonies can be easily assessed.

8. Testing FOA resistant transformants for uracil dependence is important because we frequently obtain a small number of transformants that are FOA resistant and not uracil-dependent. These clones probably have an alteration at some other locus and should be discarded.

9. The selection for FOA resistance can also be done in soft agarose. Plate cells in soft agar on base agar both containing 100 µg/mL uracil and 100 µg/mL FOA. Colonies will appear after several weeks.

10. HL5 can be used for the uracil selection instead of FM. However, untransformed cells will have a tendency to grow significantly in HL5 owing presumably to a small amount of uracil in the medium.

11. A number of *pyr5-6⁻* cell lines have been created that are available for general use. These include DelF16-11 (AX2 parental, DelF *[31]*; Hkpo DAK, unpublished), HK50 (NC4A2 parental), and HL330 (AX3 parental, HL330 *[19]*). They are available from the author on request.

4.4. Thy Selection

12. This selection seems to be tighter in HL5 than the *pyr5-6* selection. There is no need to use FM medium because the JH10 cells do not grow significantly in HL5 medium.

References

1. Loomis, W. (1975) *Dictyostelium Discoideum: A Developmental System*. Academic, New York.
2. Dingermann, T., Reindl, N., Werner, H., Hildebrandt, M., Nellen, W., Harwood, A., Williams, J., and Nerke, K. (1989) Optimization and *in situ* detection of *Escherichia coli* beta-galactosidase gene expression in *Dictyostelium discoideum*. *Gene* **85**, 353–362.
3. Howard, P. K., Ahern, K. G., and Firtel, R. A. (1988) Establishment of a transient expression system for *Dictyostelium discoideum*. *Nucleic Acids Res.* **16**, 2613–2623.
4. Mann, S. K., Yonemoto, W. M., Taylor, S. S., and Firtel, R. A. (1992) DdPK3, which plays essential roles during Dictyostelium development, encodes the catalytic subunit of cAMP-dependent protein kinase. *Proc. Natl. Acad. Sci. USA* **89**, 10,701–10,705.
5. Johnson, R. L., Vaughan, R. A., Caterina, M. J., Van, H. P., and Devreotes, P. N. (1991) Overexpression of the cAMP receptor 1 in growing *Dictyostelium* cells. *Biochemistry* **30**, 6982–6986.
6. Hall, A. L., Franke, J., Faure, M., and Kessin, R. H. (1993) The role of the cyclic nucleotide phosphodiesterase of *Dictyostelium discoideum* during growth, aggregation, and morphogenesis: overexpression and localization studies with the separate promoters of the pde. *Dev. Biol.* **157**, 73–84.
7. Hadwiger, J. A. and Firtel, R. A. (1992) Analysis of G alpha 4, a G-protein subunit required for multicellular development in *Dictyostelium*. *Genes Dev.* **6**, 38–49.
8. Faix, J., Gerisch, G., and Noegel, A. A. (1990) Constitutive overexpression of the contact site A glycoprotein enables growth-phase cells of *Dictyostelium discoideum* to aggregate. *EMBO J.* **9**, 2709–2716.
9. Anjard, C., Pinaud, S., Kay, R. R., and Reymond, C. D. (1992) Overexpression of Dd PK2 protein kinase causes rapid development and affects the intracellular cAMP pathway of *Dictyostelium discoideum*. *Development* **115**, 785–790.
10. De Lozanne, A. and Spudich, J. A. (1987) Disruption of the Dictyostelium myosin heavy chain gene by homologous recombination. *Science* **236**, 1086–1091.
11. Knecht, D. and Loomis, W. F. (1987) Antisense RNA inactivation of myosin heavy chain gene expression in Dictyostelium discoideum. *Science* **236**, 1081–1086.
12. Kuspa, A., Maghakian, D., Bergesch, P., and Loomis, W. F. (1992) Physical mapping of genes to specific chromosomes in *Dictyostelium discoideum*. *Genomics* **13**, 49–61.
13. Nellen, W., Silan, C., and Firtel, R. A. (1984) DNA-mediated transformation in Dictyostelium discoideum: regulated expression of an actin gene fusion. *Mol. Cell. Biol.* **4**, 2890–2898.
14. Devreotes, P. N. and Zigmond, S. H. (1988) Chemotaxis in eukaryotic cells: a focus on leukocytes and Dictyostelium. *Ann. Rev. Cell Biol.* **4**, 649–686.

15. Abe, T., Maeda, Y., and Iijima, T. (1988) Transient increase of the intracellular Ca2+ concentration during chemotactic signal transduction in Dictyostelium discoideum cells. *Differentiation* **39**, 90–96.
16. Cubitt, A. B. and Firtel, R. A. (1992) Characterization of phospholipase activity in Dictyostelium discoideum. Identification of a Ca^{2+}-dependent polyphosphoinositide-specific phospholipase C. *Biochem. J.* **283**, 371–378.
17. Van Duijn, B., Vogelzang, S. A., Ypey, D. L., Van der Molen, L. G., and Van Haastert, P. J. (1990) Normal chemotaxis in *Dictyostelium discoideum* cells with a depolarized plasma membrane potential. *J. Cell Sci.* **95**, 177–183.
18. Van Haastert, P. J., De Vries, M. J., Penning, L. C., Roovers, E., Van der Kaay, J., Erneux, C., and Van Lookeren Campagne, M. M. (1989) Chemoattractant and guanosine 5'-[gamma-thio]triphosphate induce the accumulation of inositol 1,4,5-trisphosphate in Dictyostelium cells that are labelled with [^{3}H]inositol by electroporation. *Biochem. J.* **258**, 577–586.
19. Kuspa, A. and Loomis, W. F. (1992) Tagging developmental genes in Dictyostelium by restriction enzyme-mediated integration of plasmid DNA. *Proc. Natl. Acad. Sci. USA* **89**, 8803–8807.
20. Franke, J. and Kessin, R. (1977) A defined minimal medium for axenic strains of Dictyostelium discoideum. *Proc. Natl. Acad. Sci. USA* **74**, 2157–2161.
21. Knecht, D. A., Jung, J., and Matthews, L. (1990) Quantification of transformation efficiency using a new method for clonal growth and selection of axenic Dictyostelium cells. *Dev. Genet.* **11**, 403–409.
22. Nellen, W. and Firtel, R. A. (1985) High copy number transformants and co-transformation in Dictyostelium. *Gene* **39**, 155–163.
23. Knecht, D., Cohen, S., Loomis, W., and Lodish, H. (1986) Developmental regulation of *Dictyostelium discoideum* actin gene fusions carried on low-copy and high-copy transformation vectors. *Mol. Cell Biol.* **6**, 3973–3983.
24. Cohen, S. M., Knecht, D., Lodish, H. F., and Loomis, W. F. (1986) DNA sequences required for expression of a *Dictyostelium* actin gene. *EMBO J.* **5**, 3361–3366.
25. Egelhoff, T. T., Brown, S. S., Manstein, D. J., and Spudich, J. A. (1989) Hygromycin resistance as a selectable marker in Dictyostelium discoideum. *Mol. Cell Biol.* **9**, 1965–1968.
26. Chang, A. C., Hall, R. M., and Williams, K. L. (1991) Bleomycin resistance as a selectable marker for transformation of the eukaryote, *Dictyostelium discoideum*. *Gene* **107**, 165–170.
27. Leiting, B. and Noegel, A. A. (1991) The ble gene of Streptoalloteichus hindustanus as a new selectable marker for *Dictyostelium discoideum* confers resistance to phleomycin. *Biochem. Biophys. Res. Commun.* **180**, 1403–1407.
28. Sutoh, K. (1993) A transformation vector for *Dictyosteium discoideum* with a new selectable marker bsr. *Plasmid* **30**, 150–154.
29. Sussman, M. (1987) Cultivation and synchronous morphogenesis of Dictyostelium under controlled experimental conditions. *Methods Cell Biol.* **28**, 9–29.
30. Jacquet, M., Guilbaud, R., and Garreau, H. (1988) Sequence analysis of the DdPYR5-6 gene coding for UMP synthase in *Dictyostelium discoideum* and com-

parison with orotate phosphoribosyl transferases and OMP/carboxylases. *Mol. Gen. Genet.* **211,** 441–445.

31. Kalpaxis, D., Zundorf, I., Werner, H., Reindl, N., Boy, M. E., Jacquet, M., and Dingermann, T. (1991) Positive selection for Dictyostelium discoideum mutants lacking UMP synthase activity based on resistance to 5-fluoroorotic acid. *Mol. Gen. Genet.* **225,** 492–500.

32. Dynes, J. L. and Firtel, R. A. (1989) Molecular complementation of a genetic marker in Dictyostelium using a genomic DNA library. *Proc. Natl. Acad. Sci. USA* **86,** 7966–7970.

33. Leiting, B. and Noegel, A. (1988) Construction of an extrachromosomally replicating transformation vector for *Dictyostelium discoideum. Plasmid* **20,** 241–248.

34. Witke, W., Nellen, W., and Noegel, A. (1987) Homologous recombination in the Dictyostelium alpha-actinin gene leads to an altered mRNA and lack of the protein. *EMBO J.* **6,** 4143–4148.

CHAPTER 33

Gene Transfer
by Electroporation in *Tetrahymena*

Jacek Gaertig and Martin A. Gorovsky

1. Introduction

Tetrahymena thermophila is a free-living ciliated protozoan useful in cell biological and molecular genetic studies. Recently DNA-mediated transformation has been developed for this microorganism *(1)*, and used to study the mechanisms of genome rearrangement *(2–4)* and structure–function relationships of the RNA component of telomerase, an enzyme responsible for synthesis of chromosomal ends *(5,6)*. *Tetrahymena* is also an interesting eukaryotic model for studying cell biological processes, such as cell motility, exocytosis, and pattern formation. Furthermore, recent evidence showed that *Tetrahymena* is one of only a few eukaryotic organisms known in which DNA recombination activity of transforming DNA is known to be exclusively homologous *(7–10)*. Thus, gene replacement can be used in *Tetrahymena* as a tool to address fundamental problems in molecular cell biology.

Like most ciliates, *Tetrahymena* cells contain two different nuclei. The somatic functions of the cell are governed by the polygenomic, transcriptionally active macronucleus, whereas the diploid, transcriptionally silent (during vegetative growth) micronucleus is responsible for germinal functions. Thus, the phenotype is determined by the macronucleus, whereas the micronucleus is used to transmit the genotype during the sexual reproductive process (conjugation). The recently developed methods genetically transform the macronucleus. The first successful DNA-mediated transformation in *Tetrahymena* was performed by direct

From: *Methods in Molecular Biology, Vol. 47: Electroporation Protocols for Microorganisms*
Edited by: J. A. Nickoloff Humana Press Inc., Totowa, NJ

injection of a drug-resistant ribosomal RNA gene (rDNA) into the macronucleus *(1)*. Although microinjection-mediated DNA transformation methods are reliable, the technique is tedious, requires expensive equipment, and is suitable only if one or very few transformants are needed. Despite development of new types of plasmid transformation vectors, until recently microinjection-mediated DNA transformation remained the method of choice for gene transfer in *Tetrahymena*.

As for many other cell types, electric-field-based approaches for cell permeabilization and gene transfer were explored in *Tetrahymena*. The ciliate cell body is unusual among unicellular eukaryotes owing to the existence of a rather complex cortical structure. Underneath the plasma membrane, there is a layer of cortical membrane vesicles (alveolar system), a dense sheet of the plasma-membrane-associated cytoskeleton (epiplasm), and arrays of cortical microtubules. Thus, the cortical cytoskeleton may provide additional physical barriers to cell permeabilization. Despite these natural obstacles, in *Tetrahymena*, high-voltage electric field pulses induces cell membrane phenomena that are observed in many other cell types. Mass electrofusion followed by cytoplasmic integration was induced when *Tetrahymena* cells were brought into close contact by an alternating electric field in a nonconductive medium and fused by high-voltage pulses *(11,12)*. Electroporation-mediated stable DNA transformation using a ribosomal gene rDNA vector was also achieved in vegetative *Tetrahymena (13,14)*. However, this method was very inefficient and even more labor-intensive than microinjection; it produced very few transformants and, more importantly, it required a prolonged growth and expansion period to isolate transformants. The low efficiency of DNA-mediated electrotransformation of vegetative cells could be owing to insufficient DNA uptake or to insufficient levels of expression of the transformed gene relative to the large number of endogenous genes in the polycopy macronucleus.

We recently described an electroporation-mediated transformation method for *Tetrahymena* that generates transformants at high efficiency *(15)*. In our protocol, instead of vegetative cells, conjugating cells were used. Conjugation in *Tetrahymena* can be induced by starvation, in a reasonably synchronous fashion *(16)*. During conjugation, two cells with different mating types form a pair. The micronuclei undergo meiosis followed by a haploid mitosis of one of the four meiotic products to produce two pronuclei in each cell. The subsequent pronuclear crossexchange between partner

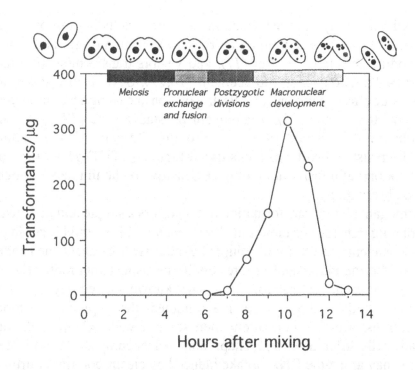

Fig. 1. Effect of conjugation timing on transformation efficiency mediated by an rDNA vector pD5H8 (data from ref. *15*). Approximate timing of occurrence of major microscopically identifiable stages of conjugation is illustrated at the top of figure.

cells followed by pronuclear fusion leads to formation of the zygotic micronucleus (synkaryon). The zygotic nucleus undergoes two mitotic divisions and gives rise to new macro- and micronuclei, while the preexisting macronucleus is resorbed. In each partner, two nuclei that are localized to the anterior cytoplasmic pole become the macronuclei, whereas the two posterior ones remain the micronuclei. During macronuclear development, the zygotic micronuclear genome is reorganized by excision of the micronucleus-specific sequences, endoreplication of most of the gene copies to ~45 times the haploid (C-) value, and amplification of a single copy of the ribosomal RNA gene to ~10,000 palindromic dimers.

In contrast to somatic cells, conjugating *Tetrahymena* exhibited a transient stage of high competence for electrotransformation. The period of electrotransformation competence overlaps with the period of macronuclear development (Fig. 1). Electrotransformation of conjugants was used with a variety of transformation strategies, including use of rDNA vectors *(15)*, vectors containing an rDNA replication origin and a protein-coding selectable marker *(17)*, and vectors transforming by functional or knockout gene replacement *(10,17)*. Several drug-resistant markers are effective in conjugant electroporation, including a paromomycin resistant rRNA gene *(15)*, a neomycin resistance gene *(17)*, a mutant cycloheximide-resistant ribosomal protein L29 gene *(15)*, a mutant oryzalin resistant/taxol hypersensitive β-tubulin (BTU1) gene *(10)*, and a mutant oryzalin resistant/taxol hypersensitive α-tubulin gene (unpublished observations).

Conjugant electrotransformation (CET) offers a simple and rapid way of isolating transformed clones in *Tetrahymena*. This method is conjugation-dependent; starved (nonconjugating) cells are not transformed under similar electroporation and culture conditions, using either replicative or gene replacement vectors *(15,17)*. It is not known why conjugating *Tetrahymena* cells at the stage of macronuclear development are more electrotransformation competent then somatic cells. Conjugants and somatic cells differ in cell size, shape, and nuclear composition, and these factors may affect the DNA uptake induced by electroporation. Furthermore, in conjugating cells, replicative vector DNA is subjected to a series of DNA endoreplications or to rDNA amplification (if the transforming vector contains an rDNA replication origin), which may result in a higher level of expression enabling more efficient selection. Also, developing macronuclei undergoing processes of genomic rearrangement and actively replicating DNA may have higher levels of enzymatic activities involved in homologous recombination, required for transformation by gene replacement vectors.

Two types of transformation vectors can be used to transform *Tetrahymena* by CET: (1) vectors transforming by homologous recombination (and lacking a replication origin), which can be used for functional gene replacement or gene knockout, and (2) replicative vectors (containing the *Tetrahymena* rDNA replication origin), which are maintained as an extrachromosomal DNA. Some of the basic transformation vectors and selectable markers for gene replacement and gene knockout that can be used with CET are listed in Table 1. Replicative vectors are of two types: rDNA vectors (containing a drug resistant rRNA gene) and plasmids containing the replication origin from rDNA and a protein coding selectable marker (for example the H4-I/*neo*/BTU2 gene [as in pH4T2]).

Table 1
Transformation Vectors and Selectable Genes Used with the CET Method

Plasmid name	Selectable marker gene	Mechanism of transformation	Cotransformation recommended?[b]	Selection	Vector reference	CET reference
pD5H8	rRNA gene	Processing replicative	No	Pmr	25	15
prD4-1	rRNA gene	Replicative	No	Pmr	7	15
pH4T2	H4-I/neo/BTU2	Replicative	No	Pmr	9,17	17
TtL29A3	rpL29	Replacement	No	Cy	8	15
pBTU1M350	BTU1 (β-tubulin 1)	Replacement	Yes	Ory or Vb	10	10
pTUB65	α-tubulin	Replacement	Yes	Ory or Vb	Unpub	Unpub
p4T21[a] (universal disruption cassette)	H4-I/neo/BTU2	Replacement	No	Pmr	17	Unpub

Abbreviations: Pmr, 120 μg/mL paromomycin; Cy, 15 μg/mL cycloheximide; Ory, 30 μM oryzalin; Vb, 30 μM vinblastine.

[a]Note that the p4T21 plasmid does not transform *Tetrahymena* by itself. The insert from this plasmid (the H4-I/neo/BTU2 gene) must be subcloned within the coding sequence of a targeted gene prior to transformation.

[b]Cotransformation is recommended when the phenotype of the transformed gene cannot be easily selected because the gene is present in the macronucleus in low number.

335

Because of the presence of an rDNA replication origin fragment, these vectors are amplified to a very high copy number in *Tetrahymena* conjugants, leading to overexpression of cloned genes. It should be noted that, owing to the presence of the rDNA origin, replicative vectors frequently integrate into the rDNA.

In the *Tetrahymena* genome, transformed DNA is replicated if it contains a replication origin fragment, or it is integrated into the endogenous locus if it contains a homologous targeting sequence. Using homologous recombination activity in *Tetrahymena*, a gene can be inactivated by disruptive replacement. An artificial gene was constructed consisting of the aminophosphotransferase coding sequence (*neo* gene) placed under the control of short regulatory sequences of the H4-I gene promoter and the BTU2 gene terminator *(9,17)*. By itself, this hybrid H4-I/*neo*/BTU2 gene does not transform *Tetrahymena*, most likely because of the absence of a replication origin and its inability to undergo homologous integration (homologous sequences are short and are from two different genes). However, the H4-I/*neo*/BTU2 gene provides a convenient "disruption cassette" that can inactivate *Tetrahymena* genes in vivo by CET. To perform a gene knockout, the H4-I/*neo*/BTU2 cassette is simply subcloned within the coding sequence of a cloned gene and the whole construct targeted to a homologous locus by CET. So far, 10 *Tetrahymena* genes have been disrupted using this method (unpublished results). Various lengths of homologous flanking sequences were used in gene knockout experiments ranging from one to several kilobases of DNA. The shortest flanking regions sufficient for correct targeting contained only ~500 bp of homologous sequence on each side of the hybrid *neo* gene, in the case of a knockout of a gene encoding a small RNA induced by starvation or heat shock (P. Fung and R. Hallberg, personal communication). In most cases, the coding sequence of the targeted gene was not eliminated, but was simply interrupted by the disruption cassette. In most constructs, transcription of the *neo* gene is in the same direction as the disrupted gene's transcription. However, in one case, efficient gene disruption occurred when the selectable H4-I/*neo*/BTU2 marker was subcloned so it was transcribed in the opposite transcriptional direction than the disrupted gene (X. Liu and M. Gorovsky, unpublished results).

Functional gene replacement can be performed using a cloned *Tetrahymena* gene that provides a selectable phenotype. For example, a mutant cycloheximide resistant ribosomal protein L29 (rpL29) gene can be used for a functional replacement of the endogenous drug-sensitive gene *(8)*. The BTU1 (β-tubulin) gene can be used both as a positive and negative selectable marker because it confers a dual phenotype of resistance to microtubule depolymerizing agents (oryzalin or vinblastine) and hypersensitivity to the microtubule-stabilizing drug taxol *(10)*. A mutant version of the BTU1 gene can be used to replace the wild-type BTU1 gene using oryzalin or vinblastine selection. The mutant tubulin gene can then be replaced in turn by the wild-type BTU1 gene by selecting for taxol resistance. Recently, a selectable α-tubulin gene replacement marker similar to the mutant BTU1 has been constructed (unpublished results).

It should be noted that owing to the polygenomic character of the macronucleus, in the typical gene replacement experiment, only partial replacement of some of the ~45 endogenous copies is obtained initially. However, replacement can be completed during subsequent vegetative divisions. The macronucleus divides amitotically, and alleles are randomly distributed at division *(18,19)*. Under selective pressure, assortment of the endogenous allele can occur within 50–80 generations if the function of the targeted gene is not essential. Partial replacement of essential genes also can be obtained (X. Liu and M. Gorovsky, unpublished results).

Here, we describe a general protocol for the conjugant electrotransformation method that can be applied to transform *Tetrahymena* with both replicative or gene replacement/gene knockout vectors.

2. Materials

2.1. Reagents and Instrumentation

1. *Tetrahymena* SPP growth medium: 1% proteose-peptone, 0.1% yeast extract, 0.2% dextrose, and 0.003% sequestrine *(20)*. Sterilize by autoclaving.
2. Starvation buffer: 10 m*M* Tris-HCl, pH 7.5. Sterilize by autoclaving.
3. Electroporation buffer: 10 m*M* HEPES buffer, pH 7.5. Sterilize by autoclaving.
4. Plating growth medium (SPPA): SPP and 1X antibiotic-antimycotic mix.
5. Antibiotic-antimycotic mix (100X): Add 1 mL of fungizone (Gibco-BRL #600-5295AE) to 100 mL of penicillin/streptomycin (Gibco-BRL #600-5140PG). Store at –20°C.
6. Paromomycin sulfate (Sigma, St. Louis, MO): 100 mg/mL stock solution in water. Store at –20°C.

7. Cycloheximide (Sigma): 12.5 mg/mL solution in ethanol. Store at –20°C.
8. 6-Methylpurine (Sigma): 15 mg/mL solution in water. Store at –20°C.
9. Oryzalin (Chem Services, West Chester, PA): 100 mM solution in DMSO. Store at –20°C.
10. Vinblastine (Sigma): 50 mM solution in DMSO. Store at –20°C.
11. Dissection scope: Low magnification is sufficient to monitor *Tetrahymena* during transformation steps and to screen for transformants.
12. Laminar flow hood: required for all plating procedures to avoid contamination.
13. Cell electroporator: to deliver an exponential electric pulse of ~4 ms length and field strength of ~1125 V/cm.
14. Small laboratory shaker set at 30°C.

2.2. Cell Lines

Suitable strains must be highly fertile. One of them should be a heterokaryon containing a dominant drug resistance gene in the micronucleus and the other containing a recessive, sensitive allele in the macronucleus. A suitable pair is the heterokaryon strain CU428.1-*Mpr/Mpr* (mp-s, mt VII) and the wild-type strain B2086 *Mpr⁺/Mpr⁺* (mp-s, mt I), available from P. J. Bruns (Cornell University, Ithaca).

3. Methods

The conjugant electrotransformation method consists of four steps.

1. Conjugation is induced between two fertile *Tetrahymena* strains of different mating types.
2. At the macronuclear development stage, the cells are concentrated, mixed with transforming DNA, and electroporated.
3. The electroporated cells are plated on a nonselective medium to recover from electric shock, complete conjugation, and grow for a few generations.
4. Finally, transformants are selected using appropriate drugs depending on the type of the transformation vector.

In most cases, the last two steps are performed using the same microtiter plates without need for time-consuming replica transfers.

3.1. Preparation of Electrotransformation Competent Tetrahymena

1. Use two strains with different mating types. One of the strains is a heterokaryon having drug resistance marker in the micronucleus (e.g., CU428). Grow both strains in 50 mL of SPP, with shaking, in 300-mL Erlenmeyer flasks. Cells should be transferred several times before the transformation

experiment. To have cells ready on a daily basis, transfer about 10^5 cells each day to 50 mL of fresh SPP in 300-mL Erlenmeyer flasks.

To slow down the aging process (*see* Note 9d), *Tetrahymena* strains should not be transferred frequently during periods when transformation experiments are not performed. Cells can be maintained without transfers in soybean cultures for several months (*21*). However, for long-term storage, *Tetrahymena* strains should be stored frozen in liquid nitrogen (*22*).

2. To starve the day-old culture, wash cells with 50 mL of 10 mM Tris, pH 7.5, by centrifugation at 1100g for 2 min. Resuspend each strain in 50 mL of Tris buffer in a 250-mL Erlenmeyer flask, and incubate at 30°C.
3. After 6–12 h of starvation, count the starved cells and adjust to 3×10^5 cells/mL.
4. Mix 50 mL of each strain of different mating type in a 2-L Erlenmeyer flask. Add 1 mL of 100X antibiotic/antimycotic solution, and incubate at 30°C with shaking at 160 rpm. Use a programmable timer to turn the shaker off 10 h before you want to perform electroporation. Cells should be starved for 18–24 h before the shaker is turned off. Fast shaking prevents formation of pairs (*23*). The cells will begin to pair when the shaker stops. If the shaking method for initiation of pairing cannot be used, then after counting, incubate 50 mL of each strain separately in a 300-mL Erlenmeyer flask at 30°C for 18–24 h (without shaking). Ten hours before electroporation, mix them in a 2-L flask and continue the incubation at 30°C.
5. Four hours after the shaker is turned off (or cells mixed), check pairing efficiency. Proceed only if more than 80% of cells are in pairs (*see* Note 9). Prepare transforming DNA by resuspending in 125 µL of the electroporation buffer (*see* Section 3.2.1.).
6. At 10–11 h after turning the shaker off (or mixing strains), transfer cells to 50-mL Corning plastic conical tubes and centrifuge for 5 min at 1100g. Discard the supernatant, and resuspend cells in 100 mL of the electroporation buffer. Centrifuge for 4 min as above.
7. Resuspend cells in 1 mL of electroporation buffer at about 3×10^7 cells/mL. Use cells immediately for electroporation.

3.2. Electroporation

We have performed three types of experiments using CET:

1. Transformation with high efficiency replicative vectors.
2. Gene replacements.
3. Gene replacements with a cotransforming high-efficiency vector.

Cotransformation is recommended for tubulin gene replacement (*10*, and unpublished results using a mutant α-tubulin gene).

1. In high-frequency transformation experiments with replicative vectors (such as rDNA vector pD5H8), use 15–20 μg of high-quality RNA-free plasmid DNA. In gene replacement experiments, release the insert from the vector to create homologous ends (or ends as close to the homologous sequence as possible) by using the appropriate restriction endonucleases. There is no need to separate insert DNA from the vector. Purify the digested DNA by extracting once with phenol/chloroform/isoamyl alcohol (25:24:1), and once with chloroform/isoamyl alcohol (24:1), then precipitate with ethanol, and resuspend in 125 μL of electroporation buffer. Use 50 μg of digested DNA/electroporation. To cotransform, add 15–20 μg of a high-efficiency vector (pD5H8) to the digested gene replacement vector DNA. All transforming DNA should be resuspended in 125 μL of electroporation buffer (*see* Note 1).

2. Mix 125 μL of concentrated cells with plasmid DNA, and transfer to a cuvet with a 0.2-cm gap. Pulse once at 250 V, 275-μF capacitance, and 13-Ω resistance. Under these conditions, the voltage peak is about 225 V, pulse length 4 ms, and the field strength is 1125 V/cm (*see* Note 2).

3. Incubate electroporated cells at room temperature for 1 min, and then resuspend in 20-mL of SPPA in a Petri dish (100-mm diameter) (*see* Note 3).

4. Dilute cells using an array of Petri plates. The dilution factor is defined as the number of 96-well plates that would be needed to plate all cells (~2.4 mL of cells/plate). Thus, the original 20-mL sample of electroporated cells represent an 8X dilution. Make further dilutions (10–20 mL for each dilution) depending on the type of transformation vector (*see below*). For all dilutions, plate 25 μL of cells into wells already filled with 175 μL of SPPA. Depending on the type of transformation experiment, use the following dilutions (*see* Notes 4–7):

   a. For high-efficiency transformation vectors (pD5H8, prD4-1, pH4T2, and derivatives), prepare one 96-well microtiter plate with 80, 160, 320, and 8000X dilutions.

   b. For gene replacement experiments, plate all the remaining 8X dilution and prepare one plate at 8000X (*see* Note 8).

   c. For gene replacement experiments in which a cotransforming vector was included, plate all remaining 8X dilution. Also plate one each at the 80, 160, 320, and 8000X dilutions.

5. Incubate cells at 30°C. Note that plates with low plating dilution (8X) in experiments in which the H4-I/*neo*/BTU2 marker is used may be incubated at room temperature for more convenient timing of subsequent selection (*see* Note 7b).

6. For all transformations, use the single plate at the 8000X dilution to test overall survival and mating efficiency (*see* Note 6). After 12–18 h, add 50 µL of SPPA medium containing 75 µg/mL of 6-methylpurine (if the strain CU428 was used) to each well to obtain a final concentration of 15 µg/mL of 6-methylpurine. Nonconjugants with 6-methylpurine-sensitive macro-nuclei will be killed. Conjugants will have developed a new 6-methylpurine resistant macronucleus from the micronucleus of the heterokaryon, which contains this gene in an unexpressed form.

7. Start selection of transformants as follows (depending on the type of trans-formation).

   a. For experiments involving high-efficiency vectors (plated at dilutions >80X), after 12–18 h, add 50 µL/well of 600 µg/mL paromomycin in SPPA to a final concentration of 120 µg/mL (in the case of rDNA and *neo* vectors) or cycloheximide at 75 µg/mL (final concentration 15 µg/mL) for *rpL29* vectors.

   b. For disruptive gene replacement experiments in which the H4-I/*neo*/BTU2 marker was used and cells were plated at low dilution (8X), paromomycin should be added either 6 h after electroporation follow-ing incubation of cells at 30°C, or 12–18 h after electroporation follow-ing incubation at room temperature prior to selection (*see* Note 10).

   c. In cotransformation experiments (dilution 8X) involving the pD5H8 rDNA vector, begin the initial selection with paromomycin 12–18 h after electroporation (with cells incubated at 30°C prior to selection). In cotransformation experiments involving the high-efficiency vector pH4T2 (with the H4-I/*neo*/BTU2 marker), the initial paromomycin selection should be performed as described above after 6–9 h if cells are incubated at 30°C or after 12–18 h if cells are left at room tempera-ture (*see* Note 10).

8. Incubate at 30°C for 4–5 d, and then calculate the transformation yield (8–320X plates) and the survival number (8000X plates) using the Poisson distribution *(24)*.

9. In the case of cotransformation experiments, make replicas of the primary paromomycin resistant transformant clones (8 or 16X plates) by transfer-ring cells into medium containing an appropriate selective drug for the gene replacement event. Secondary selection plates usually contain 150–200 µL of SPPA with an appropriate drug. Use 30 μM oryzalin or vinblastine to select for the mutant BTU1 and mutant α-tubulin gene replacement transformants and 15 µg/mL of cycloheximide in the case of mutant *rpL29* gene transformants. Note that late-stationary-phase cells are extremely sensitive to oryzalin, and even cells containing an oryzalin resistant mutant tubulin gene die under these conditions. Thus, the status

of the culture should be monitored using the dissecting scope and cells transferred onto secondary selection plates when the late-logarithmic or early stationary phase of growth is reached in most wells. In cotransformation experiments, plates at higher dilutions (80–320X) are used to calculate the frequency of the primary paromomycin resistant clones.

4. Notes
4.1. Electroporation Conditions

1. Transformation was not detectable when performed in *Tetrahymena* culture medium (SPP) instead of the standard HEPES electroporation buffer.
2. Using the ECM 600 electroporator (BTX Inc., La Jolla, CA), we determined the optimal range for a limited number of the electrotransformation parameters *(17)* using stable transformation to paromomycin resistance mediated by the high-efficiency rDNA vector (pD5H8) as an assay for electroporation efficiency. The optimal field strength was measured for two pulse length values. For 4 ms (obtained using 275-μF capacitance and 13 Ω resistance), the optimal transformation yield was obtained at 1125 V/cm (using 250-V setting). For 1.6 ms (50-μF capacitance and 24-Ω resistance), the maximal yield was observed at 1380 V/cm (using 325 V).
3. CET efficiency is very sensitive to cold. Short (2-min) incubation of cells on ice before electroporation completely abolished transformation without affecting cell survival. No significant effect was observed if cells were incubated on ice for 2 min after electroporation. Because cell survival remained at the same level, cold pretreatment likely affects DNA uptake and not cell viability, possibly by affecting membrane fluidity. Thus, cells should not be kept on ice before electroporation.
4. The frequency of transformation strongly depends on the type of vector. Replicative vectors (pD5H8, prD4-1, pH4T2) transform at high frequency, usually between 10,000 and 30,000 transformants in a single experiment *(15,17)*. Vectors transforming by gene replacement require integrative replacement within the homologous locus based on homologous recombination, which occurs at low frequency; 1–10 transformants/experiment were obtained in functional gene replacement experiments *(10,15* and unpublished results). For knockout gene replacement using the universal disruption cassette H4-1/*neo*/BTU2, up to 100 transformants were obtained in a single experiment. Tubulin gene replacement vectors, such as the mutant BTU1 *(10)* and the mutant α-tubulin gene (unpublished observation), should be cotransformed with a high-frequency replicative vector (pD5H8) to obtain the highest transformation frequency.
5. The conformation of transforming DNA plays a significant role in transformation efficiency. For the replicative vector pD5H8, subjecting the plas-

mid DNA to multiple cycles of freezing and thawing dramatically reduced transformation yield *(15)*, suggesting that relaxed DNA is less efficient than supercoiled. Also, linearization of the same vector reduced the transformation efficiency significantly *(15)*. Because a linear replicative (rDNA) vector is efficient in transformation when microinjected into the macronucleus *(25)*, the effect of linearization on electroporation-mediated transformation is likely the result of decreased DNA uptake, and not altered gene expression or replication of the transformed DNA. However, for a replacement-type vector *(17)*, linearization of plasmid stimulated transformation efficiency based on homologous recombination. In this case, the linearization of vector DNA created homologous ends that probably stimulated homologous recombination within the targeted locus to a degree that more than compensated for the reduced uptake of linear DNA. Several types of carrier DNA from animal cells were found either to have no effect or actually decrease transformation frequency using the pD5H8 vector.

4.2. Plating and Selection Conditions

6. The cells subjected to electroporation need to be diluted appropriately prior to plating. The optimal plating cell concentration depends on the type of transformation vector. In experiments in which the frequency of transformation is expected to be low (gene replacement or gene knockout), nearly all treated cells should be plated at high cell concentration, except for the small fraction of cells plated at very high dilution to measure the conjugant survival rate (*see* Note 8). For other experiments in which high-frequency transformation is expected (using replicative vectors) and quantification of transformation yield is needed or in which it is important to isolate independent transformant clones, several dilutions of cells are used to ensure that at least one plate has the optimal transformant number. The frequency of transformation is calculated using the Poisson distribution *(24)*.

7. It is very useful to know how many cells survive in a given electroporation experiment. Final transformation yield is a result of both survival efficiency and transforming vector efficiency. The cell-survival yield is dependent on cell concentration, quality of DNA, and fertility of strains. Thus, the cell-survival data help to compare transformation results obtained using different vectors or data from independent experiments. Also, if transformation efficiency is too low, knowing the cell-survival values may help to eliminate some of the potential technical problems (*see* Section 4.3.). Cell survival is defined here as the number of conjugation progeny clones per experiment. If one of the strains used for transformation is a drug resistant heterokaryon (contains the dominant 6-methylpurine resistance gene *Mpr* in the micronucleus, and a macronucleus with the sensitive phenotype),

only conjugation progeny and not parental cells or cells that aborted conjugation will become resistant to 6-methylpurine. Thus, the cell survival is simply the total number of 6-methylpurine resistant clones per experiment. To obtain cell survival number, some cells should be plated at very high dilution (8000X) and selected with 6-methylpurine.

8. In gene replacement experiments, only partial replacement of the 45 macronuclear copies occurs initially. The original transformant clones need to be analyzed (e.g., using Southern Blotting) to verify if the desired gene replacement event occurred since an occasional outcome is an integration of the insert DNA into one of the flanking regions of the targeted locus *(9,17)*. To complete gene replacement (completely eliminate the endogenous gene copies through assortment), transformed clones need to be grown in a selective medium for another 50–80 generations (generation time is ~3 h at 30°C) by daily transfers. Under drug selection, the endogenous allele is lost if it is not essential.

4.3. Troubleshooting

9. If there are very few transformants and overall survival is very low (<50,000 conjugation progeny resistant to 6-methylpurine), one or more of the problems listed below may be responsible.

    a. RNA is present in the DNA preparation. This problem appears to be more serious in case of large-size plasmids (>10 kbp). Plasmid DNA needs to be checked for the presence of RNA using an agarose gel. If a low-mol-wt diffuse RNA band is visible, plasmid DNA needs to be digested with a combination of RNase A and T1, and repurified using standard organic extractions and ethanol precipitation.

    b. Ions in the final electroporation mixture: If the electroporation medium is too conductive (indicated by a shorter pulse length, <3.7 ms), owing to an inadequate washing of cells or DNA, the survival rate may decrease, presumably because of increased heating produced during electric pulse application. Either reprecipitate DNA or wash cells more carefully.

    c. Bacterial or fungal contamination: Contaminating microorganisms grow well during mating since the cells do not feed. We observed a dramatic decline in the survival rate that correlated with the presence of unidentified microorganisms in *Tetrahymena* cultures. This effect was observed only if cells were subjected to electroporation. Check conjugating cells under a microscope for the presence of contaminants. Add 1 mL of antibiotic mix (*see* Section 2.) when cells are mixed. If survival is still low, use a fresh stock of cells or purify cells by serial drop transfers in SPPA medium *(24)*.

d. The strains used for transformation are old and give low frequency of conjugation progeny. Although *Tetrahymena* strains can be grown vegetatively indefinitely, clones undergo aging and show a gradual decrease in ability to perform germinal functions. Even though old strains are still able to form pairs when mixed with cells of a different mating type, conjugation in aged strains usually is terminated prematurely without developing new macronuclei. Also, conjugation involving an aged clone tends to be less synchronous. Decrease in fertility is associated with a decrease in competence for transformation by CET (unpublished data). Strains were observed to deteriorate in a relatively short period of 6–12 mo, even without extensive culture growth (unpublished observation). Maintenance of strains in long-term soybean medium *(19)* is not sufficient to prevent germinal aging of *Tetrahymena* strains. If other potential problems (listed above) are eliminated, check the strains as follows: Mix cells of two different mating types, one of which is a functional heterokaryon. After 6 h, add an equal volume of 2X SPPA. At 8 h isolate 48 pairs individually into single drops of SPPA. Grow cells for 3 d at 30°C. Replica-transfer into media containing 15 µg/mL 6-methylpurine (if the CU428 strain is used) or other appropriate medium to select for true conjugants. Efficient strains give ≥90% drug resistant conjugation progeny. If too many drug-sensitive progeny are produced (macronuclear retention), obtain fresh cells from frozen stocks.

e. Pairing of mixed strains is not efficient. If mixed cells do not reach 80–95% level of pairing at 4 h after mixing, the population of conjugants in the mating reaction may not be sufficiently synchronous to yield mass transformation. Poor mating efficiency may result from mixing unequal numbers of cells of each mating strain. Furthermore, even trace amounts of chemicals (including dishwasher detergent) left on glassware may inhibit mating. Finally, a slow mating reaction may be caused by insufficient aeration. Use a flask large enough to ensure that fluid depth does not exceed 1 cm.

10. If there are very few transformants, but overall survival is high (>100,000 of drug resistant conjugation progeny cells/experiment), examine the possible causes listed below.

a. DNA of high-frequency replicative vectors is not supercoiled. Replicative vectors (i.e., pD5H8) transform more efficiently when supercoiled. Check DNA on a minigel. Avoid freezing and thawing DNA used for transformations by making small aliquots and freezing; use only once.

b. DNA is not completely digested in a gene replacement experiment. DNA used in gene replacement experiments must be completely digested to stimulate homologous recombination by releasing the insert at homologous ends.

c. Too little DNA is used for a gene replacement experiment. DNA may be lost during purification following digestion. Use more DNA for digestion than the final amount needed for transformation. Check amount of digested and purified DNA before electroporation by agarose minigel separation using standards of known concentration for comparison.

d. Plating dilution too low (initial cell concentration too high): Vectors based on the neomycin resistance gene placed under control of the H4-I gene promoter display decreased efficiency of selection when plated at low dilutions (high cell concentrations) *(17)*. This problem can be avoided by using greater dilutions or by incubating cells at room temperature prior to selection. For high-efficiency (replicative vector) experiments, dilutions ≥80X are recommended, and the drug should be added no later than 12–18 h after electroporation. Because of the large number of cells that must be screened in low-efficiency (gene replacement) experiments, low plating dilutions are recommended. The drug can be added as early as 6–9 h after electroporation in the case of vectors based on the H4-I/*neo*/BTU2 marker.

e. Electroporation delayed after conjugant cells are washed and concentrated in the electroporation buffer: Conjugant cells prepared for electroporation quickly lose high competence for transformation. Usually, several electroporations can be performed in a row using a batch of cells from the same mating. If several electroporations are planned with the same batch of cells, all steps should be performed as quickly as possible.

f. Electroporation of cells delayed after addition of DNA: Electroporation must be performed immediately after mixing cells with DNA. A decrease in transformation rate is observed if electroporation is delayed, presumably because of the presence of secreted DNase activity by concentrated cells *(13)*.

g. Optimal timing of transformation was missed. This is probably the most common cause of low efficiency of transformation. The electroporation-mediated transformation is more sensitive to timing than microinjection-mediated conjugant transformation, probably because lower amounts of transformed DNA are delivered per nucleus (transforming DNA competes less efficiently with endogenous gene markers for replication and expression). The maximal transformation efficiency is usually observed between 10 and 11 h after mixing (or after shaker is turned off). Missing the optimum time-point by as little as 30 min may result in a significant reduction in transformation frequency. The more synchronous the mating, the higher the transformation yield, but also the greater chance of missing the optimal transformation time window. Unfortunately, mating kinetics vary slightly between experiments, and it is difficult to predict when the exact period for maximal efficiency occurs. There is little to be done except repeat the experiment, make sure the temperature is well controlled, and perform electroporation 10 h after mixing the cells (or turning off the shaker). If possible, cells can be electroporated at two time-points (10 and 11 h after mixing) to increase the chance for having highly competent conjugants.

References

1. Tondravi, M. M. and Yao, M.-C. (1986) Transformation of *Tetrahymena thermophila* by microinjection of ribosomal RNA genes. *Proc. Natl. Acad. Sci. USA* **83**, 4369–4373.
2. Godiska, R. and Yao, M.-C. (1990) A programmed site-specific DNA rearrangement in *Tetrahymena thermophila* requires flanking polypurine tracts. *Cell* **61**, 1237–1246.
3. Yao, M.-C., Yao, C.-H., and Monks, B. (1990) The controlling sequence for site-specific chromosome breakage in *Tetrahymena*. *Cell* **63**, 763–772.
4. Yasuda, L. F. and Yao, M.-C. (1991) Short inverted repeats at a free end signal large palindromic DNA formation in Tetrahymena. *Cell* **67**, 505–516.
5. Yu, G.-L., Bradley, J. D., Attardi, L. D., and Blackburn, E. H. (1990) *In vivo* alteration of telomere sequences and senescence caused by mutated *Tetrahymena* telomerase RNAs. *Nature* **344**, 126–132.
6. Yu, G.-L. and Blackburn, E. H. (1991) Developmentally programmed healing of chromosomes by telomerase in *Tetrahymena*. *Cell* **67**, 823–832.
7. Yu, G.-L., Hasson, M., and Blackburn, E. H. (1988) Circular ribosomal DNA plasmids transform *Tetrahymena thermophila* by homologous recombination with endogenous macronuclear ribosomal DNA. *Proc. Natl. Acad. Sci. USA* **85**, 5151–5155.
8. Yao, M.-C. and Yao, C.-H. (1991) Transformation of *Tetrahymena* to cycloheximide resistance with a ribosomal protein gene through sequence replacement. *Proc. Natl. Acad. Sci. USA* **88**, 9493–9497.
9. Kahn, R. W., Andersen, B. H., and Brunk, C. F. (1993) Transformation of *Tetrahymena thermophila* by microinjection of a foreign gene. *Proc. Natl. Acad. Sci. USA* **90**, 9295–9299.
10. Gaertig, J., Thatcher, T. H., Gu, L., and Gorovsky, M. A. (1994) Electroporation-mediated replacement of a positively and negatively selectable β-tubulin gene in *Tetrahymena thermophila*. *Proc. Natl. Acad. Sci. USA* **91**, 4549–4553.
11. Gaertig, J., Kiersnowska, M., and Iftode, F. (1988) Induction of cybrid strains of Tetrahymena thermophila by electrofusion. *J. Cell Sci.* **89**, 253–261.
12. Gaertig, J. and Iftode, F. (1989) Rearrangement of the cytoskeleton and nuclear transfer in *Tetrahymena thermophila* cells fused by electric field. *J. Cell Sci.* **93**, 691–703.

13. Brunk, C. F. and Navas, P. (1988) Transformation of *Tetrahymena thermophila* by electroporation and parameters effecting cell survival. *Exp. Cell Res.* **174**, 525–532.

14. Orias, E., Larson, D., Hu, Y.-F., Yu, G.-L., Karttunen, J., Lovlie, A., Haller, B., and Blackburn, E. H. (1988) Replacement of the macronuclear ribosomal RNA genes of a mutant *Tetrahymena* using electroporation. *Gene* **70**, 295–301.

15. Gaertig, J. and Gorovsky, M. A. (1992) Efficient mass transformation of *Tetrahymena thermophila* by electroporation of conjugants. *Proc. Natl. Acad. Sci. USA* **89**, 9196–9200.

16. Martindale, D. W., Allis, C. D., and Bruns, P. J. (1982) Conjugation in *Tetrahymena thermophila*. A temporal analysis of cytological stages. *Exp. Cell Res.* **140**, 227–236.

17. Gaertig, J., Gu, L., Hai, B., and Gorovsky, M. A. (1994) Gene replacement and high frequency vector-mediated electrotransformation in *Tetrahymena*. *Nucleic Acids Res.* **22**, 5391–5398.

18. Nanney, D. L. (1980) *Experimental Ciliatology*. John Wiley, New York.

19. Doerder, F. P., Deak, J. C., and Lief, J. H. (1992) Rate of phenotypic assortment in *Tetrahymena thermophila*. *Dev. Genet.* **13**, 126–132.

20. Gorovsky, M. A., Yao, M.-C., Keevert, J. B., and Pleger, G. L. (1975) Isolation of micro- and macronuclei of Tetrahymena pyriformis, in *Methods in Cell Biology*, vol. 9 (Prescott, D. M., ed.) Academic, New York, pp. 311–327.

21. Williams, N. E., Wolfe, J., and Bleyman, L. K. (1980) Long term maintenance of *Tetrahymena* spp. *J. Protozool.* **27**, 327.

22. Flacks, M. (1979) Axenic storage of small volumes of *Tetrahymena* cultures under liquid nitrogen: a miniaturized procedure. *Cryobiology* **16**, 287–291.

23. Bruns, P. J. and Brussard, T. B. (1974) Pair formation in *Tetrahymena pyriformis*, an inducible developmental system. *J. Exp. Zool.* **188**, 337–344.

24. Orias, E. and Bruns, P. J. (1975) Induction and isolation of mutants in Tetrahymena, in *Methods in Cell Biology* (Prescott, D. M., ed.) Academic, New York, pp. 247–282.

25. Yao, M.-C. and Yao, C.-H. (1989) Accurate processing and amplification of cloned germ line copies of ribosomal DNA injected into developing nuclei of *Tetrahymena thermophila*. *Mol. Cell. Biol.* **9**, 1092–1099.

CHAPTER 34

Transfection of the African and American Trypanosomes

John M. Kelly, Martin C. Taylor, Gloria Rudenko, and Pat A. Blundell

1. Introduction

The protozoan genus *Trypanosoma* contains several pathogenic parasites of major medical and veterinary importance. These include *Trypanosoma brucei* subspecies, the causative agents of African trypanosomiasis, and *Trypanosoma cruzi*, the causative agent of American trypanosomiasis or Chagas' disease. The development of transformation procedures for these organisms has been a major technical advance. Applications include investigation of the molecular determinants of virulence, functional analysis of potential chemotherapeutic targets, and studies on the mechanisms of gene expression, drug resistance, and developmental regulation. Electroporation is the predominant method used to transfect trypanosomes. The protocols described here are in routine use in our respective laboratories in Amsterdam *(T. brucei)* and London *(T. cruzi)*, and have proven to be straightforward and reliable.

The transformation of *T. brucei* by electroporation differs in several respects from transformation of other kinetoplastids, such as *Leishmania* and *T. cruzi*. First, introduced DNA appears to integrate into the genome almost exclusively by homologous recombination *(1–3)*; the formation of episomes is very rare *(4)*. Second, in transient assays, transcription seems to be less promiscuous than in *Leishmania,* and a genuine trypanosome promoter must be present. Currently, only three of these have been extensively characterized: the ribosomal DNA promoter *(5),*

From: *Methods in Molecular Biology, Vol. 47: Electroporation Protocols for Microorganisms*
Edited by: J. A. Nickoloff Humana Press Inc., Totowa, NJ

the variant-specific surface glycoprotein (VSG) promoter *(6,7)*, and the procyclic acidic repetitive protein (PARP) promoter *(8–10)*. All of these promoters are highly active on episomes. Stable transformation in *T. brucei* generally requires linearized DNA; supercoiled DNA is very inefficient by comparison. Recently, transfection of the previously intractable bloodstream-form was developed by Carruthers and Cross *(11)*. The protocol presented here has slight modifications from that original protocol. However, transformation of the bloodstream-form remains at least 1000-fold less efficient than transformation of the cultured insect form for reasons that are as yet unclear.

Transformation of *T. cruzi* can be mediated by integration *(12,13)*, which always occurs by homologous recombination, or by episomal vectors *(14,15)*. The protocol described here has been used to introduce both plasmid and cosmid shuttle vectors into *T. cruzi*. A plasmid vector (pTEX) was constructed using flanking sequences derived from the *T. cruzi* glycosomal glyceraldehyde 3-phosphate dehydrogenase (gGAPDH) genes *(16)* with the neomycin phosphotransferase gene as a selectable marker. Transformed cells contain multiple copies of the vector predominantly as large extrachromosomal elements composed of head-to-tail tandem repeats. A cosmid vector (pcosTL), a derivative of pTEX, mediates the introduction of large DNA fragments (up to 40 kb) into *T. cruzi*, where the recombinant molecules can replicate as extrachromosomal circles present in multiple copies *(15)*. The protocol can also be used to facilitate transformation of *Leishmania* species using the same vectors. The sequences involved in autonomous replication and transcriptional regulation within the vectors have yet to be unequivocally identified. However, the same splice leader addition site upstream of the drug-selectable marker gene is effective in both *T. cruzi* and *Leishmania*.

2. Materials

2.1. T. brucei *Electroporation*

1. SDM-79 medium for insect-stage *T. brucei*: modified from Brun and Schönenberger *(17)*. For 1L: 7.0 g Dulbecco's Modified Eagle medium powder (Gibco, Paisley, Scotland), 2.0 g Medium 199 powder (Gibco), 8.0 mL MEM amino acid solution (Gibco), 6.0 mL MEM nonessential amino acids (Gibco), 1.0 g glucose (anhydrous), 8.0 g HEPES, 5.0 g MOPS, 2.0 g NaHCO$_3$, 200 mg L-alanine, 100 mg L-arginine HCl, 70 mg L-methionine, 80 mg L-phenylalanine, 600 mg L-proline, 60 mg L-serine, 160 mg taurine, 350 mg L-threonine, 100 mg L-tyrosine, 10 mg adenosine, 10 mg gua-

nosine, 50 mg D (+)-glucosamine HCl, 4 mg folic acid, 2 mg *p*-aminobenzoic acid, 0.2 mg biotin, 60 mg penicillin.

Dissolve mixed powders in approx 700 mL double-distilled water, and then add both MEM amino acid solutions. Adjust the pH to 7.3 with NaOH. Make up to a final volume and sterilize by filtration. Store at 4°C. Before use, add 10% v/v heat-inactivated fetal calf serum (FCS) (heat-inactivate for 30 min at 56°C) and 2 mL/L sterile hemin (2.5 mg/mL in 50 mM NaOH, filter-sterilized). The complete medium can be stored at 4°C for 1 mo or frozen at –20°C for at least 1 yr.

2. HMI-9 for bloodstream-form trypanosomes *(18)*: 365 mL Iscove's MDM (Gibco), 1 mM hypoxanthine, 50 µM bathocuproine sulfonate, 1 mM cysteine, 2 mM β-mercaptoethanol, 1 mM pyruvate, 160 µM thymidine, 50 U/mL penicillin, 50 µg/mL streptomycin penicillin/streptomycin, 30 µg/mL kanamycin. Mix well, sterilize by filtration, and add 50 mL of heat-inactivated FCS and 50 mL of Serum Plus (Hazleton Biologics).

The solutions for preparation of HMI-9 should be stored as concentrated stock solutions, and the medium made just prior to use. Store HMI-9 medium at 4°C for no longer than a week.

Plates consist of 9 parts HMI-9 medium with 1 part 6.5% LMP agarose (Ultrapure, BRL). For 20 plates, add 2.6 g agarose to 40 mL water in a 500-mL bottle. Autoclave, and then add 360 mL HMI-9 to the liquid agarose. Mix thoroughly, and pour 20 mL each into 100-mm culture plates. For selective plates, the drug should be included with the HMI-9 before mixing with agarose.

3. Zimmerman Postfusion Buffer (ZM): 132 mM NaCl, 8 mM KCl, 8 mM Na$_2$HPO$_4$, 1.5 mM KH$_2$PO$_4$, 0.5 mM magnesium acetate, 0.09 mM CaCl$_2$.
4. ZMG: 1X ZM supplemented with 0.5% glucose.
5. TE buffer: 10 mM Tris-HCl, pH 7.5, 1 mM EDTA.
6. Glycerol freezing medium: 80% SDM-79 (procyclic) or 80% HMI-9 (bloodstream-form), 10% heat-inactivated FCS, 10% glycerol. After mixing, sterilize by filtration and store at –20°C.

2.2. T. cruzi *Electroporation*

1. Growth medium for *T. cruzi* epimastigotes: RPMI 1640 (Gibco BRL) supplemented with 0.5% (w/v) trypticase (BBL, Cockeysville, MD), 0.5% (w/v) HEPES, 0.03M hemin, 10% (v/v) FCS (heat-inactivated), 2 mM sodium glutamate, 2 mM sodium pyruvate and antibiotics. Prepare as follows: Make sterile stocks (×100) of trypticase (0.175 g/mL, filter-sterilized), HEPES (1M, filter-sterilized), and hemin (2.5 mg/mL in 0.01M NaOH, autoclaved). Add 2.8 mL of trypticase solution, 2 mL of HEPES, and 0.8 mL of hemin to each 100 mL of RPMI stock, together with 10 mL of FCS, 1 mL of 200 mM sodium glutamate/200 mM sodium pyruvate

(with penicillin and streptomycin to give final concentrations of 250 U/mL and 250 µg/mL, respectively). The glutamine/pyruvate/antibiotic solution should be filter-sterilized before addition.

2. PBS: 70 mM NaCl, 8 mM Na$_2$HPO$_4$, 2 mM NaH$_2$PO$_4$, pH 7.2.
3. Electroporation buffer: 272 mM sucrose, 7 mM sodium phosphate pH 7.2. Filter-sterilize, and store aliquots at −20°C.
4. G418 (Sigma): Prepare stocks of 50 mg/mL in T.E. buffer; filter-sterilize, and store at −20°C.

3. Methods

3.1. Electroporation of Procyclic T. brucei

1. Grow culture form procyclic trypanosomes in 5 mL of SDM-79 medium until midlogarithmic growth phase.
2. Count the number of cells (i.e., with hemocytometer).
3. Pellet the cells at 3000g for 5 min in a bench-top centrifuge.
4. Wash the cells by resuspending the pellet in 10 mL PBS. Resuspend the trypanosomes at 2×10^7 cells/mL in ZM buffer.
5. Pipet 5 µg DNA (*see* Note 1) into a 0.2-cm gap electroporation cuvet as a drop on the side.
6. Add 500 µL of the trypanosome suspension onto the DNA drop, ensuring that this mixes well—do not create air bubbles.
7. Pulse with 1500 V, 25-µF capacitance. The time constant should be 0.2 ms. Strain 427-60 requires two pulses for optimal efficiency. All other strains so far tested can only withstand one pulse. Trypanosome killing should be between 50 and 80%.
8. Immediately pipet 500 µL of SDM-79 medium into the cuvet. Mix thoroughly, and transfer the cell suspension into a culture flask containing 4 mL of SDM-79. At this stage, no selective drug should be present. Allow the trypanosomes to recover by incubating the flask at 27°C for 36–48 h.
9. Count the number of surviving cells. We usually set up a dilution series during selection, since this gives some idea of the efficiency of transformation and allows the rapid characterization of both (semi-)clonal (lowest outgrowing flask) and polyclonal (top flask of series) populations (*see* Note 3).
10. Outgrowth of the transformants should be visible in the more dilute flasks after 14 d. We generally passage the cells 1:5 for the first passage after selection, and then at higher dilutions until they grow readily at 1:1000. This ensures that all nontransformed cells are dead.

3.2. Electroporation of Bloodstream-Form T. brucei

1. Isolate trypanosomes from the blood of an infected rodent by centrifuging 3-mL aliquots of whole blood in a 15-mL Falcon tube at 3000g for 10 min.

The trypanosomes will appear as a thick white layer on top of the "buffy coat." Pipet the white layer into 10 mL of HMI-9 medium at room temperature. Dilute an aliquot, as appropriate, and count the trypanosomes. Each cuvet will require 5×10^7 parasites. Do not use a DE52 column to purify trypanosomes for electroporation, since this seems to reduce the efficiency and/or survival.

2. Resuspend the cells at 1×10^8 cells/mL in ZMG.
3. Add 5 µg DNA (*see* Note 1) and then 500 µL of cell suspension to a 0.2-cm gap cuvet.
4. Pulse the cells once at 1.5 kV, 25 µF; the optimal parameters may vary from strain to strain.
5. Immediately rinse the cuvet with 500 µL of HMI-9 medium, and add the cells to a flask containing 4.5 mL of HMI-9 medium.
6. After each series, count the surviving trypanosomes. This will give an indication of how well they have tolerated the conditions used. Incubate the flasks at 37°C for 18 h to allow recovery of the transfectants.
7. Count the number of motile cells in each flask.
8. Resuspend the cells at 2×10^7 cells/mL in HMI-9 medium. Pipet 500 µL of this suspension onto the center of each selective plate. Close the plate, and swirl **gently** to spread the trypanosomes across the surface.
9. Allow the plates to dry for 15 min in sterile hood (until the liquid has just soaked into the plate), with the lids off.
10. Equilibrate the plates overnight in a 37°C CO_2 incubator, and then seal the plates with parafilm. Do not tip the plates because the agarose will break.
11. Colonies should be visible after 6–8 d. Transformed colonies are removed from the plate by dropping 10–50 µL of HMI-9 onto the colony and pipeting gently up and down. Increase the volume to 200 µL, and transfer to another selective plate to expand the population. After this plate is full, the trypanosomes can be washed off and expanded by replating or by growth in an animal if selection is no longer required. In our experience, most integrations, even into minichromosomes, appear to be stable for long periods in the absence of selection. It is important to note that once the plate begins to turn orange, indicating acidification, the trypanosomes will die within hours if not washed off and replated. The bloodstream-form cannot regulate its internal pH.

3.3. Electroporation of T. cruzi

Work with *T. cruzi* involves risk of contracting a serious infection. Care should be taken to conform rigorously to the designated codes of practice for handling this organism (*see* Note 11). Most *T. cruzi* electroporation experiments in our laboratory have been carried out using the

Hoefer Progenitor electroporator (Model PG1). In contrast to apparatus based on the capacitor discharge method (e.g., Bio-Rad Gene Pulser), the machine uses a timed switch on the output of a DC power supply. The unit generates a square wave for the duration of the pulse. The pulse is administered from a circular electrode probe that fits into the wells of a 24-well tissue-culture tray (e.g., Nunclon). The distance between the ring electrode and the center electrode is 5.5 mm; thus, at 400 V, the voltage gradient is 728 V/cm. The conditions described have been optimized for the *T. cruzi* Sylvio-X10.6 clone and may vary between different strains.

1. Pellet logarithmically growing *T. cruzi* epimastigotes ($3000g$, 10 min) from a 50–100 mL culture, wash once in 10 mL of PBS, and resuspend at 10^8 cells/mL in electroporation buffer. This generally represents a 10-fold concentration of the parasites.
2. Add supercoiled vector DNA (5–100 µg) to the chamber of a 24-well microtiter plate, and add a 1-mL suspension of cells. Incubate on ice for 10 min.
3. Electroporate the cells using 400 V, 99 ms square-wave pulses; pulse 8–10 times. This results in approx 50% cell killing.
4. Incubate on ice for 10 min, and then add 200-µL aliquots to 1 mL of normal growth medium, and incubate at 27°C overnight. This procedure can be carried out in the same microtiter plate (*see* Note 12).
5. Add 1 mL from each well to a flask containing 10 mL of growth medium plus 100 µg/mL G418, and incubate the cells at 27°C. Different strains of *T. cruzi* have different susceptibility to G418; use a range of concentrations (10–200 µg/mL) until the optimal dose for selection has been established.
6. After 3 d, add a further 20 mL of growth medium (including G418) to each of the flasks.
7. Five days later, remove 1.5 mL of cells to a fresh flask containing 15 mL of growth medium + G418.
8. Drug-resistant cells can usually be observed 7–11 d later. Leave the cells for a further week before subculturing (1:10).

3.4. Storage of Transformed Trypanosomes

Cryopreservation of trypanosomes is a straightforward procedure that should be used to maintain backup stocks of each transformed line.

1. Pellet actively growing trypanosomes ($3000g$), and resuspend in the appropriate freezing mix (growth medium + 10% v/v glycerol) at 3×10^7–1×10^8 trypanosomes/mL.

2. Aliquot into cryotubes (liquid nitrogen-resistant vials), and freeze at about 1°C/min to –70°C using a programmable freezing unit. Alternatively, wrap tubes in toweling, and place in a box in –70°C freezer overnight. Then transfer tubes to liquid nitrogen for long-term storage.
3. To recover trypanosomes from cryopreservation, thaw for 5 min in a 25°C water bath, and transfer organisms to a tube containing 5 mL of growth medium. Pellet cells (3000*g*), and then resuspend in appropriate growth medium and incubate as usual.

4. Notes

For successful transformation, the quality of DNA used is paramount, since low efficiencies can result from poor-quality DNA. If transfecting with a linearized plasmid, check an aliquot on an agarose gel for complete digestion. Extract the digest with 1:1 phenol:chloroform, and precipitate the DNA with ethanol. Resuspend the DNA in up to 50 μL sterile TE for each electroporation. If the transfected DNA is a fragment from a digest, purify the appropriate fragment on agarose gel in the absence of ethidium bromide. After running the gel, cut off the markers, and stain them in ethidium bromide. Use these as a guide to isolate the gel slice containing the fragment. Ethidium bromide is trypanocidal and must be removed using isoamyl alcohol extraction followed by ethanol precipitation. CsCl may also be toxic, so for transfection of supercoiled DNA, it is important to use DNA free of CsCl.

For purification of the DNA from agarose, we find that the glass-milk method (e.g., Geneclean, Bio101 Inc.) gives the best-quality DNA for transfection, although electroelution may also be used with good results. If using electroelution, the DNA must be cleaned afterward by extraction with 1:1 phenol:chloroform followed by ethanol precipitation. In DNA manipulations, it is important to bear in mind that this DNA is to be transfected into cells maintained in tissue-culture conditions and that the DNA must therefore also be sterile.

4.1. Transfection of Procyclic T. brucei

1. Procyclic survival will be drastically reduced if the trypanosomes are left in ZM for more than 1 h.
2. Optimal electrical parameters may vary somewhat from strain to strain and must be determined empirically for each new strain used. For instance, we find that freshly differentiated procyclics from the 221 bloodstream-form are best transformed with one pulse, whereas procyclics of strain 427-60

that have been maintained in long-term culture require two pulses for optimum efficiency.

3. In general, a dilution series should be set up with flasks containing 10^2–10^7 cells; the rest of the transformed population can be immediately frozen down as a stabilate in case outgrowing populations become contaminated. For all but the flask with 10^7 cells, it is necessary to add 1×10^6–1×10^7 wild-type cells/flask to allow the transfectants to grow out (below a certain population density, *T. brucei* will tend to die regardless of the drug effect).

4. To calculate the amount of drug required for selection, carry out an IC_{50} test on the strain of interest, since drug susceptibility varies between strains. The procyclic form is much more resistant to neomycin, hygromycin, and phleomycin than the bloodstream-form; hence, when using freshly differentiated procyclics, the IC_{50} of the parent bloodstream-form is not a good guide. For strain 427-60 procyclics, we find that 20 µg/mL G418 or 25 µg/mL hygromycin give outgrowth only of transformants. We have never observed natural resistance occurring at these antibiotic concentrations. There is sometimes batch variation in the drugs used. When starting a new batch, check the IC_{50} again if problems are encountered.

5. Suitable controls for transfection and selection should always include trypanosomes electroporated in the absence of DNA, since selection of these should result in complete cell death. Also, if there is a much greater survival in mock-transfected cells compared to DNA transfectants, it may indicate problems with DNA preparations. Transfection of a well-characterized construct in parallel with new constructs is a good control for the efficiency of electroporation itself, since many constructs give reasonably reproducible efficiencies of transfection.

4.2. Transfection of Bloodstream-Form **T. brucei**

6. Only prepare trypanosomes for four cuvets at a time, since long incubation periods in electroporation buffer result in lower survival. As a general rule, trypanosomes prepared from rats seem to be better able to survive electroporation than in vitro cultured trypanosomes.

7. Motile cell counts together with counts of immediate survival allow one to calculate the number of doublings occurring during the recovery period. This information is essential when estimating the efficiency of transfection.

8. Plates must be dry, since wet plates allow free migration of trypanosomes and prevent formation of tight colonies. However, it is vitally important not to overdry the plates, since this will result in poor survival and hence loss of transformants.

9. We recommend that stabilates be made as early as possible, since the plating method appears to be very prone to bacterial infection.

10. For selection of bloodstream-form transformants, it is important to note that the bloodstream-form is, in general, much more susceptible to the standard antibiotics used (neomycin and hygromycin) than is the procyclic form. We usually select at 2 µg/mL of either drug.

4.3. Transfection of T. cruzi

11. At least 50 cases of laboratory-acquired infections with *T. cruzi* have been identified *(19)*. Strict adherence to good microbiological codes of practice is essential at all times when working with this organism. Wear close-fitting rubber gloves for all procedures involving live parasites. Use 70% ethanol for decontamination of gloves and working surfaces. Contaminated material should be disposed of immediately by immersion in 70% ethanol or bleach (sodium hypochlorite). Whenever possible, work (including electroporation) should be carried out within a safety cabinet (Class II BS5726). Avoid any procedures that might lead to droplet suspensions, and wear safety glasses and a breathing mask.

12. When carrying out multiple transfections of *T. cruzi* using the Hoefer Progenitor apparatus, care should be taken to avoid crosscontamination. Do each transfection in a separate microtiter plate, and rinse the electrode thoroughly in sterile distilled water and 70% ethanol between samples. It is possible to perform between 10 and 20 consecutive transfections without crosscontamination.

13. To determine the efficiency of transfection, a dilution series can be performed. Add diluted aliquots of electroporated cells to a well containing 10^7 nontransfected parasites (in 1 mL of growth medium) and select as described above (*see* Section 3.3., steps 5–8). Transfection efficiencies are usually in the range of 10^{-5}–10^{-6}/µg DNA for both plasmid and cosmid vectors. This protocol can also be used with slight modification to transform *Leishmania donovani, L. mexicana,* and *L. major.* We have found *L. donovani* to be more sensitive to G418, and we select with 25 µg/mL of G418. In addition, it is unnecessary to dilute or passage the parasites (*see* Section 3.3., steps 6 and 7) to select drug-resistant cells.

Acknowledgments

We thank Vern Carruthers and George Cross for generously giving us their *T. brucei* bloodstream-form transformation protocol several months prior to publication, and P. Borst, in whose laboratory all our work on *T. brucei* transformation has been carried out.

References

1. ten Asbroek, A. L. M. A., Ouellette, M., and Borst, P. (1990) Targeted insertion of the neomycin phosphotransferase gene into the tubulin gene cluster of *Trypanosoma brucei. Nature* **348,** 174,175.

2. Lee, M. G.-S. and Van der Ploeg, L. H. T. (1990) Homologous recombination and stable transfection in the parasitic protozoan *Trypanosoma brucei*. *Science* **250,** 1583–1587.

3. Eid, J. and Sollner-Webb, B. (1991) Stable, integrative transformation of *Trypanosoma brucei* that occurs exclusively by homologous recombination. *Proc. Natl. Acad. Sci. USA* **88,** 864–868.

4. ten Asbroek, A. L. M. A., Mol, C. A. A. M., Kieft, R., and Borst P. (1993) Stable transformation of *Trypanosoma brucei*. *Mol. Biochem. Parasit.* **59,** 133–142.

5. Zomerdijk, J. C. B. M., Kieft, R., Shiels, P. G., and Borst, P. (1991) Alpha-amanitin-resistant transcription units in trypanosomes: a comparison of promoter sequences for a VSG gene expression site and for the ribosomal RNA genes. *Nucleic Acids Res.* **19,** 5153–5158.

6. Zomerdijk, J. C. B. M., Oullette, M., ten Asbroek, A. L. M. A., Kieft, R., Bommer, A. M. M., Clayton, C. E., and Borst P. (1990) The promoter for a variant surface glycoprotein gene expression site in *Trypanosoma brucei*. *EMBO J.* **9,** 2791–2801.

7. Jefferies, D., Tebabi, P., and Pays, E. (1991) Transient activity assays of the *Trypanosoma brucei* variant surface glycoprotein gene promoter: control of gene expression at the posttranscriptional level. *Mol. Cell. Biol.* **11,** 338–343.

8. Rudenko, G., Le Blancq, S., Smith, J., Lee, M. G.-S., Rattray, A., and Van der Ploeg, L. H. T. (1990) Procyclic acidic repetitive protein (PARP) genes located in an unusually small α-amanitin resistant transcription unit: PARP promoter activity assayed by transient DNA transfection of *Trypanosoma brucei*. *Mol. Cell. Biol.* **10,** 3492–3504.

9. Sherman, D. R., Janz, L., Hug, M., and Clayton, C. (1991) Anatomy of the parp gene promoter of *Trypanosoma brucei*. *EMBO J.* **10,** 3379–3386.

10. Brown, S. D., Huang, J., and Van der Ploeg, L. H. T. (1992) The promoter for the procyclic acidic repetitive protein (PARP) genes of *Trypanosoma brucei* shares features with RNA polymerase I promoters. *Mol. Cell. Biol.* **12,** 2644–2652.

11. Carruthers, V. and Cross, G. A. M. (1992) DNA-mediated transformation of bloodstream-form *Trypanosoma brucei*. *Nucleic Acids Res.* **21,** 2537,2538.

12. Otsu, K., Donelson, J. E., and Kirchhoff, L. V. (1993) Interruption of a Trypanosoma cruzi gene encoding a protein encoding 14-amino acid repeats by targeted insertion of the neomycin phosphotransferase gene. *Mol. Biochem. Parasitol.* **57,** 317–330.

13. Hariharan, S., Ajioka, J., and Swindle, J. (1993) Stable transformation of *Trypanosoma cruzi:* inactivation of the PUB 12.5 polyubiquitin gene by targeted gene disruption. *Mol. Biochem. Parasitol.* **57,** 15–30.

14. Kelly, J. M., Ward, H. M., Miles, M. A., and Kendall, G. (1992) A shuttle vector which facilitates the expression of transfected genes in *Trypanosoma cruzi* and *Leishmania*. *Nucleic Acids Res.* **20,** 3963–3969.

15. Kelly, J. M., Das, P., and Tomás, A. M. (1994) An approach to function complementation by introduction of large DNA fragments into *Trypanosoma cruzi* and *Leishmania donovani* using a cosmid shuttle vector. *Mol. Biochem. Parasit.* **65,** 51–62.

16. Kendall, G., Wilderspin, A. F., Ashall, F., Miles, M. A., and Kelly, J. M. (1990) *Trypanosoma cruzi* glycosomal glyceraldehyde-3-phosphate dehydrogenase does not conform to the "hotspot" topogenic signal model. *EMBO J.* **9**, 2751–2758.
17. Brun, G. A. and Schönenberger, M. (1979) Cultivation and *in vitro* cloning of procyclic culture forms of *T. brucei* in semi-defined medium. *Acta Trop.* **36**, 289–292.
18. Hirumi, H. and Hirumi, K. (1989) Continuous cultivation of *Trypanosoma brucei* bloodstream-forms in a medium containing a low concentration of serum proteins without feeder layers. *J. Parasit.* **75**, 985–989.
19. Brener, Z. (1984) Laboratory-acquired Chagas' disease: an endemic disease among parasitologists? in *Genes and Antigens of Parasites—A Laboratory Manual*, 2nd ed. (Morel, C. M., ed.), Fundação Oswaldo Cruz, Rio de Janeiro, pp. 3–10.

CHAPTER 35

Electroporation in *Giardia lamblia*

A. L. Wang, Tiina Sepp, and C. C. Wang

1. Introduction

Electroporation was first applied in *Giardia lamblia* in 1990 by Furfine and Wang *(1)* to introduce a full-length giardiavirus (GLV) single-stranded RNA (ssRNA) into uninfected *G. lamblia* cells. After electroporation, the cells were grown to confluency and passed serially at 1:13 dilutions. It was found that by the eighth passage, ssRNA had replicated to form the viral genomic dsRNA in detectable amounts. Fully assembled and infectious virus particles could be recovered from the culture medium supernatant as well. This is, to the best of our knowledge, the first successful example that the genomic transcript of a dsRNA virus can be introduced into the host cell to generate a mature virus.

More recently, the same technique was applied to transfect the *G. lamblia* isolates that were otherwise highly resistant to GLV infection *(2)*. The resistant cells were electroporated with the total RNA isolated from virus-infected *G. lamblia* cells under the same conditions as described in *(1)*. Once again, the proliferation of GLV genome and the expression of viral capsid protein inside the electroporated GLV resistant cells were evident after the electroporation (Fig. 1), and virus particles infectious to the sensitive *G. lamblia* isolates were found to be present in the culture medium supernatant *(3)*. The intracellular presence of viral dsRNA was monitored up to the ninth passage and was found to show no sign of diminishing. The viral genome continued to proliferate and could apparently be maintained indefinitely in the infected cells, even though the virus particles thus released remained incapable of infecting

From: *Methods in Molecular Biology, Vol. 47: Electroporation Protocols for Microorganisms*
Edited by: J. A. Nickoloff Humana Press Inc., Totowa, NJ

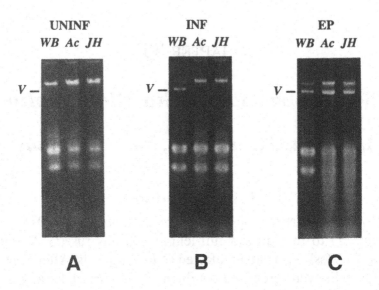

Fig. 1. Electroporation of virus-free *G. lamblia* cells with GLV RNA. Agar-ose-gel electrophoretic analysis of the total RNA from 1.5×10^6 uninfected (A), GLV-infected (B), or GLV-RNA electroporated *G. lamblia* cells (C). The ethidium bromide-stained GLV dsRNA is marked as V. The band above V is contaminating DNA, and two bands below V are large and small subunit ribosomal RNAs, respectively. The multiplicity of infection for experiments in (B) is 10^4 virus particles/cell. WB, *G. lamblia* GLV-sensitive WB strain; Ac, *G. lamblia* GLV-resistant Ac strain; JH, *G. lamblia* GLV-resistant JH strain.

naive untreated resistant cells. It turned out that the resistance to GLV infection in these *G. lamblia* strains that were refractory to viral infection could be directly attributed to the lack of specific receptors for GLV on the cell surface *(3)*. When the initial blockage of viral entry is bypassed by introducing GLV ssRNA into these resistant cells *via* electroporation, all other cellular machineries needed for the proliferation of giardiavirus are present in these resistant cells to complete the viral replication cycle.

To date, these two studies remain the only published reports in which an exogenous gene was successfully introduced into and expressed in *Giardia* cells. This apparent failure to generalize the application of electroporation for the introduction of other foreign genes into *Giardia* is largely because of a gap in our understanding as to what constitutes the promoters or ribosomal binding sequences in this protozoon that diverted very early from the other eukaryotes in its evolutionary path *(4)*. Of the approximately handful of *G. lamblia* genes that have been cloned and sequenced, all of them show extremely short (as short as 1–3 nucleotides) upstream untranslated sequences in their messenger RNAs (mRNA) *(5)*. Thus, it is the lack of knowledge regarding the *cis*-regulating elements that control the expression of genes and the ribosomal binding region in mRNA from *Giardia* that limits the application of this powerful transfection technique in this parasitic protozoon. It is believed that once this knowledge becomes available, electroporation will become a very important means for the transformation of *G. lamblia* cells.

2. Materials

1. Medium for in vitro cultivation of *G. lamblia* trophozoites *(6)*: For each 100 mL of medium, add: 0.1 g K_2HPO_4, 60 mg KH_2PO_4, 2.0 g casein digest (BBL, Cockeysville, MD) *(see* Note 1), 1.0 g yeast extract (BBL), 1.0 g glucose, 0.2 g NaCl, 0.2 g cysteine HCl, 20 mg ascorbic acid, and 75 mg bile (Sigma, St. Louis, MO) *(see* Note 2). Adjust pH to 7.05 with 1*N* NaOH, and then add: 10 mL, heat-inactivated calf serum (Gibco [Gaithersburg, MD] or Hyclone [Logan, UT]) *(see* Note 3), 1 mL penicillin-streptomycin (Gibco), 0.1 mL fungizone (Gibco) *(see* Note 4), and 1 mL BME vitamins (Gibco). Filter-sterilize. Store the filtered medium at 4°C for no more than 10 d to 2 wk. Discard any unused medium if crystalline cysteine sediments to the bottom of tube within this period. Presence of cysteine sediments means that the pH of the medium has decreased drastically on storage, and this medium will promptly kill *Giardia* cells.
2. Phosphate-buffered saline (PBS): 0.15*M* NaCl and 10 m*M* sodium phosphate, pH 7.2. Sterilize by autoclaving *(see* Note 8).
3. Chemicals: Common chemicals and water used for the preparation of RNAs must be of the highest level of purity available and treated to be free of RNase contamination. Molecular biology laboratory manuals provide instructions regarding RNase-free manipulation.
4. Special equipment for the handling of *Giardia* culture: Although the trophozoites stage of *G. lamblia* we work with is noninfectious, the organism is nonetheless a human pathogen. Safety precautions and P2 biological containment as stipulated by NIH, such as wearing protective gloves, should be strictly adhered to. Special equipment for biosafety includes the laminar flow hood for the transferring and inoculation of G. *lamblia* culture *(see* Note 6) and the autoclave for sterilization. All contaminated

pipets, vessels, or spent culture medium must be autoclaved or treated with disinfectant before disposal.

3. Methods

1. Prepare *G. lamblia* total RNA from early log-phase cultures ($1.2–1.5 \times 10^6$ cells/mL) of GLV-infected trophozoites by the hot phenol method *(7)* (*see* Notes 5, 6, and 10).
2. Electrophorese 2 mg of total RNA in 0.8% agarose gel in Tris-borate-EDTA buffer. Excise gel slices containing the ssRNA, and elute ssRNA from the gel by electroelution (Elutrap by Schleicher & Schuell, Keene, NH) (*see* Note 7). Determine the yield of ssRNA by UV absorption at 258 nm before adding 100 mg of tRNA from *S. cerevisiae* as carrier and precipitating the RNAs with ethanol.
3. Grow the recipient *G. lamblia* WB cells to 95% confluence ($1.2–1.5 \times 10^6$ cells/mL), and harvest by centrifugation at 5000*g* for 10 min at 4°C (*see* Note 5). Wash the cells twice with 10 mL of ice-cold PBS, each time pelleting by centrifugation as before, and resuspend to 10^7 cells/mL with the same buffer.
4. Pipet an estimated 25 ng of gel-purified ssRNA plus 25 mg of tRNA carrier into a 0.2-cm electroporation cuvet with 0.5 mL of the cell suspension, and incubate on ice for 10 min before applying the electric current. Pulse at 2.5 kV, 25 µF, and 400 Ω. The time constant should be 0.6–0.7 ms under these conditions (*see* Notes 9 and 10).
5. Incubate the cells on ice for 15 min, and then gently dilute into 15 mL of fresh culture medium prewarmed to 37°C, leaving little air space inside the tube to maintain a relatively anaerobic condition (*see* Note 12). Subculture the electroporated cells after they reach confluency in 3–4 d by assuming 1:15 dilution for each passage. After the first passage, the cells begin to gain in vigor and will grow to confluency in 2 d. Extract total RNA from the cells, and examine on agarose gel electrophoresis for the presence of GLV dsRNA as described earlier. GLV dsRNA is usually detectable (i.e., presence of 2–5 ng GLV dsRNA in 10^5 cells) after the second or third passage.
6. To test for the presence of infectious virus particles in the culture medium supernatant, chill the confluent cells on ice for 10 min to detach from culture tubes, and centrifuge the cell suspension at 5000*g* for 10 min at 4°C. Pass the culture supernatant through a 0.2-µm Millipore filter to remove contaminating cells and cell debris. Add 1 mL of this filtered spent medium to 14 mL of culture freshly inoculated with the GLV-sensitive *G. lamblia* WB strain at 1×10^5 cells/mL. After the culture reaches confluency, extract total RNA from these cells, and assay for GLV dsRNA by agarose gel electrophoresis as described before (*see* Notes 12 and 13).

4. Notes

4.1. In Vitro Cultivation of G. lamblia Trophozoites

1. Not all lots of casein digest will adequately support the growth of *Giardia*. We routinely use BBL 97023. If a certain lot does not work, request casein digest from a different lot.
2. Bile used for preparation of the medium should not be confused with bile salts, which are detergents and will kill *Giardia*.
3. Heat-inactivated fetal calf serum or horse serum can substitute satisfactorily for calf serum.
4. The concentration of fungizone used in the *Giardia* culture medium is only one-tenth of that customarily used for mammalian cells. Too much fungizone will kill this protozoon.

4.2. Isolation of GLV RNA

5. The general procedure to harvest *Giardia* cells is to chill the culture tubes in ice water for 10 min to allow the flagellate to detach from the surface of culture vessel. Transfer cell suspensions or washings only in the laminar flow hood to avoid inhaling the contaminated aerosol. After phenol has been added, all subsequent manipulations can be performed outside the laminar flow hood.
6. Total RNA from *G. lamblia* is prepared as follows *(6)*: Prepare pH 5 phenol by equilibrating redistilled phenol with double-distilled water. Discard the water layer, and aliquot phenol into 5–10 mL vials. Keep one vial at 65°C for immediate use, and store the rest at –20°C in the dark. Resuspend washed *G. lamblia* cell pellet (*see* Section 3., step 3) in 1.5-mL microfuge tube with 20 vol of $0.1M$ sodium acetate, pH 4.8. Add 1/20 vol of 20% sodium lauryl sulfate (SDS). Quickly vortex the mixture to lyse the cells. Immediately add an equal volume of phenol prewarmed to 65°C, and vortex vigorously for 1 min. Centrifuge at $11,000g$ for 10 min at 4°C. Transfer the top aqueous layer to a clean 1.5-mL tube, and re-extract with an equal volume of hot phenol. Transfer the aqueous layer to a clean tube. Repeat the phenol extraction until the interphase is clean. RNA may be precipitated from the aqueous phase of the last extraction by adding 2.5 vol of cold ethanol. Alternatively, an aliquot of the aqueous phase may be directly applied to an agarose gel for electrophoresis.
7. GLV ds or ssRNA can also be recovered from agarose gel slices by RNaid resin (Bio 101, La Jolla, CA) or by centrifuging through a Spinex cellulose acetate column (Costar, Cambridge, MA). Recovery methods using low-melting-point agarose or silica powder give poor results, since RNA stains poorly in the former and does not bind well to the latter.

4.3. Electroporation

8. It is a good idea to prepare a batch of PBS large enough for multiple tests in order to achieve more reproducible results. This is because a small variation in the ionic strength of this buffer will result in relatively large change of electric current under high voltage.

9. We use only 0.2-cm cuvets from Bio-Rad (Richmond, CA). Cuvets of 0.4-cm width are not suitable because they are unable to attain the same electric field (V/cm). Cuvets may be rinsed with water, soaked in 75% ethanol, and air-dried in laminar flow hood for reuse. If arcing occurs during electroporation, the cuvet should be discarded.

10. Total RNA isolated from GLV-infected *G. lamblia* cells can also be used to transfect the virus-free *Giardia* cells. In this case, there is no need to add exogenous carrier RNA. Although this total RNA contains GLV dsRNA in addition to ssRNA, dsRNA is not infectious by electroporation *(1)*. The success of transfecting *G. lamblia* with the GLV RNA depends largely on the purity, as well as the abundance and integrity of the ssRNA. The input RNA must be thoroughly washed with 75% ethanol to remove contaminating salts or phenol. Depending on the quality of the input RNA, successful transfection can be achieved with 10–250 μg total RNA/mL cell suspension.

11. Use two to four pulses if necessary. Pulsing more than once increases the efficiency of transfection, but decreases the survival rate. We find a survival rate of 5–10% to be optimal for successful transfection.

12. To assure an adequate survival rate of the electroporated cells, it is essential to start with vigorously growing cells, and to maintain cells at 4°C during the electroporation experiment as much as possible. After the electric field has been applied, leave the cells in the same cuvet undisturbed on ice for at least 15 min. Transfer the cells into prewarmed culture medium by gently pipeting once. Do not stress the cells by repeated pipeting.

13. In control experiments, where the purified GLV ssRNA was replaced by purified GLV dsRNA, there was no sign of GLV replication in the electroporated cells. Also, when ssRNA was simply mixed with the recipient cells without applying the electric current, no GLV could be detected in the treated cells. The absolute requirement for the electric field in this experiment suggested that: (a) electroporation was needed for GLV ssRNA to enter into the *Giardia* cells, and (b) there was no viral contamination in our purified GLV ssRNA samples.

References

1. Furfine, E. S. and Wang, C. C. (1990) Transfection of the *Giardia lamblia* double-stranded RNA virus into *Giardia lamblia* by electroporation of a single-stranded RNA copy of the viral genome. *Mol. Cell. Biol.* **10**, 3659–3663.

2. Miller, R. L., Wang, A. L., and Wang, C. C. (1988) Identification of *Giardia lamblia* strains susceptible and resistant to infection by the double-stranded RNA virus. *Exp. Parasitol.* **66,** 118–123.

3. Sepp, T., Wang, A. L., and Wang, C. C. (1994) Giardiavirus-resistant *Giardia lamblia* lacks a virus receptor on the cell membrane surface. *J. Virol.* **68,** 1426–1431.

4. Sogin, M. L., Gunderson, J. H., Elwood, H. J., Alonso, R. A., and Peattie, D. A. (1989) Phylogenetic meaning of the kingdom concept: an unusual ribosomal RNA from *Giardia lamblia. Science* **243,** 75–77.

5. Alonso, R. A. and Peattie, D. A. (1992) Nucleotide sequence of a second alpha giardin gene and molecular analysis of the alpha giardin genes and transcripts in *Giardia lamblia. Mol. Biochem. Parasitol.* **50,** 95–104.

6. Keister, D. B. (1983) Axenic culture of *Giardia lamblia* in TY1-S-33 medium supplemented with bile. *Trans. R. Soc. Trop. Med. Hyg.* **77,** 487,488.

7. Wang, A. L. and Wang, C. C. (1986) Discovery of a specific double-stranded RNA virus in *Giardia lamblia. Mol. Biochem. Parasitol.* **21,** 259–276.

Index

A

Agrobacterium,
  A. rhizogenes, 171
  A. tumefaciens, 171–178
  binary vector, 173
  tumor-inducing (Ti) plasmid, 171
Aspergillus,
  A. nidulans, 279, 281, 282, 284
  A. oryzae, 281
  selectable markers, 282

B

Bacillus subtilus, 205, 209, 228
Bacteroides, 161–169
  B. fragilis, 162, 163, 166
  B. ovatus, 163
  B. thetaiotomicron, 162
  B. unifomis, 163
  conjugative transfer from *E. coli,*
    161, 162
  plasmids, 162
β-galactosidase (lacZ), 73, 108, 137,
  321
β-glucuronidase, 89
Beuveria,
  B. bassiana, 282
  B. sulfurescens, 282
  hyphal fragment electroporation,
    285, 286, 288
Borrelia burgdorferi, 253–259
  selectable markers, 256
Brucella, 143–148
  B. abortus, 143–147
  B. melitensis, 146
  plasmid vectors, 144

C

Camplyobacter jejuni, 128, 180
Candida, 291–302
  C. albicans, 291
  C. maltosa, 291–300
  C. utilis, 291
  shuttle vectors, 294
Capacitance,
  effect on electroporation, 88
  effect on pulse length, 32, 225
Caulobacter crescentus, 52
cDNA,
  libraries, 67, 74–77
  saturation capacity, 75
  synthesis, 68, 71
Cell stress, 20
Cell viability, 20
Cell wall weakening agents (glycine,
  threonine, lysozyme), 185, 191,
  195–198, 204, 205, 219, 222,
  223, 231, 247, 287
Chloramphenicol acetyl transferase
  (cat), 304, 306, 310
Clostridium, 161, 227–235
  C. acetobutylicum, 227–229, 231
  C. botulinum, 227–229, 231
  C. perfringens, 227–229, 231, 232
  DNase, 228, 231
  oxygen tolerance, 232
  shuttle plasmids, 227–229

D

Dictyostelium, 310, 321–330
  selectable markers, 324–328
  vectors, 323
Dielectrophoresis, 27, 94